OSIRIS

A RESEARCH JOURNAL
DEVOTED TO THE
HISTORY OF SCIENCE
AND ITS
CULTURAL INFLUENCES

PUBLICATION AND EDITORIAL OFFICE
DEPARTMENT OF HISTORY AND SOCIOLOGY OF SCIENCE
UNIVERSITY OF PENNSYLVANIA
215 SOUTH 34TH STREET
PHILADELPHIA, PENNSYLVANIA, 19104-6310, USA

SUGGESTIONS FOR CONTRIBUTORS TO OSIRIS

1. **Manuscripts** (original plus two copies) may be submitted to any member of the Editorial Board, or to the OSIRIS Editorial Office, 215 South 34th Street, Philadelphia, PA 19104-6310, USA. Please include an **abstract** of approximately 150 words. Contributors are advised to retain a copy for reference. If return of submitted material is desired, include return postage or international reply coupons. OSIRIS is now devoted primarily to thematic issues, conceived and compiled by guest editors.

2. Manuscripts should be **typewritten** or processed on a **letter-quality** printer and **double-spaced** throughout, including quotations and notes, on paper of standard size or weight. Margins should be wider than usual to allow space for instructions to the typesetter. The right-hand margin should be left ragged (not justified) to maintain even spacing and readability.

3. OSIRIS normally uses double-blind refereeing; authors should therefore identify themselves only on a detachable cover sheet.

4. Bibliographic information should be given in **footnotes** (not parenthetically in the text), typed separately from the main body of the manuscript, **double-** or even **triple-spaced,** numbered consecutively throughout the article, and keyed to reference numbers typed above the line in the text.

a. References to **books** should include author's full name; complete title of the book, underlined (italics); place of publication and publisher's name for books published after 1900; date of publication, including the original date when a reprint is being cited; page numbers cited. *Example:*

Joseph Needham, *Science and Civilisation in China,* 5 vols., Vol. I: *Introductory Orientations* (Cambridge: Cambridge Univ. Press, 1954), p. 7.

b. References to articles in **periodicals** should include author's name; title of article, in quotes; title of periodical, underlined; year; volume number, Arabic and underlined; number of issue if pagination requires it; page numbers of article; number of particular page cited. Journal titles are spelled out in full on first citation and abbreviated subsequently. *Example:*

John C. Greene, "Reflections on the Progress of Darwin Studies," *Journal of the History of Biology,* 1975, *8*:243–273, on p. 270; Dov Ospovat, "God and Natural Selection: The Darwinian Idea of Design," *J. Hist. Biol.,* 1980, *13*:169–174, on p. 171.

c. When first citing a reference, please give the title in full. For succeeding citations, please use an abbreviated version of the title with the author's last name. *Example:*

Greene, "Reflections," p. 250.

5. Please mark clearly for the typesetter all unusual alphabets, special characters, mathematics, and chemical formulae, and include all diacritical marks.

6. A small number of **figures** may be used to illustrate an article. Line drawings should be directly reproducible; glossy prints should be furnished for all halftone illustrations.

7. Manuscripts should be submitted to OSIRIS with the understanding that upon publication **copyright** will be transferred to the History of Science Society. That understanding precludes OSIRIS from considering material that has been submitted or accepted for publication elsewhere.

OSIRIS (ISSN 0369-7827) is published once a year.

A subscription to OSIRIS is $39 for institutions. Individual subscriptions are $29 (hardcover) and $20 (paperback).

Address advertising inquiries, single-issue orders, and new subscriptions to HSS Publications Office, 215 South 34th Street, Philadelphia, PA 19104-6310, U.S.A. Address renewal orders, claims for missing issues, and changes of address to HSS Business Office, P.O. Box 529, Canton, MA 02021, U.S.A. Claims for issues not received should be sent within four months of publication of the issue in question.

OSIRIS is indexed in major scientific and historical indexing services, including *Biological Abstracts, Current Contents, Historical Abstracts,* and *America: History and Life.*

Second-class postage paid at Canton, MA, and additional mailing offices.

Typeset and printed at The Sheridan Press, Inc.

Hardcover edition, ISBN 0-934235-18-X.
Paperback edition, ISBN 0-934235-17-1.

RENAISSANCE MEDICAL LEARNING

EVOLUTION OF A TRADITION

EDITED BY

MICHAEL R. McVAUGH

AND

NANCY G. SIRAISI

A RESEARCH JOURNAL
DEVOTED TO THE HISTORY OF SCIENCE
AND ITS CULTURAL INFLUENCES
SECOND SERIES VOLUME 6 1990

OSIRIS 1990 SECOND SERIES VOLUME 6

RENAISSANCE MEDICAL LEARNING: EVOLUTION OF A TRADITION

Edited by MICHAEL R. McVAUGH and NANCY G. SIRAISI

Cover: An exchange between master and students, from the opening of Galen's *Tegni* or *Ars parva* as found in Vienna, Osterreiches Nationalbibliothek, MS 2315, folio 145r. Courtesy of the Loren C. MacKinney Collection, University of North Carolina.

A medical master and students. From Avicenna's Canon *in Glasgow, MS Hunter 9, folio 39. Courtesy of the Loren C. MacKinney Collection, University of North Carolina.*

Introduction

By Michael R. McVaugh and Nancy G. Siraisi

THE ESSAYS in this volume of *Osiris* address the theme of medical knowledge in western Europe between the twelfth and the sixteenth centuries. They trace developments in the ways in which the specialized knowledge appropriate to the medical profession was conceived, articulated, and put to use. Their topic is thus by no means coextensive with the history of medicine in medieval and Renaissance Europe. Recent scholarship on this, as on other periods of medical history, encompasses a range of topics broader and more diversified than was once traditional. Today, historians of medieval and Renaissance medicine, like their colleagues who specialize in more recent periods, include among their current interests such subjects as the history of morbidity and epidemiology; the social and cultural history of the management of public health and health-related institutions; the experiences, attitudes, and medical beliefs of patients and of non-elite practitioners; and the cultural assumptions embodied in medical or physiological theories. Indeed, a number of the authors represented in this volume have published or are at work on such studies in what is broadly speaking the social history of medicine and medicine in popular culture.[1]

But among this diversity of topics the nature of medical knowledge holds a place of central importance. Among medieval branches of knowledge, medicine had the distinctive feature of bridging academic learning on the one hand and

[1] It is not possible to provide a full bibliography within the confines of this introduction. The following is a representative sampling of work published since 1975. Jean-Noël Biraben, *Les hommes et la peste en France et dans les pays méditerranéens,* 2 vols. (Paris: Mouton, 1975–1976); Ann G. Carmichael, *Plague and the Poor in Renaissance Florence* (Cambridge: Cambridge Univ. Press, 1986); William Eamon and Gundolf Keil, " 'Plebs amat empirica': Nicholas of Poland and His Critique of the Medieval Medical Establishment," *Sudhoffs Archiv,* 1987, *71*:180–196; Luis García-Ballester, *Los Moriscos y la medicina: Un capítulo de la medicina y la ciencia marginadas en la España del siglo XVI* (Barcelona: Editorial Labor, 1984); Monica H. Green, "Women's Medical Practice and Medical Care in Medieval Europe," *Signs,* 1989, *14*:434–473; Danielle Jacquart, *Le milieu médical en France du XIIe au XVe siècle* (Geneva: Librarie Droz, 1981); Jacquart and Claude Thomasset, *Sexuality and Medicine in the Middle Ages* (Princeton, N.J.: Princeton Univ. Press, 1988; original French ed., 1985); Michael MacDonald, *Mystical Bedlam: Madness, Anxiety, and Healing in Seventeenth-Century England* (Cambridge: Cambridge Univ. Press, 1981); Irma Naso, *Medici e strutture sanitarie nella società tardo-medievale: Il Piemonte dei secoli XIV e XV* (Milan: F. Angeli, 1982); Katharine Park, *Doctors and Medicine in Early Renaissance Florence* (Princeton, N.J.: Princeton Univ. Press, 1985); Carol Rawcliffe, "The Profits of Practice: The Wealth and Status of Medical Men in Later Medieval England," *Social History of Medicine,* 1988, *1*:61–78; Andrew W. Russell, ed., *The Town and State Physician in Europe from the Middle Ages to the Enlightenment* (Wolfenbüttel: Herzog August Bibliothek, 1981); Paul Slack, *The Impact of Plague in Tudor and Stuart England* (London/Boston: Routledge & Kegan Paul, 1985); Charles Webster, ed., *Health, Medicine, and Mortality in the Sixteenth Century* (Cambridge: Cambridge Univ. Press, 1979). Michael McVaugh and Luis García-Ballester have in preparation books on the social history of medicine in the Crown of Aragon during the fourteenth century. For additional bibliography on medieval and Renaissance medicine in general see Faye M. Getz, "Western Medieval Medicine," *Trends in History,* 1988, *4*:37–54; and Nancy G. Siraisi, "Some Current Trends in the Study of Renaissance Medicine," *Renaissance Quarterly,* 1984, *37*:585–600.

crafts, trades, and professions on the other. In western Europe social, institutional, and intellectual developments beginning in the late eleventh century combined to produce a state of affairs in which medicine was both an accepted learned (later, university) discipline, with ties to natural philosophy and an authoritative literature of its own, and an occupation involving technical skills pursued for gain. And in medicine, unlike the other "lucrative science," law, practice involved the performance—or at the very least the supervision—of physical activity; while medical techniques, again unlike legal ones, had aspects that were essentially craftsmanlike in nature. Medicine was thus simultaneously text-based and practice-based. As a result, between the thirteenth and the sixteenth centuries the abilities demanded of the most highly esteemed members of the medical profession, namely, university professors of medicine, included both the explication of ancient texts and the supervision of patients.[2]

Hence, although the assimilation and putting to use of Greek and Arabic medicine in the medieval West in many respects paralleled similar developments in other disciplines, medicine presented special difficulties or, perhaps, offered special opportunities. The elite minority among medical practitioners who participated fully in the world of Latin academic learning had contact with transmitted texts of Greek and Arabic philosophy as well as medicine, but unlike scholastic natural philosophers they were also involved in a practical, technical activity. Perhaps as a result they were notably self-conscious about the relations in their discipline between authority and experience, theory and practice, the universal and the particular, and speculative philosophy and technical mastery. Medieval and Renaissance medical knowledge and its place in the history of learning is therefore a subject that crosses boundaries between the traditionally delineated subject matter of intellectual history, history of science, history of medicine, and history of technology.

Nonetheless, our subject has by now acquired a distinct identity, an identity that is of relatively recent origin. Twenty-five years ago a collection of essays that took seriously European medical thought in the five centuries preceding what may still for present purposes be termed the Scientific Revolution would have been unthinkable. Historians tended to view most of the academic Latin medical writings of the period between Constantine the African and Vesalius as mere book learning and verbal games, as products of an abstruse system cut off from reality—and from the world of medieval medical practice—and hence as sterile and uninteresting: "medieval" and "scholastic," in fact, with all the scornful associations those words can possess. Only insofar as medicine had apparently escaped an academic setting did it seem to hold any interest: anatomy and surgery had alone been the subjects of serious historical scholarship.

Today, however, there is a growing recognition that the intellectual activity of the academic medical community of the later Middle Ages and Renaissance is not satisfactorily characterized by these stereotypes. Far from distancing themselves

[2] There is no doubt that learned university professors of medicine in medieval and Renaissance universities were consulted by patients and supervised their treatment. Michael R. McVaugh, "The Two Faces of a Medical Career: Jordanus de Turre of Montpellier," in *Mathematics and Its Applications to Science and Natural Philosophy in the Middle Ages: Essays in Honor of Marshall Clagett*, ed. Edward Grant and John E. Murdoch (Cambridge: Cambridge Univ. Press, 1987), pp. 301–324, analyzes one particularly well-documented fourteenth-century example.

from experience and practice, medical masters displayed a continual preoccupation with understanding how to express that experience and how to fit it into the body of natural knowledge they were gradually reacquiring from the classical past. Far from being a period of sterility ended by a burst of radical innovation in the late Renaissance, their intellectual world was a time of continuing growth and development—shaped on the one side by new institutions and expanding scientific horizons and on the other by a new disease environment (created by the spread of plague from the mid-fourteenth century, for example, or by that of syphilis from the 1490s on). We are coming to see the twelfth century as the starting point for a long medical Renaissance,[3] for an intellectual tradition that, evolving over four hundred years, would eventually identify, formulate, and begin to resolve a broad range of issues defining a new science of medicine.

This new understanding of a "long Renaissance" of medical learning has grown out of a range of equally new areas of scholarly investigation.[4] The organization and evolving curriculum of most of the principal medical faculties—Salerno, Montpellier, Bologna—are now more clearly understood; scholars have begun to map out the stages by which Greek and Arabic texts were recovered by the schools, and by which this textbook knowledge was applied and articulated.[5] We

[3] Danielle Jacquart speaks of a medical Renaissance of the eleventh and twelfth centuries in her "A l'aube de la renaissance médicale des XIe–XIIe siècles: L'*Isagoge Johannitii* et son traducteur," *Bibliothèque de l'Ecole des Chartes,* 1986, *144*:209–240.

[4] The highly selective references in the remainder of the introduction are supplemented by the footnotes in the individual articles that follow. These notes constitute an extensive bibliography of recent work on medical knowledge in the medieval and Renaissance centuries. A general overview is also provided by Siraisi, "Some Current Trends" (cit. n. 1). Valuable current bibliographies of texts and studies on medieval and Renaissance as well as ancient medicine are included in the *Society for Ancient Medicine and Pharmacy Newsletter* (editor, Wesley D. Smith, University of Pennsylvania).

[5] Once again, only some principal items of work appearing since 1975 can be indicated here. Medical faculties: Paul Oskar Kristeller's seminal articles on Salerno are collected, with new appendixes and lists of manuscripts, in his *Studi sulla Scuola medica salernitana* (Naples: Istituto Italiano per gli Studi Filosofici, 1986); Luis García-Ballester, "Arnau de Vilanova (c. 1240–1311) y la reforma de los estudios médicos en Montpellier (1309): El Hipócrates latino y la introducción del nuevo Galeno," *Dynamis,* 1982, *2*:97–158; Nancy G. Siraisi, *Taddeo Alderotti and His Pupils: Two Generations of Italian Medical Learning* (Princeton, N.J.: Princeton Univ. Press, 1981); Jerome J. Bylebyl, "The School of Padua: Humanistic Medicine in the Sixteenth Century," in *Health, Medicine, and Mortality,* ed. Webster (cit. n. 1), pp. 355–370; Tiziana Pesenti, *Professori e promotori di medicina nello Studio di Padova dal 1405 al 1509: Repertorio bio-bibliografico* (Padua/Trieste: Edizioni Lint, 1984). Texts of Greek and Arabic origin: Pearl Kibre, *Hippocrates Latinus: Repertorium of Hippocratic Writings in the Latin Middle Ages* (New York: Fordham Univ. Press, 1985); John M. Riddle, "Dioscorides," in *Catalogus translationum et commentariorum: Medieval and Renaissance Latin Translations and Commentaries,* Vol. IV, ed. F. Edward Cranz and Paul Oskar Kristeller (Washington, D.C.: Catholic Univ. America Press, 1980), pp. 1–143; Eugene F. Rice, Jr., "Paulus Aegineta," *ibid.,* pp. 145–191 (articles on the *fortuna* of other ancient medical authors will appear in future volumes); Richard J. Durling has in preparation a *Galenus Latinus,* which will include text editions as well as a repertory of manuscripts and early editions (portions of this project have appeared as journal articles and separately published editions); Roger French, "*De iuvamentis membrorum* and the Reception of Galenic Physiological Anatomy," *Isis,* 1979, *70*:96–109; Vivian Nutton, *John Caius and the Manuscripts of Galen* (Cambridge: Cambridge Philological Society, 1987); Nancy G. Siraisi, *Avicenna in Renaissance Italy* (Princeton, N.J.: Princeton Univ. Press, 1987). A useful guide to bibliographies of current work on manuscript studies and to printed catalogues of manuscripts in the field of medieval and early Renaissance medicine is contained in Peter Murray Jones, "Medical Books before the Invention of Printing," in *Thornton's Medical Books, Libraries, and Collectors,* ed. Alain Besson (3rd rev. ed., Aldershot/Brookfield, Vt.: Gower, 1990), pp. 1–29. The texts and translations appearing in the revived Corpus Medicorum Graecorum have included a number of items of importance for the study of sixteenth-century medicine. Finally, Paul Oskar Kristeller, *Iter Italicum: A Finding List of Uncatalogued or Incompletely Catalogued Humanistic Manuscripts of the Renaissance in Italian and Other Libraries* (London: Warburg Institute, 1963–), contains a good number of medical items.

have careful studies—in a few cases critical editions[6]—of some writings (commentaries or independent texts) by the leading figures of this medical Renaissance, which show us their authors attempting to integrate their medical learning with other intellectual movements, first Aristotelian natural philosophy and then humanism as well. And we are beginning to see attempts to offer a synthetic interpretation of medical method or ideas in particular professional or intellectual milieus.[7]

The essays presented in this volume have been designed to convey a sense of the intellectual unity and coherence of the learned tradition within this "long Renaissance" of medicine, while introducing readers to some of the new themes being explored in current scholarship: how such things as textual traditions, pedagogical techniques, institutional frameworks, and relations with other disciplines and with the extra-academic world conditioned and shaped the subject as it evolved. Broadly speaking, we can divide this renaissance into two phases: first, a period of interaction (from the twelfth to the early fourteenth century) between medicine and intellectual developments in other areas that led to rapid theoretical consolidation and disciplinary formation; second, the articulation of this framework and the exploration of intellectual innovation from the mid-fourteenth century to the end of the sixteenth.

In the essays that follow, the authors work out in detail some of the implications of this periodization (without pretending to an exhaustive treatment). The period 1100–1350 was shaped by the need to make sense of the newly available Greek medical literature, organized and extended with understanding and often originality by Arabic-language authors.[8] It is convenient to view this literature as entering into Latin in two principal waves of translation: one associated with Constantine the African (d. ca. 1087), communicating, among other works, the general work on medicine of Haly Abbas (ᶜAlī ibn al-ᶜAbbās al-Majūsī, fl. 10th century) known in Constantine's version as the *Pantegni;* the other, a hundred years later, dominated by the translations ascribed to Gerard of Cremona, including not only numerous major writings of Galen, but also those of Rhazes (al-Rāzī, d. 925) and Avicenna (al-Ḥusain ibn ᶜAbdallāh Ibn Sīnā, d. 1037). Latin medical writers had somehow to assimilate this continually expanding body of material, to establish a conceptual framework for it, and to develop techniques for instruction and communication.

Thus Jerome Bylebyl shows how the Platonizing philosophers of the twelfth century, convinced that natural science (*physica*) should give first attention to

[6] *Arnaldi de Villanova Opera medica omnia,* ed. Luis García-Ballester, J. A. Paniagua, and Michael R. McVaugh (Granada/Barcelona: Univ. Barcelona, 1975–), is in the process of providing critical editions of all the writings on medicine of a leading academic physician and intellectual figure of the late thirteenth and early fourteenth centuries. This modern enterprise thus extends to a different area of medieval medical writing the task pioneered by Julius Leopold Pagel and Karl Sudhoff in their work on surgical texts. The need for, and lasting value of, text editing is certainly demonstrated by the extent to which historians of medieval surgery still make use of the editions produced by Pagel and Sudhoff in the late nineteenth and early twentieth centuries.

[7] Siraisi, *Taddeo Alderotti* (cit. n. 5), is intended as an endeavor of this kind.

[8] The medical texts that circulated in western Europe during the early Middle Ages (which were by no means negligible, though fewer in number, slighter, and less theoretically oriented in content than those that became available from the late eleventh century) of course also derived from the medicine of Greco-Roman antiquity. No attempt is made here either to give an account of early medieval medicine or to discuss the social and economic transformations underlying the rise of university faculties of medicine, medical professionalization, and so on.

man, to physiological explanation, helped confer the name and thereby a lofty status on the new theoretical medicine. It was at this time that Salerno, in southern Italy, became the first important center to revive medical learning: Mark Jordan illustrates some of the pedagogical and authorial devices used by Salernitan medical authors to transform medicine into a subject elaborated by scholastic treatment and embedded in a new matrix, that of Aristotelian natural philosophy.

During the thirteenth century the setting of the new universities allowed medical faculties of importance to take root at Bologna, Paris, and Montpellier, but also led to the emergence of tensions between natural-philosophical and medical knowledge; Michael McVaugh examines the attempt at Montpellier in the years around 1300 to identify and define a specifically medical knowledge dependent, in part, on the empirical results of practice. The strength of this new type of institutional learned medicine is illustrated in the essay by Luis García-Ballester and his collaborators showing that in the fourteenth century some Jewish practitioners, heirs to their own rich medical tradition (based in large part on essentially the same Greco-Arabic medicine as that of the Latins), recognized the intellectual appeal—and doubtless the social and professional usefulness—of the new Latin scholastic forms of medical learning.

Between the later fourteenth and the sixteenth centuries development took place within an established framework, as innovations in textual analysis and attention to clinical experience both served as stimuli. Chiara Crisciani analyzes the extent to which different genres of medical writing (commentaries, guides to practice, works on surgery, and pharmacological treatises) produced during the thirteenth to the fifteenth centuries reveal a consciousness of the past as past, and a recognition of the possibility—or desirability—of change. In investigating the attitudes of fifteenth-century physicians to plague and to medical astrology (an ancillary discipline the practice of which became much more widespread after 1300), Danielle Jacquart explores the connections between new issues, individual judgment, and the study of the particular; by juxtaposing one French and two Italian authors, she offers insights on the relation between different medical traditions (a theme also touched on by McVaugh and García-Ballester) and suggests by implication an emerging predominance of Italian over French medicine.

In this same period after 1350 an enlarged vernacular medical literature began to emerge. In addition to compilations of remedies, it included translations of Latin works that embodied the new scholastic medicine and coined a new technical vocabulary; compendiums and expositions based on Latin sources or sources translated from Latin; and treatises composed directly in the vernacular. In this way the assimilation and reformulation of medical knowledge in the learned (academic) world—the theme of the present volume—began to be recapitulated in extra-academic texts. Vernacular authors and translators selected, adapted, and reworked texts and information from the Latin tradition according to the needs of a diversified world of medical and surgical practice, a process that (although not treated here) is also receiving considerable scholarly attention.[9] Medical teachings had diffused widely in Latin during the prior stage

[9] Again, only a small selection of pertinent titles published since 1975 can be listed here, but see, e.g., Gerhard Baader and Gundolf Keil, eds., *Medizin im mittelalterlichen Abendland* (Darmstadt: Wissenschaftliche Buchgesellschaft, 1982); Keil et al., eds., *Fachprosa-Studien: Beiträge zur mittelalterlichen Wissenschafts- und Geistesgeschichte* (Berlin: E. Schmidt, 1982); the monograph series

of the "long medical Renaissance": Salernitan concepts stimulated the philosophical thought of northern France in the twelfth century; Pietro d'Abano brought a Parisian formation to Padua early in the fourteenth. The vernacular writings of the late fourteenth and the fifteenth centuries ensured that this international medical culture would enjoy a still wider audience, until linguistic drift helped evolve distinct traditions of popular medicine in the various territories and regions of early modern Europe. Studies of this process of transmission are accumulating more and more telling evidence to invalidate the stereotype of completely separate medical worlds of abstract, scholastic theorizing and empirical or folkloric practice. It becomes clear that practitioners at all levels of medical activity about which we have any information drew on essentially common traditions of theory and practice, though with highly varying degrees of expertise, sophistication, and intellectualization. Furthermore, it is also becoming evident that the quantities of vernacular and more or less "popular" medical publications that appeared in the sixteenth century expanded a genre already in existence in the age of manuscript books and were not an innovation due to the introduction of printing.

In general, recent work on sixteenth-century medicine is characterized by a lack of insistence on a sharp break with the medieval or early Renaissance past, and by an interest in looking at a broad picture of medical knowledge and practice. Vesalius and Paracelsus are by no means exhausted as subjects of study,[10] but their towering figures—and the subjects of anatomy and of Paracelsianism in all its guises—should not be allowed to obscure the rest of the view. The final three articles in this volume explore the ambiguous relation of medical knowledge in the sixteenth-century university world to its past. Humanistic insistence on the importance of the direct study of Greek medical texts, in evidence from the late fifteenth century, represented a heightening of dependence on the authority of the most ancient past; yet ironically the intensive study of ancient authorities soon led with increasing frequency to the self-conscious repudiation of one or more aspects of established medical tradition. Nancy Siraisi's study of Giovanni Argenterio examines the career of a radical critic of Galen who, like Vesalius, received an important part of his training in the atmosphere of intense enthusiasm for a new and improved Galenism that characterized the medical faculty of the University of Paris during the 1530s. Siraisi endeavors to show how Argenterio's institutional setting as a university professor, his need for patronage, and his own formation in Galenist theory and text exposition, as well as his personal experiences as a practitioner, interacted to shape his critique. In the relatively flexible and open academic medical world of the second half of the sixteenth century Argenterio's were among various more or less radical ideas

Würzburger medizinhistorische Forschungen, ed. Keil (1975–), which includes editions of a number of German vernacular texts; Faye M. Getz, "Gilbertus Anglicus Anglicized," *Medical History*, 1982, *26*:436–442; Linda E. Voigts, "Scientific and Medical Books," in *Book Production and Publishing in Britain, 1375–1475*, ed. Jeremy Griffiths and Derek Pearsall (Cambridge: Cambridge Univ. Press, 1989), pp. 345–402; Voigts and Michael R. McVaugh, *A Latin Technical Phlebotomy and Its Middle English Translation* (Transactions of the American Philosophical Society, 74.2) (Philadelphia, 1984); and Tony Hunt, *Popular Medicine in Thirteenth-Century England* (Cambridge: D. S. Brewer, 1990). The publication of Linda E. Voigts and Patricia Kurtz's "Catalogue of Incipits of Scientific and Medical Writings in Old and Middle English" (forthcoming) will ensure still further study of one particular vernacular tradition.

[10] We make no attempt here to indicate bibliography on the large subfields of Renaissance anatomy and Paracelsianism.

that became part of the common currency of debate. Yet despite such debates, and despite widespread and well-known innovations in some aspects of teaching (for example, in anatomy, medical botany, and, in some settings, the beginnings of clinical instruction), medical pedagogy in the second half of the century remained profoundly conservative in important ways.

Richard Durling's presentation of a propaedeutic text by Girolamo Mercuriale is highly instructive in this respect. Mercuriale was a celebrated professor at Padua, in his time the most important medical school in Europe; students were attracted to Padua especially by its reputation for up-to-date teaching in anatomy and practical medicine. Mercuriale himself engaged vigorously in current medical debates and wrote pioneering—at least in their choice of subject matter—studies of gymnastic medicine and skin diseases. But much of the advice Mercuriale had to give young students would have been appropriate three or four hundred years before. In the text presented by Durling the impact of humanism has chiefly been to extend the list of books and authors that the neophyte must master—not only ancient medical writers in new translations and editions and new commentaries, but poets and historians as well. Yet medieval authorities and commentators have not disappeared from the list. Good medical education, for Mercuriale, still involved overwhelming feats of bookish study of texts and commentaries. Mercuriale's mnemonic tricks belong recognizably to the same tradition as the exegeses of the Salernitans or the concordances of the thirteenth-century schools; but his recommendation that students keep commonplace books shows that a method of controlling information taught in Renaissance schools had penetrated medical training.[11]

Durling's contribution to this volume also highlights the continuing need to edit and publish the sources of the medical thought of the period, so many of which remain in manuscript, and, where desirable, to translate them for a wider audience. Only in this way can we hope eventually to have a fully satisfactory understanding of the "long medical Renaissance."

By contrast, the final study in the volume, Vivian Nutton's paper on the reception of Fracastoro's ideas about contagion, deals with a figure who has traditionally been accorded the status of a major medical innovator who was overlooked and ignored by his contemporaries. Nutton dismisses this stereotype in favor of an analysis of the fate of Fracastoro's theories of disease transmission that illustrates the complex interaction of innovation and tradition in the intellectual world of sixteenth-century medicine. The article provides abundant evidence that Fracastoro's theories, far from being ignored by his contemporaries, were widely disseminated and eagerly debated by learned medical men in many parts of Europe, both in and outside the university milieu. But the very flexibility, diversity, and openness of the debate over the causes and transmission of diseases such as plague and syphilis, the more or less uncontrolled proliferation of medical vocabulary and suggested causal mechanisms, and the underlying confidence in Galenic physiology and pathology all served to make it possible for many sixteenth-century physicians to absorb Fracastoro's account of "contagion" and "seeds" into a generally Galenic account of disease transmission.

These final papers deal with individuals who have been assigned not only to

[11] See Ann M. Blair, "Restaging Jean Bodin: The *Universae Naturae Theatrum* (1596) in Its Cultural Context" (Ph.D. diss., Princeton Univ., 1990), pp. 4–5.

the "long medical Renaissance" of this volume but also to a distinct "medical Renaissance of the sixteenth century."[12] We certainly do not dispute the existence of the latter renaissance. But did the sixteenth century mark a real break with medicine's past? The question evades an easy answer. Rather, it calls for the painstaking examination of the details of particular debates, careers, and milieus, an undertaking to which we hope this volume may contribute. But to speak in general terms, the pace and scale of intellectual development certainly seemed to contemporaries (as they do to us) to have greatly increased. Many physicians undoubtedly perceived themselves as engaged in a movement of renewal that involved challenging texts and authors long accorded unquestioned acceptance. Much fuller knowledge of the surviving textual heritage of Greek medicine was unquestionably achieved. And in due course, challenges first addressed to medieval authorities were put to the revered Greeks as well. But in intellectual as well as institutional terms, the actual framework of medical knowledge remained largely intact, even at the century's end: "The most obvious signs of change," it has recently been concluded, "[lie] in the rhetoric and the ideological positions adopted by the medical men of the sixteenth century."[13]

Of course there were important, and ultimately revolutionary, changes. The antecedents of a new physiological understanding must be sought in sixteenth-century anatomy and, indeed, sixteenth-century studies of Greek anatomical, physiological, and biological texts. Yet Vesalius's repudiation of aspects of Galenic anatomy was initiated as an endeavor to perfect an existing system of knowledge. Only the attacks of Paracelsus and some of his followers attempted a wholesale repudiation of that system, and their ideas supplemented or provided an alternative to but were not able to supplant orthodox medical learning. Moreover, Paracelsianism itself incorporated many elements from earlier traditions. Rather than regard the medicine of the sixteenth century as representing a sharp break with the past, then, it seems—as the title of this collection is intended to imply—preferable to see it as constituting simultaneously the final stage in the formation and the first stage in the dissolution of the medical Renaissance.

In conclusion, first a few notes as to editorial policy. We have chosen arbitrarily to refer through this volume to medieval and Renaissance authors by vernacular names (Arnau de Vilanova, Pietro d'Abano), except in a few rare cases in which an English or Latin name is much more familiar. We have not, however, tried to standardize completely all references to citations from these authors, and therefore different contributors to this collection may cite different early printed editions of the same work. We have consciously tried to employ as little Latin terminology as possible, even though English translation of technical terms is bound to be anachronistic and somewhat inaccurate; in only a few cases where the Latin term means something significantly different from the English cognate (notably *scientia*/science) have we let the Latin term stand. The Latin orthography in quotations and titles may also vary, reflecting that in the original works

[12] Andrew Wear, Roger K. French, and Iain M. Lonie, eds., *The Medical Renaissance of the Sixteenth Century* (Cambridge: Cambridge Univ. Press, 1985).

[13] *Ibid.*, p. xi.

and the habitual spelling of the periods from which they come. Finally, space limitations have required us to omit Latin texts from most footnotes.

This issue of *Osiris* owes much to the following people, whom we would like to thank: Arnold Thackray, for encouraging us to undertake the project and for giving helpful suggestions about its formation; our authors, all of whom have been patient collaborators in the preparation of this volume; the referees of the collection, for much constructive criticism; Julia A. McVaugh, for compiling the index; and Frances Coulborn Kohler and the History of Science Society Publications Office staff, for tireless assistance and meticulous attention to detail that have enhanced the scholarly quality and the presentation of these essays.

The Medical Meaning of *Physica*

*By Jerome J. Bylebyl**

DURING THE LONG INTERVAL from the fifth century B.C. to the early modern period, the classical traditions of medicine and natural philosophy underwent repeated interactions, resulting in significant overlaps of content and method. The two disciplines nevertheless retained distinct identities through most of this coexistence, with the notable exception of a medieval phase in which scholars tended to conflate them.[1] By the ninth century the word *physica,* without having lost its classical meaning of "natural philosophy," was beginning to displace *medicina* as the designation of medical learning (as distinct from practice), and during the twelfth century *physica* came to refer to both the learning and the practices associated with rational medicine. Similarly, by the early Middle Ages *physici* (natural philosophers) were being cited as literary authorities on specifically medical issues,[2] and during the twelfth century the singular *physicus* became a preferred alternative to *medicus* (healer) for designating an individual medical expert. After prevailing for several additional centuries, these usages passed out of currency in continental Europe, stranding the English words "physic" and "physician" as major vestiges.[3]

The demise of medical *physica* in Latin and the Romance vernaculars is consistent with a general return to classical usage during the Renaissance, but as Paul Oskar Kristeller has emphasized, the medieval terminology is of great inter-

* Institute of the History of Medicine, The Johns Hopkins University, 1900 East Monument Street, Baltimore, Maryland 21205.

[1] Any discussion of this issue must begin with the seminal passage in Paul Oskar Kristeller, "The School of Salerno: Its Development and Contribution to the History of Learning," *Bulletin of the History of Medicine,* 1945, *17*:138–194, on pp. 159–160. Of the earlier literature see Louis Dubreuil-Chambardel, *Les médecins dans l'ouest de la France, aux XIe et XIIe siècles* (Paris: Société Française d'Histoire de Médecine, 1914); and Loren C. MacKinney, *Early Medieval Medicine, with Special Reference to France and Chartres* (Baltimore: Johns Hopkins Press, 1937). Of more recent literature, the publications of Brian Lawn (cited below, n. 6) are crucial. Darrel W. Amundsen has surveyed the place of medicine in traditional classifications of knowledge in "Medicine and Surgery as Art or Craft: The Role of Schematic Literature in the Separation of Medicine and Surgery in the Late Middle Ages," *Transactions and Studies of the College of Physicians of Philadelphia,* 5th Ser., 1979, *1*:43–57. The essays in Peter Dronke, ed., *A History of Twelfth-Century Western Philosophy* (Cambridge: Cambridge Univ. Press, 1988), epitomize and extend a vast body of pertinent scholarship; individual essays from this volume are cited below.

[2] MacKinney, *Early Medieval Medicine,* p. 131, takes such references to mean that by the ninth century living practitioners were already being called *physici.* Kristeller, "School of Salerno," p. 159, n. 77, maintains, more plausibly, that *physici* in these early instances are rather a class of authors.

[3] Faye M. Getz is investigating the later medieval tradition of English physic; see "Charity, Translation, and the Language of Medical Learning in Medieval England," *Bull. Hist. Med.,* 1990, *64*:1–17. Harold J. Cook has traced this ideology into the early modern period; see "Physicians and the New Philosophy: Henry Stubbe and the Virtuosi-Physicians," in *The Medical Revolution of the Seventeenth Century,* ed. Roger French and Andrew Wear (Cambridge: Cambridge Univ. Press, 1989), pp. 246–271. For the prevalence of this terminology outside England see Katharine Park, *Doctors and Medicine in Early Renaissance Florence* (Princeton, N.J.: Princeton Univ. Press, 1985), esp. pp. 58–59.

est precisely as a departure from such norms.[4] My present aims are to consider how *physica* gravitated toward a medical meaning during the early Middle Ages, and how this meaning was understood during the twelfth century, a period that Kristeller has identified as pivotal in this regard. This was a time when new translations were greatly enriching the tradition of theoretical medicine in western Europe. It was also the first of the medieval centuries in which natural science loomed large among scholarly interests. But it was also the last such period in which Aristotle's *libri naturales*—his wide-ranging works on natural science—were not generally available as possible models.

Scholars were thus challenged to create their own *physica* out of the less comprehensive resources at their disposal. Lacking a definitive taxonomy, they tended to construe *physica* in generic terms, as the rational explanation of "natures" by means of underlying principles, notably the four elements and the four primary qualities.[5] Brian Lawn has shown that such explanatory *physica* was represented by a substantial literature dealing with a mélange of specific questions, but the best examples of *physica* known to twelfth-century scholars were the newly available textbooks of rational medicine.[6] This coincidence of method and subject was incorporated into a distinct ideology according to which *physica* had a moral imperative to devote its explanatory powers first of all to the human body and its health-related needs. Personification and allegory, so vital to medieval discussions of higher learning, were also important in maintaining a balance between the generality of *physica* and its application to medical ends.

This was arguably the mainstream view of natural philosophy at mid century, but gradually European scholars became familiar with the *libri naturales*. These represented a definitive natural philosophy in which medicine per se did not loom large and whose principal end was the attainment of truth. Equally important, these works arrived in tandem with a Greco-Arabic tradition that, although it closely linked the study of natural science with that of rational medicine, positively exaggerated the differences between these two disciplines.[7] These conflicting intellectual and moral values produced polarization, as *physica* long survived as a distinctly medical ideology, while its close equivalent *scientia naturalis* took on a more austere meaning in relation to Aristotelian science.[8]

[4] Kristeller, "School of Salerno" (cit. n. 1), pp. 159–160. By "medical *physica*" I mean the use of the word *physica* alone to convey both a philosophical and a medical meaning. Rarely, the modifier *medicinalis* may be added to *physica*. See the ninth-century commentator on Martianus Capella discussed below; for the twelfth century see *Dominicus Gundissalinus De divisione philosophiae*, ed. Ludwig Baur (Beiträge zur Geschichte der Philosophie des Mittelalters, 4.2–3) (Münster: Aschendorff, 1903), p. 83.

[5] Richard McKeon, "Medicine and Philosophy in the Eleventh and Twelfth Centuries: The Problem of Elements," *Thomist*, 1961, *24*:211–256; and Charles Burnett, "Scientific Speculations," in *Twelfth-Century Philosophy*, ed. Dronke (cit. n. 1), pp. 151–176, esp. pp. 166–176.

[6] Brian Lawn, *The Salernitan Questions: An Introduction to the History of Medieval and Renaissance Problem Literature* (Oxford: Clarendon Press, 1963), esp. pp. 1–72; and Lawn, ed., *The Prose Salernitan Questions* (London: Oxford Univ. Press for the British Academy, 1979). See also Burnett, "Scientific Speculations," p. 168.

[7] Nancy G. Siraisi, *Arts and Sciences at Padua: The Studium of Padua before 1350* (Toronto: Pontifical Institute of Mediaeval Studies, 1973), pp. 109–112, 152; Siraisi, "Taddeo Alderotti and Bartolomeo da Varignana on the Nature of Medical Learning," *Isis*, 1977, *68*:27–39; and Siraisi, *Taddeo Alderotti and His Pupils: Two Generations of Italian Medical Learning* (Princeton, N.J.: Princeton Univ. Press, 1981), pp. 118–146.

[8] E.g., Gundissalinus, who was one of the first scholars in Europe to be familiar with the full taxonomy of Aristotelian natural science, used the word *physica* twice with reference to natural

There is much to be learned about this later interplay, but, as noted, my discussion will focus on the earlier, formative period. I shall begin by considering certain large features of ancient natural philosophy and medicine that facilitated the later reorientation of *physica* along medical lines. Then I shall examine a series of Latin authors from late antiquity to the twelfth century to trace some of the stages by which the word *physica* came to compete with *medicina* as the designation of medical learning. Finally, I shall consider several twelfth-century authors to see how knowledge drawn from works on rational medicine was incorporated into the evolving framework of general natural philosophy and, conversely, how *physica* emerged as the accepted name for the new rational medicine itself.

I. ANCIENT ROOTS OF MEDICAL *PHYSICA*

In classical Greek, the consideration of nature (*physis*) could be designated by various terms, including *peri physeōs* (concerning nature), *physiologia* (the study of nature), and several forms of the adjective *physikos* (natural). In the *Physics* Aristotle used the neuter plural *ta physika* (natural things) to designate his general account of movement and change. But in referring to natural science in a more comprehensive sense, he favored the feminine singular *hē physikē,* a contraction of *hē physikē epistēmē* (natural knowledge) or *hē physikē philosophia* (natural philosophy).[9] These phrases could be rendered into Latin as *scientia naturalis* or *philosophia naturalis,* and Latin authors could also refer to *rerum natura* or *naturae* (the nature or natures of things). But like many other terms for higher learning, the feminine singular *physica* was directly borrowed into classical Latin, and it was this form that later acquired the medical connotations under discussion here.

This word is frequently translated into modern English as "physics," but the present subject will seem less strange if its broader association with nature is kept in mind. Indeed, over an extended formative period Greek medical theory could not be sharply distinguished from the philosophers' inquiry "on nature." Thus the Hippocratic treatise *On the Nature of Man* sets forth an avowedly medical theory of the four humors but goes on to relate this to the cosmic qualities of hot, cold, wet, and dry, which determine the "natures" of all existing things. Conversely, Plato's *Timaeus* is a philosopher's account of the cosmos and the four elements which goes on to a consideration of man and medicine.[10]

---

science and once with reference to medicine but otherwise adhered to *scientia naturalis* and *medicina,* respectively; see his *De divisione philosophiae* (cit. n. 4), pp. 15, 27, 83; pp. 19–28, 83–89. See also how gingerly Aquinas avoided using the word *physici* when commenting on the very passage (at the beginning of *De sensu et sensato*) where Aristotle discusses *physici* and *medici:* Thomas Aquinas, *Opera omnia* (Parma, 1852–1873; rpt. New York: Musurgia, 1949), Vol. XX, pp. 145–148. See also Nancy G. Siraisi, "The Medical Learning of Albertus Magnus," in *Albertus Magnus and the Sciences: Commemorative Essays, 1980,* ed. James A. Weisheipl (Toronto: Pontifical Institute of Mediaeval Studies, 1980), pp. 379–404, esp. pp. 379–384; and Heinrich Schipperges, "Zur Unterscheidung des 'Physicus' vom 'Medicus' bei Petrus Hispanus," *Asclepio,* 1970, *22*:321–327.

[9] Cf., e.g., *Physics* 8.1 with *Metaphysics* 6.1. See also Hermann Bonitz, *Index Aristotelicus,* in *Aristotelis Opera,* ed. Immanuel Becker, 5 vols. (Berlin, 1831–1870), Vol. V, pp. 834–835.

[10] G. E. R. Lloyd, intro. to *Hippocratic Writings,* ed. Lloyd, trans. J. Chadwick and W. N. Mann (New York: Penguin, 1978), pp. 37–45; Hippocrates, *On the Nature of Man,* esp. 3, 4; and Plato, *Timaeus,* esp. 69–92.

Plato conceived of the cosmos and man as modeled upon one another, necessitating an explanation of why the human body is subject to disease and death, and this in turn led to advice on maintaining health and curing disease. Thus in absolute terms, human physiology and medicine might loom rather large within an anthropocentric natural philosophy.

But from the time of Aristotle onward, natural science and medicine became more clearly demarcated. To be sure, Aristotle himself pointed out that the more enlightened medical experts have much to say about natural science, while the more accomplished students of nature extend their inquiries to medical issues. However, even in acknowledging this congruence, he presupposed his own division of philosophy into theoretical and practical branches.[11] Theoretical philosophy has the attainment of truth as its primary object, and Aristotle placed natural philosophy on this side, rather than among the practical disciplines, notably ethics, which provide guidance for the general conduct of life. Aristotle assigned medicine to yet a third group, namely, the "productive arts," which aim at a specific result, such as health. Natural philosophers are therefore free to consider medical issues, but they ought not to be in the business of offering medical advice. Furthermore, while the *libri naturales* do contain much that is relevant to medicine, Aristotle's own promised account of health and disease failed to appear.[12] Thus while Aristotelian natural philosophy was by no means antimedical, it was massively nonmedical inasmuch as it dealt with so much else. Conversely, as Hellenistic medical authors went on to develop their own subject systematically, there was progressively less incentive for philosophers to conduct similar inquiries.[13]

But medicine and natural philosophy never ceased to have important topics in common, and so the lines were kept open for a closer integration in the postancient period. In the Greek-speaking East and, later, in the Islamic world, this synthesis included Aristotelian philosophy, which retained its nonmedical identity even as it entered into a close alliance with medicine. In the Latin West, however, the synthesis was initially based upon non-Aristotelian sources that were more permissive of an outright coalescence, so as to produce the medical *physica* under consideration here.[14]

[11] Aristotle, *On Sense and Sensible Objects* 1, 436a17–436b2; and *On Respiration* 21, 480b22–31. For discussions of these two passages in the thirteenth century see the references to Albertus (Siraisi, "Medical Learning of Albertus Magnus") and Aquinas in n. 8 above. Charles Schmitt examines Renaissance interpretations in "Aristotle among the Physicians," in *The Medical Renaissance of the Sixteenth Century,* ed. Andrew Wear, R. K. French, and I. M. Lonie (Cambridge: Cambridge Univ. Press, 1985), pp. 1–15. For Aristotle's broader views on the divisions of learning see James A. Weisheipl, "The Nature, Scope, and Classification of the Sciences," in *Science in the Middle Ages,* ed. David C. Lindberg (Chicago: Univ. Chicago Press, 1978), pp. 461–482, esp. pp. 464–468.

[12] Schmitt, "Aristotle among the Physicians" (cit. n. 11), pp. 2–3.

[13] See Heinrich von Staden, *Herophilus: The Art of Medicine in Early Alexandria* (Cambridge: Cambridge Univ. Press, 1989), esp. pp. 89–114. See also Ludwig Edelstein, "The Relation of Ancient Philosophy to Medicine," in his *Ancient Medicine,* ed. Owsei Temkin and C. Lilian Temkin (Baltimore: Johns Hopkins Press, 1967), pp. 349–366; and Owsei Temkin, "Greek Medicine as Science and Craft," in his *The Double Face of Janus and Other Essays in the History of Medicine* (Baltimore: Johns Hopkins Univ. Press, 1977), pp. 137–153.

[14] On the East see Schmitt, "Aristotle among the Physicians," pp. 2–3; and Owsei Temkin, *Galenism: Rise and Decline of a Medical Philosophy* (Ithaca, N.Y.: Cornell Univ. Press, 1973), pp. 51–94. Lawn, *Salernitan Questions* (cit. n. 6), pp. 1–72, has a thorough account of the sources of European *physica* down through the twelfth century. My present concern is to identify the large factors that made the terminology itself seem appropriate to twelfth-century scholars. In this section I refer primarily to the ancient sources; for corroboration see Sections II and III.

One of these sources was the tradition of natural history, which, in contrast to classical natural philosophy, was concerned with description rather than explanation. This tradition found a consummate exponent in the Roman author Pliny (ca. A.D. 23–79), whose *Historia naturalis* was available throughout the Middle Ages. Altogether, Pliny encompassed a vast array of subjects, but his largest single topic was the medicinal properties of plants, animals, and minerals. He himself did not use the word *physica* in relation to his subject, but his references to a beneficent *Natura* are all-pervasive, thus making it evident to later readers that *physica,* the study of nature, could be a rich trove of medicinal lore.[15] However, Pliny idealized simple herbal medicine and had little use for the theories of Greek doctors, whereas for medieval scholars the word *physica* came to stand for theoretical medicine in the Greek tradition.[16] Thus while it had a distinctly Plinian strand, medical *physica* was not an exclusively Plinian ideology.[17]

Allusions to nature were also pervasive within rational medicine itself, and this permitted medieval scholars to make direct and substantial connections with rational *physica.* We have seen that the Hippocratic author could discuss "the nature [*physis*] of man" as the foundation of medicine, while Galen (ca. A.D. 129–199) could similarly refer to "the *physiologia* of man" as knowledge shared in common by philosophers and medical authors.[18] In more formal schemata, other Greek authors placed *to physiologikon* (the natural scientific) first among the parts of rational medicine.[19] Several early medieval medical texts incorporated Latin versions of such outlines, which called this division variously *physiologicon, physiologica,* and *physica,* while in works that were newly translated in the late eleventh century this first part of medicine was identified both as *res naturales* (natural things) and as *physica.*[20]

[15] See esp. the summaries of Books 20–33 as given in Pliny, *Natural History,* Book 1.

[16] *Ibid.,* 25.1–6, 26.6–9, and 29.1–8. For an overview see William Stahl, *Roman Science: Origins, Development, and Influence to the Later Middle Ages* (Madison: Univ. Wisconsin Press, 1962), pp. 101–119; on medicine, p. 103.

[17] *A fortiori,* it seems implausible that medical *physica* owed a great deal to the ancient use of the neuter plural *physica* to refer to occult practices such as charms and amulets. Cf. John of Salisbury's attitude, in the *Policraticus,* toward occult *physica* with his attitude toward rational *physica: Ioannis Saresberiensis episcopi Carnotensis Policratici sive De nugis curialium et vestigiis philosophorum, Libri VIII,* ed. Clement C. I. Webb (Oxford: Clarendon Press, 1904), Vol. I, 415c–d, 417c–d, 475c–476a. On the general background to such occult *physica* see R. P. Festugière, *La révélation d'Hermes Trismégiste,* Vol. I: *L'astrologie et les sciences occultes* (3rd ed., Paris: Gabalda, 1950), pp. 187–216. The case for this derivation of medical *physica* is made by Alf Önnerfors, *In medicinam Plinii studia philologica* (Lund: CWK Gleerup, 1963), pp. 11–19. Önnerfors also presupposes a general Plinian background to medical *physica.*

[18] Galen, *Methodus medendi* 2.5 (*Claudii Galeni Opera omnia,* ed. K. G. Kühn [Leipzig, 1821–1833], Vol. 10, p. 107); cf. Galen, *De constitutione artis medicae* (Kühn, Vol. 1, p. 227) on *physiologia* as general natural philosophy.

[19] Pseudo-Galen, *Introductio* 7 (Kühn, Vol. 14, pp. 689–690); pseudo-Galen, *Definitiones medicae* 11 (Kühn, Vol. 19, pp. 351–352).

[20] See, e.g., a commentary on *De sectis,* discussed by Owsei Temkin in "Studies on Late Alexandrian Medicine, I: Alexandrian Commentaries on Galen's *De sectis ad introducendos,*" in *The Double Face of Janus* (cit. n. 13), pp. 178–197, esp. p. 189. Codex Ambrosianus G 108 inf., fol. 26r, shows the form *fisiologicon.* Temkin also cites Salvatore De Renzi, *Collectio Salernitana,* Vol. I (Naples, 1852), pp. 87–88, where a similar scheme is reproduced from a commentary on the *Aphorisms.* The form is *physiologica.* A third text is pseudo-Soranus, *Questiones medicinales,* discussed by Lawn, *Salernitan Questions* (cit. n. 6), p. 12. The text is printed in V. Rose, *Anecdota Graeca et Graecolatina,* 2 vols. (Berlin, 1864–1870), Vol. II, pp. 243–274, where the form (p. 251) is initially *physiologica* but then repeatedly *physica.* Pre-Salernitan codices containing these texts are listed by A. Beccaria, *I codici di medicina del periodo presalernitano* (Rome: Edizione di Storia e Letteratura, 1956), pp. 456, 462, 486. The eleventh-century translations include the *Isagoge, Pantegni,* and *Premnon physicon,* all of which are discussed in Section III, below.

Such surveys were concerned with the humors, parts, and functions of the human body in their "natural" or healthy state. These topics might seem to be clearly circumscribed in relation both to the rest of medicine and to general natural philosophy, but both boundaries could readily be breached. The word *physiologica* contains the element *logica,* which was known to be the defining characteristic of rational medicine. Hence "natural reasoning" (*physica ratio*) could be conceived as applying throughout rational medicine, including the understanding of the unnatural or diseased state, as well as the restoration and preservation of the natural state.[21] On the other hand, the surveys of *physiologica* began with the four elements and the four primary qualities, topics that belonged to general natural philosophy. Thus both terminology and substance might suggest that the whole first division of medicine, or, by extension, the whole of medical reasoning, belonged to natural philosophy *rather than* to medicine itself.

Since the medical tradition offered no sure grounds for resolving the latter ambiguity, it is highly relevant that medieval scholars had early access to the other side of Aristotle's symmetry, namely, the tradition of medicine within philosophy. Plato's *Timaeus* had gone on to exert a unique influence on both the philosophy and the medicine of later antiquity, and a substantial portion of the dialogue was translated into Latin by Calcidius in the fourth or fifth century A.D. Calcidius omitted the last part, including the more specifically medical sections, but in his accompanying commentary he made it clear that Plato had given a comprehensive account of the human body, just as he had done for the cosmos. Calcidius also claimed that *physici* as well as *medici* had investigated the body through dissection, and he gave priority to the *physici* in this regard. Furthermore, early on he enumerated all the topics discussed by Plato in the *Timaeus,* including, with regard to the human body, its humors, its parts and their functions, and its diseases and their cure.[22] Thus there was ample room for much or even all of rational medicine within a Platonic natural philosophy, and the doctrine of the microcosm could even make such knowledge seem crucial to the whole discipline.

The medieval legacy of Stoicism, conveyed by familiar authors such as Cicero, also contributed to medical *physica* by weakening some of the disciplinary boundaries that Aristotle had erected. Aristotle had determined that logic was not a part of philosophy per se, and as noted, he divided philosophy into theoretical and practical branches. The Stoics, however, reverted to a more integral view of philosophy, which they divided into *logica, ethica,* and *physica,* thus placing natural philosophy on an equal basis with two other disciplines that should be applied in daily life.[23] Furthermore, the Stoics held that *physica* itself had important practical implications, in that such knowledge was essential for choosing a way of life in harmony with nature.[24] They did not understand such choices

[21] See pseudo-Soranus, *Questiones medicinales,* p. 251.

[22] *Timaeus a Calcidio translatus commentarioque instructus,* ed. J. H. Waszink, in *Plato Latinus,* Vol. IV (London: Warburg Institute; Leiden: E. J. Brill, 1962), pp. 228, 256–258, and 60–61.

[23] The Stoic strands in twelfth-century thought are surveyed by Michael Lapidge, "The Stoic Inheritance," in *Twelfth-Century Philosophy,* ed. Dronke (cit. n. 1), pp. 81–112, esp. pp. 99–112. For the tripartite scheme see *ibid.,* p. 84. Heinrich von Staden has pointed out to me that this scheme is attributed to Xenocrates, the third head of the Platonic Academy. See W. K. C. Guthrie, *A History of Greek Philosophy,* Vol. V (Cambridge: Cambridge Univ. Press, 1978), p. 478. Guthrie also notes its apparent acceptance by Aristotle in *Topics* 105b20.

[24] Cicero, *De finibus* 3.72–73.

primarily in relation to bodily health, but medieval scholars found it possible to regard *ethica* and *physica* as two normative parts of philosophy, of which the first was responsible for the health of the soul, while the second oversaw the health of the body.

We can thus find in ancient learning a natural history tradition preoccupied with practical medicine, a medical tradition grounded in natural philosophy and human physiology, a Platonic tradition deeply concerned with the microcosm, and a Stoic tradition of practical or ethical natural philosophy. And these were also principal motifs of medical *physica* as it had evolved by the mid-twelfth century.

But this brief overview should not obscure the magnitude of the change that took place from ancient to medieval usage. The Stoics themselves, like their pre-Socratic forebears, would have identified natural philosophy first of all with cosmology, while even Pliny's *Natural History* began with the latter topic. Furthermore, Calcidius clearly decided to omit the medical sections of the *Timaeus* and thus to accentuate the cosmological. Conversely, in his dialogue *Saturnalia,* Macrobius (early fifth century) included as an interlocutor an expert on "human bodies" and "the theory of medicine [*medicinae ratio*]." At one point this character cites an anatomical misconception from the *Timaeus* as typical of what happens when a philosopher oversteps his bounds, because "anatomy belongs properly to medicine."[25] In his angry rejoinder, a Platonic philosopher reminds him that philosophy is the source of all other disciplines. More specifically, medicine belongs to the natural part of philosophy (*physica pars philosophiae*), although in comparison with the study of the heavens it represents "the lowest dregs of *physica*." Thus the medical expert is portrayed as upholding the complete autonomy of medicine, while the philosopher associates *physica* primarily with the heavens, granting to *medicina* (properly so-called) only marginal status within this larger domain. The latter view reflected a certain contempt. But Augustine (354–430) could regard *medici,* as experts on the human body, more favorably than *physici,* who study the heavens and the earth.[26]

Cosmology, the hallmark of classical *physica,* by no means disappeared from late ancient and early medieval learning, but its identity was profoundly changed. Secular learning survived, not primarily under the rubric of philosophy and its branches, including *physica,* but under that of the seven liberal arts, of which astronomy and geometry could be associated with the heavens and the earth, respectively. Indeed, not only did the whole of the mathematical quadrivium take on a cosmological aura, but all seven of the liberal arts were enveloped in cosmic allegory.[27] And while theologians might or might not look askance at the non-Christian origins of such speculations, they had their own format for cosmological discussions in the tradition of the *Hexaemeron,* or commentary on the six days of creation in Genesis.[28]

[25] Macrobius, *Saturnalia* 7.3.3, 7.15.1–15, ed. J. Willis (Leipzig: Teubner, 1963), pp. 409, 452–454 (the latter passage discussed by Lawn, *Salernitan Questions* [cit. n. 6], pp. 4, 14–15).

[26] Augustine, *Enchiridion,* Chs. 9, 16, in *Patrologiae cursus completus, series Latina,* ed. J. P. Migne (Paris, 1845–) (hereafter ***Patrologia Latina***), Vol. 40, cols. 235, 238–239.

[27] William H. Stahl and Richard Johnson with E. L. Burge, *Martianus Capella and the Seven Liberal Arts,* 2 vols. (New York: Columbia Univ. Press, 1971–1977), Vol. I, esp. pp. 3–40, 55–71, 81–98, 125–148, 171–201.

[28] David C. Lindberg, "Science and the Early Church," in *God and Nature: Historical Essays on*

This diffusion of cosmology into other disciplines did not prevent its continued association with the word *physica,* but the relationship was inevitably weakened. The arts of the quadrivium may have been interpreted along qualitative, cosmological lines, but this did not negate their traditional identification with mathematics. More important, although the biblical account of creation might be regarded as a kind of *physica,* the issue could not be pressed too far without coming up against the more potent claims of Christian theology. *Physica* did, however, retain an unambiguous claim to represent the rational explanation of natures at the subcosmic level. But while this approach might in principle encompass many subdivisions of knowledge, the one that was in fact most readily accessible to early medieval scholars was medicine.

Medicine itself entered into late ancient learning with its traditional identity intact. Influential theologians such as Augustine and Jerome (ca. 331–419/20) made pervasive reference to the analogical "medicine of the soul," but for both authors this comparison presupposed the existence of a learned and morally legitimate medicine of the body.[29] Furthermore, the Christian obligation to comfort the sick was extended to include the practice of secular medicine, opening the way for a long tradition of medical learning within the Church itself. Indeed, insofar as natural science was perceived as relevant to medicine, it could share in this moral sanction. Jerome urged Christians to study the works of *physici* such as Pliny and Aristotle in order to appreciate God's providence in creating medicinal substances, although even in doing so he acknowledged the coequal expertise of *medici* such as Galen. Similarly, Cassiodorus (ca. 480–ca. 575) urged the monastic *medici* to "learn the natures of herbs," but without neglecting *medicina* in their literary curriculum.[30]

But if medicine held a respectable place in the theological tradition, its status was more ambiguous in relation to the liberal arts. To some ancient authors the medical art itself had seemed worthy of the designation "liberal," but even in giving a cosmological orientation to the liberal arts, Martianus Capella (fl. after 410) formally excluded medicine on the grounds of its noncelestial character.[31] However, the first view also survived into medieval learning,[32] and medical *physica* seems to have emerged as the synthesis of these two positions: *physica* could be regarded as a body of medically relevant learning having the prestige to stand among the liberal arts, but thereby making *medicina* seem inferior by comparison.

---

the Encounter between Christianity and Science, ed. Lindberg and Ronald L. Numbers (Berkeley/Los Angeles: Univ. California Press, 1986), pp. 19–48.

[29] Arthur Stanley Pease, "Medical Allusions in the Works of St. Jerome," *Harvard Studies in Classical Philology,* 1914, *25*:73–86; Sister Mary Emily Keenan, "Augustine and the Medical Profession," *Transactions and Proceedings of the American Philological Association,* 1936, *67*:168–190; and Rudolph Arbesmann, "The Concept of 'Christus Medicus' in St. Augustine," *Traditio,* 1954, *10*:1–28. For a broader overview see Darrel W. Amundsen, "Medicine and Faith in Early Christianity," *Bull. Hist. Med.,* 1982, *61*:326–350.

[30] Jerome, *Adversus Jovinianum* 2.331–332 (*Patrologia Latina,* Vol. 23, cols. 305–306); and Cassiodorus, *Institutiones* 1.31, ed. R. A. B. Mynors (Oxford: Clarendon Press, 1937), pp. 78–79.

[31] Stahl, Johnson, and Burge, *Martianus Capella* (cit. n. 27), Vol. I, p. 93; Vol. II, p. 346. Other late ancient authors dropped medicine tacitly. On earlier discussions of medicine vis-à-vis the liberal arts see Fridolf Kudlien, "Medicine as a 'Liberal Art' and the Question of the Physician's Income," *Journal of the History of Medicine,* 1976, *31*:448–459; on later views see Amundsen, "Medicine and Surgery" (cit. n. 1), esp. pp. 48–52.

[32] MacKinney, *Early Medieval Medicine* (cit. n. 1), pp. 95–96, 121–134.

The word *medicina* was vulnerable in this regard because its relevance went considerably beyond the domain of learning. It could certainly serve in the title of a literary text whose learned character spoke for itself, but *medicina* could also refer concretely to an herb or remedy, as well as to a practical art that was not necessarily based upon learning. Medieval scholars were made sensitive to the latter distinction by the lingering effects of the most fundamental medical controversy of antiquity, that between empiricists and rationalists. Medical empiricism had arisen as a philosophically sophisticated reaction to the growth of medical theory, and its exponents had advocated the critical accumulation of experience as an alternative form of medical learning.[33] But surviving rationalist polemic seems to have persuaded medieval scholars that empiricism was a benighted *medicina* whose practitioners were deficient in their learning.[34] Consequently, if the word *medicina* were to convey an unambiguously learned meaning, it needed auxiliaries such as *littera, scientia, logica,* or *ratio. Physica* can be regarded as another such auxiliary, but one that proved uniquely potent to displace *medicina,* first in laying claim to all learning relevant to medical practice, and eventually as the name for all of rational medicine, including its practices.

But the continued use of *medicina* for these same purposes suggests that medieval scholars were not so desperate for an alternative that they would have grasped at straws. It therefore seems unlikely that *physica* could have competed with *medicina* in this regard if it had not seemed to have an inherently plausible claim to do so, based upon the large and positive associations that had existed between natural science and medicine long before the emergence of a distinctly medical conception of *physica.*

II. *PHYSICA:* FROM COSMOLOGY TO MEDICAL LEARNING

In the works of Isidore of Seville (ca. 560–636), the encyclopedist who was so influential during the early Middle Ages, a number of the trends discussed in the previous section are already discernible. Perhaps most striking, among the twenty books of Isidore's *Etymologies* pride of place goes to the seven liberal arts, while all of philosophy is reduced to a single short chapter within this account.[35] Isidore defines philosophy as "knowledge of human and divine matters joined with a zeal for proper living" and lists *logica, ethica,* and *physica* as its principal branches. He identifies *physica* as "the investigation of nature" and "the inquiry concerning causes," as well as "the causes of heaven and the power of natural things." Listing as its four branches the liberal arts of arithmetic, geometry, music, and astronomy, Isidore a few lines later cites Genesis as an instance of *physica.*

But Isidore also wrote a short treatise, *De rerum natura,* that helped preserve a less shadowy tradition of cosmological *physica.* This treatise contains little of direct medical relevance other than a chapter on the meteorological causes of

[33] Ludwig Edelstein, "Empiricism and Skepticism in the Teaching of the Greek Empiricist School," in *Ancient Medicine* (cit. n. 13), pp. 195–203.

[34] See Lawn, *Salernitan Questions* (cit. n. 6), p. 10 and n. 5.

[35] *Isidori Etymologiarum sive originum libri XX* 2.24.1–16, ed. W. M. Lindsay, 2 vols. (Oxford: Clarendon Press, 1911). Cf. Cora E. Lutz, "Remigius' Ideas on the Classification of the Seven Liberal Arts," *Traditio,* 1956, *12*:65–86, esp. p. 69. On Isidore's towering influence on medieval learning see Stahl, *Roman Science* (cit. n. 16), pp. 212–223.

pestilences. In addition, when introducing the cosmos, Isidore referred to the doctrine of the microcosm and to the analogy between the four bodily humors and the four cosmic elements, but the brevity of his remarks implied that these topics were not highly germane to the present subject. Conversely, his tract *De medicina,* part of the *Etymologies,* features a fuller account of the humoral theory which also mentions the analogy with the elements.[36]

Isidore's emphasis on the humors reflects a broader effort to present medicine as a discipline that is, ideally, both learned and theoretical. He prefaced his account of the humors with an overview of the ancient medical sects in which he awarded superiority to the rationalists because of their reliance on rational causality. Within the *Etymologies,* this book on medicine is preceded only by those on the seven liberal arts. Isidore noted that medicine is not formally included among "the other liberal disciplines" because it is more comprehensive than they are, in the sense that the *medicus* must be familiar with each of the liberal arts.[37] His arguments for grammar and rhetoric are quite general, but he equates the relevance of dialectic with the superiority of rationalist medicine. Furthermore, the *medicus* must know arithmetic to comprehend the periodicity of diseases; geometry to understand the various influences of locales on health; music because of its therapeutic properties; and astronomy because of the health-related influences of the stars and the seasons. Isidore had previously shown that philosophy also embraces all the liberal arts, and so he concluded *De medicina* by designating medicine as "the second philosophy," that is, the philosophy of the body, just as philosophy is regarded as the medicine of the soul.

Isidore could scarcely have done more to uphold the learned status of medicine, and yet his account is not without its problematic aspects. In discussing the ancient medical sects, he portrayed empiricism not as a learned reaction against rationalism, but as a primitive form of *medicina* whose exponents were ignorant of causal reasoning.[38] But even though he regarded this medicine based only on *experientia* as inferior, he did not reject it as invalid, showing that learning was not an inherent attribute of *medicina* as such. Furthermore, by Isidore's account even rationalist medicine was a very slender discipline indeed. Apart from the theory of the four humors, *De medicina* consists primarily of a catalogue of diseases, giving the Greek names and humoral causes of each. The book does not discuss therapeutics except in general terms, nor does it include the human body beyond the humoral theory.

But such knowledge is not missing from the *Etymologies.* The last ten books comprise a survey of natural history in a Plinian mold, and here we find various catalogues of medicinal substances.[39] More important, whereas Pliny began his *Natural History* with an account of the cosmos, Isidore gives first place to "man and his parts," and it is here, in Book 11, that we actually find the fuller account of the microcosm that was omitted from *De rerum natura.* Furthermore, Isidore

[36] Isidore of Seville, *De natura rerum liber* 39 and 9, ed. G. Becker (Berlin, 1857; Amsterdam: Adolf M. Hakkert, 1967), pp. 67–68, 21; and Isidore, *Etymologiae* 4.5 (cit. n. 35). See William D. Sharpe, *Isidore of Seville: The Medical Writings: An English Translation with an Introduction and Commentary* (Philadelphia: American Philosophical Society, 1964), esp. pp. 23–26 on macrocosm and microcosm.

[37] Isidore, *Etymologiae* 4.8.

[38] *Ibid.,* 4.4.

[39] *Ibid.,* esp. Books 16 and 17.

begins this account with a definition of *Natura* as the creator of all things, and accordingly, the unnamed experts cited with regard to the body are *physici*. And these *physici* are also knowledgeable about such things as the humoral theory and the diagnosis and prognosis of disease—the very sorts of expertise that he attributes to *medici* in Book 4.[40]

Thus if *De rerum natura* and *De medicina* represent the ancient distinction between natural philosophy and medicine, this section beginning with *Natura* seems to anticipate the later conception of a natural science that is centered on the human body. In other ways, too, Isidore tightened this web of relationships. As we have seen, in the chapter on philosophy he had divided *physica* into arithmetic, geometry, music, and astronomy, while in *De medicina* he pointed out the relevance of these same disciplines to learned medicine. But in his *De differentiis* he broadened his taxonomy of *physica* to include *medicina* as a subordinate part.[41] Thus cosmological *physica* has a special pertinence to *medicina*, while the latter also belongs to *physica* conceived more comprehensively.

Both of Isidore's classifications of *physica* were well known to subsequent scholars, and from the early Carolingian period we have two poems on the liberal arts that depict two different relationships with *physica*. In the one poem, by Theodulf of Orleans (ca. 760–821), *physica* is given distinctly cosmological attributes as the parent discipline of the quadrivial arts. But in the other, by Dungal the Irishman (ca. 820), *physica* is an eighth liberal art whose identity is unmistakably medical.[42] In this section of the poem Dungal first inquires as to the reader's need for *medicina*, with reference to medicinal herbs, but then goes on to salute the medical heroes Apollo, Asclepius, and Hippocrates as the founders of *physica* as a body of knowledge. Dungal specifically cites Hippocrates for his "famous dogmas," which is probably a reference to the humoral theory, if not to rationalist medicine more generally. And it is especially striking that he substitutes the word *physica* where his source (Isidore) uses *medicina*.[43]

By the ninth century a number of leading teachers in the monastic schools were in fact studying medical texts in close conjunction with their commentaries on the liberal arts.[44] Since medicine was not their primary interest, they probably would have had little hesitation about attributing medical learning to the higher discipline of *physica*. However, awareness of the cosmological meaning seems to have inhibited the outright equation of *physica* with such medical knowledge. Thus Martin of Laon (819–875) endorsed Isidore's broader taxonomy of *physica*, but indicated that the quadrivium had a stronger claim to inclusion under this heading than did medicine.[45] Similarly, in a commentary on Martianus Capella,

[40] *Ibid.*, *physici:* 11.1.37, 125; 11.2.27; pulse: 11.1.120; urine: 11.1.138.

[41] Isidore of Seville, *De differentiis*, in *Patrologia Latina*, Vol. 133, cols. 93–94; cited by Amundsen, "Medicine and Surgery" (cit. n. 1), pp. 50–51.

[42] Theodulf of Orleans, in *Monumenta Germaniae historica: Poetae Latini aevi Carolini*, ed. Ernest Duemmler, Vol. I (Berlin, 1881), pp. 544–547, esp. lines 65–114; and Dungal the Irishman, *ibid.*, pp. 408–410, stanza 8. Both poems are cited by MacKinney, *Early Medieval Medicine* (cit. n. 1), pp. 187–188.

[43] Dungal, in *Monumenta Germaniae historica*, p. 410, stanza 8. Duemmler provides extensive annotations referring to Isidore, in this instance to *Etymologiae* 4.3.

[44] John J. Contreni, "The Study and Practice of Medicine in Northern France during the Reign of Charles the Bald," *Studies in Medieval Culture*, 1976, *6–7*:43–54. I am grateful to my student Walt Schalick for calling my attention to this valuable article.

[45] *Ibid.*, pp. 49, 53n48.

once attributed to John Scotus Erigena (d. ca. 877–879), the unmodified word *physica* is used repeatedly to refer to cosmology and the elements, whereas knowledge of the human body is attributed to a qualified *medicinalis physica*.[46]

In a work of undoubted authenticity, Erigena himself extended an unqualified *physica* up to and perhaps beyond the limits acceptable to Christian orthodoxy. At the beginning of his *De divisione naturae* (*On the Division of Nature*), he defined *physis* or *natura* as comprehending all that exists, including God as both the beginning and the end of all else. When he later came to stipulate the divisions of philosophy, however, he placed God under the separate heading of theology, in parallel with natural philosophy, logic, and ethics. Without the deity, *physica* was defined pluralistically as the investigation of "the reasons of natures [*rationes naturarum*], whether in their causes, or in their effects."[47] This definition included not only the cosmos, the four elements, the human species, and all other corporeal species, but also the whole chain of incorporeal causes that, according to John's Neoplatonism, linked God with his creation. Indeed, he claimed to find his whole hierarchy of causes and effects within the literal account of creation in Genesis, which therefore belonged to *physica* rather than to theology.[48]

Erigena's bold vision of nature was highly controversial and may have helped create a secure place for a more modest conception of *physica*. His younger contemporary Remigius of Auxerre (ca. 841–ca. 908) could link *physica* with cosmology as well as with the quadrivium,[49] but in his influential commentary on Martianus he also slanted it toward a medical meaning. In Martianus's allegory *De nuptiis Philologiae et Mercurii,* the seven liberal arts are to serve as the handmaids of Philology, but before reaching their heavenly domain she compounds an ointment to protect her body from the celestial fire. Remigius, who identified Philology with Reason, explained her ointment as *medicina a contrario,* that is, a rational remedy that counteracts the heat by cold. However, he inferred that Philology's ointment was itself eternal, and so he also supplied an allegorical interpretation: the ointment represents "*scientia physica,* which disputes about the natures of minerals, trees, plants, animals, and all created things." Thus without being limited to what was medically relevant, *physica* could be conceived as an allegorical *medicina* when its knowledge was directed toward medical ends. Subsequently, however, Remigius explained that the literal *medicina* is to be excluded from the liberal arts because it "relies more on *experientia* than on *ratio*."[50]

[46] *Iohannis Scotti Annotationes in Marcianum,* ed. Cora E. Lutz (Cambridge, Mass.: Mediaeval Academy of America, 1939), p. 3, line 28, for *medicinalis physica;* cited by Contreni, "Study and Practice of Medicine," p. 52n31. Contreni also identifies many other medical allusions in this commentary. For nonmedical references to *physica* and *physici* see e.g., *Annotationes,* pp. 4.8, 6.31, 7.13, 11.5, 29.15, 30.14, 44.6, 63.25 (the number after the period here refers to the line on the page).

[47] John Scotus Erigena, *De divisione naturae libri quinque,* in *Patrologia Latina,* Vol. 122, cols. 441–444, 705. Cf. Lutz, "Remigius' Ideas" (cit. n. 35), pp. 70–71. On p. 70n27, Lutz notes that in *Super ierarchiam caelestem* (*Patrologia Latina,* Vol. 122, col. 521) Erigena defines *physica* as "naturalis scilicet scientiae omnium rerum, quae post Deum sunt."

[48] Erigena, *De divisione,* cols. 688–742, esp. cols. 691, 693, 705, 706. In col. 707 he declares, "In his ergo omnibus nulla allegoria, sed nuda solummodo physica consideratio tractatur."

[49] Lutz, "Remigius' Ideas" (cit. n. 35), pp. 76–77; and *Remigius Autissiodorensis Commentum in Martianum Capellam,* ed. Cora E. Lutz, 2 vols. (Leiden: E. J. Brill, 1962–1965), Vol. II, pp. 278–279.

[50] Remigius, *Commentum,* Vol. I, pp. 153–154, and Vol. II, pp. 17–18; cited by Lutz, "Remigius' Ideas," pp. 77–78, nn. 70, 75.

A century later Richer of Rheims described a lively controversy about the relationship between *physica* and the quadrivium, centering on the figure of Gerbert (ca. 945–1003). Richer wrote his *History* during the 990s after studying under Gerbert, the eminent teacher of the liberal arts who later became pope. According to Richer's account, Gerbert upheld the autonomy of mathematics and its four branches based upon a taxonomy (actually, that of Aristotle) in which theoretical philosophy has three coequal divisions, namely, *physica,* mathematics, and theology. In this passage, Richer did not indicate what would be left to *physica* if it were deprived of the cosmological quadrivium, but Loren C. MacKinney has identified numerous other passages where Richer displayed his own knowledge of medicine, including three in which he cited *physici* as his authorities on the Greek names and humoral causes of diseases.[51]

Richer also provided a lengthy anecdote whose central motif was the relationship of learning to medical skill. The story is probably fictitious, but this makes it all the more valuable as the apparent expression of an ideology. It concerns two rival doctors at a royal court, one a learned French bishop and the other an unlettered Salernitan.[52] Richer conceded that both men were highly expert in the *ars medicinae,* while also ascribing to each a distinctive attribute, namely, *littera* and *experientia,* respectively. The point of the story was that the French cleric not only trounced the Salernitan in disputations but also outwitted him in practical skill. And when Richer named the overall topic of the learned debates, it was *rerum naturae,* "the natures of things," a close equivalent of *physica.* Richer did not state in general what that phrase covered, but on one occasion he listed such practical subjects as *farmaceutica* and *chirurgica,* and he alleged that the Salernitan could not even understand the meanings of such Greek terms.

What Richer seems to convey anecdotally was stated more formally at the turn of the twelfth century by Baudri of Bourgueil (1046–1130). In an allegorical poem on the arts and sciences written no later than 1102, Baudri reserved the places of honor for the seven liberal arts and medicine.[53] After surveying various aspects of medical practice and learning, he declared that the effigy of medicine should carry this inscription: "This is the *ars medicina,* who disputes on *physica;* when she is in charge our bodies will enjoy better health."[54] Thus the personified *medicina* both heals and possesses knowledge, but all her knowledge is apparently equated to *physica.* Baudri's preceding survey had indicated that such *physica* centered on the theory of the four humors (ten lines) and the human body (ten lines), as well as medicinal substances and diseases (twenty lines).

Physica had no place in Baudri's hierarchy except as an attribute of the personified *medicina,* but the dependency could be reversed. We see this in an early

[51] Richer of Rheims, *Histoire de son temps* 3.55–60, ed. and trans. J. Gaudet (Paris: Libraires de la Société de l'Histoire de France, 1845), Vols. 40 and 43; Vol. 43, pp. 62–72. This passage is preceded by an account of Gerbert's work on arithmetic, astronomy, and geometry, under the rubric of mathematics (pp. 54–62). For Richer's views on medicine see Loren C. MacKinney, "Tenth Century Medicine as Seen in the History of Richer of Rheims," *Bull. Hist. Med.*, 1934, *2*:347–375; and MacKinney, *Early Medieval Medicine* (cit. n. 1), p. 131, re *physici.*

[52] MacKinney, "Tenth Century Medicine," pp. 367–368; and Richer, *Histoire* 2.59 (Vol. 40, pp. 214–216).

[53] *Les oeuvres poétiques de Baudri de Bourgueil (1046–1130),* ed. Phyllis Abrahams (Paris: Librairie Ancienne Honoré Champion, 1926), no. 196, lines 962–1361, pp. 221–230; on the date, p. 232. Cited by Lawn, *Salernitan Questions* (cit. n. 6), p. 13.

[54] Baudri, *Oeuvres,* p. 230, lines 1332–1333: "Haec est de physica quae disputat ars medicina,/ Quae praeeunte magis corpora nostra valent."

twelfth-century example in which *physica* is personified as the teacher of a famous *medicus*. The *medicus* in question was Faricius of Abingdon (d. 1117), a native Italian who lived most of his life in England, first as monk and *medicus* at Malmesbury, then as a doctor at the royal court, and finally as the abbot of Abingdon and serious (but unsuccessful) candidate for the archbishopric of Canterbury. In a literary work Faricius described himself (in common with St. Luke) as a *medicus,* and after his death the same term was applied to him by the Abingdon chronicler and by William of Malmesbury (ca. 1090–1143).[55] William, writing in 1125, also quoted an encomium by another monk of Malmesbury that includes the following couplets on Faricius's learning:

> Omnibus instructus quos tradit littera fructus,
> Ad decus ecclesiae vertit momenta sophiae.
> Omnibus imbutus quas monstrat phisica leges,
> Ipsos demeruit medicandi munere reges.[56]

In the first and third lines *littera* and *physica* are personified in parallel ways, so that Faricius is "provided with all the fruits that letters convey, and has turned the force of this wisdom to the adornment of the Church"; and "being imbued with all the laws that *physica* demonstrates, he has served the very kings in his office of healing." The hyperbole suggests that *physica* should be understood as natural science in a broad sense, although such knowledge is assumed to find fulfillment through the practice of healing kings.

In his history of the English kings, completed by 1125, William of Malmesbury himself showed that *physica* might be personified as a healer in its own right. The instance occurs in William's highly rhetorical apology that he, a monk, was so dedicated to secular *littera,* especially history.[57] He acknowledged that indiscriminate reading might endanger salvation but maintained that he had confined himself to those subjects having some clear justification, namely, *logica, physica, ethica,* and *historia.* There can be no doubt that *physica,* when bracketed between *logica* and *ethica,* stands for natural philosophy in an unqualified sense, but as William names each discipline he identifies its morally acceptable purpose through personification. We thus have *logica,* "which strengthens eloquence"; *physica,* "which heals the ills of bodies";[58] *ethica,* "which disposes souls for proper living"; and *historia,* "which, by examples, stimulates its readers to follow good and avoid evil." Thus William cited healing as a blanket justification for

[55] Faricius, *Vita S. Aldhelmi,* in *Patrologia Latina,* Vol. 89, p. 65B; Joseph Stevenson, ed., *Chronicon monasterii de Abingdon,* 2 vols. (London, 1856), Vol. II, p. 44; and William of Malmesbury, *De gestis pontificum Anglorum,* ed. N. E. S. A. Hamilton (Rerum Brittannicarum Medii Aevi Scriptores, 52) (London, 1870; n.p.: Kraus Reprint, 1964), p. 192. On Faricius see Charles H. Talbot and Eugene A. Hammond, *The Medical Practitioners in Medieval England: A Biographical Register* (London: Wellcome Historical Medical Library, 1965), pp. 45–46; and Edward J. Kealey, *Medieval Medicus: A Social History of Anglo-Norman Medicine* (Baltimore: Johns Hopkins Univ. Press, 1981), pp. 65–70.

[56] William of Malmesbury, *De gestis pontificum,* pp. 192–193. On the dates of William's histories see Rodney Thomson, *William of Malmesbury* (Suffolk: Boydell Press, 1987), pp. 3–4.

[57] William of Malmesbury, *De gestis regum Anglorum* 2, prologue, ed. William Stubbs (Rerum Brittannicarum Medii Aevi Scriptores, 90) (London, 1887), Vol. I, p. 103. See Thomson, *William of Malmesbury,* pp. 26–28, on William's general concern with the justification of secular learning.

[58] William, *De gestis regum,* Vol. I, p. 103: "physicam, quae medetur valitudini corporum, aliquanto pressius concepi." This is the locus classicus for medical *physica,* but one should beware of regarding it as a formal definition.

studies in *physica*. In his parallel references to body and soul we can perhaps see, in place of the ancient motif of ethics as the "medicine of the soul," a new motif of *physica* as the "ethics of the body."[59]

William was one of the most deeply learned scholars in early twelfth-century Europe, and while by no means a *medicus,* he was keenly aware of the issue of learning as it pertained to clerical *medici.* Thus he wrote of John of Villula (d. 1122), the bishop of Bath, that he was "a highly regarded *medicus* as a result of practice, but not of *scientia.*" William added, "Indeed, [John] is almost entirely lacking in letters, but he nevertheless enjoys the company of learned men in the hope that by this association some of their praise will redound to him."[60]

Thus there was a sharply polemical side to this issue, and in the *Didascalicon,* the influential survey of higher learning written during the 1120s by Hugh of St. Victor (ca. 1096–1141), this negative ideology was applied to the discipline of medicine in its entirety. In contrast to the earlier tendency to group medicine informally with the seven liberal arts, Hugh endorsed the sneering view that medicine is second from the bottom among the so-called seven mechanical arts, just below hunting and just above theatrics.[61] To Hugh this meant that medicine comprised such directly physical acts as the insertion of drugs into bodily orifices, but not the theoretical learning that determined such actions. Hugh did not explicitly refer medical theory to *physica,* but he did construe the latter generically as explanation by means of the four elements, a trait quite prominent in the medical learning known to him.

With Hugh, *medicina* reached a low point of intellectual estimation, and it was soon to be stigmatized on moral grounds as well. This attack stemmed precisely from the frequency with which monks such as Faricius used their medical expertise for worldly advancement. During the 1130s the first canons were passed seeking to curb this pattern, and these official decrees of the Church associated *medicina* with vices such as avarice and lewdness in order to make its secular practice by monks seem especially shameful.[62]

This period of polemic coincides with the time when the title *physicus* was becoming current for individual medical practitioners. By the 1150s John of Salisbury (ca. 1115–1180), himself a canon lawyer, defender of the liberal arts, and biting social critic, could hint at a relationship between these two developments. Referring to medical alumni of Salerno and Montpellier, he criticized them on medical grounds as empiricists; on intellectual grounds as insufficiently learned in natural philosophy; and on moral grounds as motivated by avarice rather than

[59] For evidence of his studies in natural science see Thomson, *William of Malmesbury* (cit. n. 56), pp. 52, 59, 204; and William of Malmesbury, *Polyhistor,* ed. Helen Testroet Ouellette (Binghamton, N.Y.: Center for Medieval and Early Renaissance Studies, 1982), pp. 45–61.

[60] William, *De gestis pontificum* (cit. n. 55), p. 195 and n. 9. On John of Villula see Talbot and Hammond, *Medical Practitioners,* pp. 192–193; and Kealey, *Medieval Medicus,* pp. 57–59 (both cit. n. 55).

[61] *The Didascalicon of Hugh of St. Victor: A Medieval Guide to the Arts,* trans. and ed. Jerome Taylor (New York: Columbia Univ. Press, 1961), 2.26, pp. 78–79; Peter Sternagel, *Die Artes mechanicae im Mittelalter: Begriffs- und Bedeutungsgeschichte bis zum Ende des 13. Jahrhunderts* (Kallmünz Opf.: Verlag Michael Lassleben, 1966), esp. Chs. 7 and 9, on Hugh and his influence; and Amundsen, "Medicine and Surgery" (cit. n. 1), pp. 51–55. For Hugh's views on *physica* see *Didascalicon,* 2.16–17, pp. 71–72. See also 3.2, p. 83, where Hugh cites Pliny as the principal literary authority on *physica.*

[62] Darrel W. Amundsen, "Medieval Canon Law on Medical and Surgical Practice by the Clergy," *Bull. Hist. Med.,* 1978, *46*:23–44, esp. pp. 28–36.

This classification of knowledge from Gregor Reisch's* Margarita philosophica *(Friborg, 1504) is similar to Hugh of Saint Victor's in that it combines Aristotle's taxonomy of philosophy with the liberal and the mechanical arts. Medicine occurs twice in Reisch's scheme: second from the very bottom among the mechanical arts, and near the center, where he refers to "*physica *or natural [philosophy], under which theoretical medicine is also contained." By the long list at right, Reisch further identifies natural philosophy with Aristotle's* libri naturales. *In the early twelfth century, the latter had been unknown even by title, making it possible to more nearly equate* physica *with its medical applications. Photograph courtesy of the Jonas Friedenwald Memorial Library, Wilmer Ophthalmological Institute, the Johns Hopkins University School of Medicine.

by the interests of their patients.[63] Furthermore, in all three respects they pretended that they were just the opposite, making them a major example of fraud, comparable to those who enter monasteries in quest of an easy life. In summary, John declared, "Either under the pretext of religion they enter into cloisters, or under the appearance of philosophizing and of public utility they take refuge in *physica*."[64] Thus he seems to imply that the very word *physica* had become a sham to disguise a money-grubbing medical career. And just a few years later a revised canon was issued in which *physica* indeed replaced *medicina* as the subject that was proscribed for the regular clergy.[65]

Given the rough treatment of *medicina*, it would not be surprising if learned practitioners, whether monks or not, sought to redefine their expertise in other terms. But, as we have seen, in the liberal arts tradition *physica* had long been available as such an alternative. Furthermore, John of Salisbury's own views refute the notion that the medical *physica* of the twelfth century was a mere pretense. For just as John did not believe that the cloister per se was a sham, so also he assumed that there was a true medical *physica*, suffused with rationality and ethics. And he gives good indications that this ideal derived its credibility from the newly available medical learning, as well as from the emergence in Europe of scholars who were indeed worthy of being called *physici*.

III. THE NEW MEDICINE AND THE NEW NATURAL SCIENCE

There are ample grounds beyond John of Salisbury's views for regarding the medical *physica* of the twelfth century in a positive light, as a creative synthesis that took full advantage of the knowledge available at the time. Indeed, twelfth-century scholars seem to have been quite familiar with the etymology, classical meaning, and high prestige of the word *physica*, and many of them, like John, also had the linguistic sensitivity to avoid deliberate abuse of such a word. Furthermore, this was also the time of a vigorous revival of cosmology, which thus provided a potential alternative to medicine as the main subject matter of *physica*. Finally, in the Latin versions of Greek and Arabic medical texts that became available during this period, the subject under discussion was routinely identified as *medicina*.

Yet scholars representing a broad spectrum of disciplines formed a strong consensus that these very texts ought more properly to be regarded as works on *physica*. This reaction is especially striking with regard to the *Isagoge* of Johannitius, an adaptation of an Arabic original (the *Questions on Medicine* of Ḥunain ibn Isḥāq) that became the main primer of rational medicine for twelfth-century scholars. Its opening words declared, "*Medicina* is divided into two parts, namely theory and practice,"[66] but these very words seem to have strengthened the conviction that the source of medical theory must be *physica*, rather than

[63] *Ioannis Saresberiensis episcopi Carnotensis Metalogicon Libri IIII*, ed. Clement C. I. Webb (Oxford: Clarendon Press, 1929), 1.4, 830c–831c. Concerning those who study at Salerno and Montpellier, John variously asserts: "suum in philosophia intuentes defectum"; "repente, quales fuerant philosophi, tales in momento medici eruperunt. Fallacibus enim referti experimentis in brevi redeunt, sedulo exercentes quod didicerunt"; "cum dolor cruciat egrotantem sibique cooperantur languentis exulceratio et avaritia medentis"; "Numquid enim nature secretos latentesque cuniculos deprehendet homo totius philosophie ignarus."

[64] John, *Metalogicon* 832a: "Nam, ut dictum est, aut sub pretextu religionis mergebantur in claustris, aut sub imagine philosophandi et utilitatis publice confugiebant ad physicam."

[65] Amundsen, "Medieval Canon Law" (cit. n. 62), pp. 28–36.

[66] Johannitius, *Isagoge*, in *Articella* (Venice, 1491), fol. 2r.

medicine itself. This conviction perhaps stemmed in part from familiarity with the Aristotelian taxonomy in which an individual discipline can stand on only one side of the dichotomy between theory and practice. Indeed, Hugh of St. Victor played a major role in the revival of this taxonomy, and in denying that *medicina* possessed its own theory he was also deliberately contradicting the *Isagoge*.[67]

But besides the *Isagoge* we must also consider two other influential texts, the *Pantegni* and the *Premnon physicon*, that gave an even more direct impetus to medical *physica*. The *Pantegni* is an encyclopedia of rational medicine that was also based upon an Arabic model; the *Premnon physicon* is a treatise on the human body and soul that derived from a Greek original. Like the *Isagoge*, both were translated in southern Italy in the late eleventh century, but in these two instances the translators also added prefaces in which they commented on the relationship between *physica* and *medicina*.

The translator of the *Pantegni*, Constantine the African (d. ca. 1087), was regarded during the twelfth century as the author of the whole work. In his preface he identified his subject as *medicina litteralis* (learned medicine) and inquired which branch of higher learning medicine is subordinated to.[68] He named *logica*, *ethica*, and *physica* as the possibilities and argued that medicine is not congruent with any one of these but does have important connections with all three. Thus Constantine seemed to repudiate a unique relationship between *medicina* and *physica*, but in doing so he also declared that the *medicus* ought to be *rationabilis* as well as conversant with *moralia* and *res naturales*. These terms are offered as Latin equivalents for the three divisions of learning, but while logic and ethics could easily be identified as autonomous disciplines, *res naturales* is the very name of the *Pantegni's* first major division, which deals with the healthy human body, beginning with the general doctrine of the four elements and the four qualities. The *Isagoge* begins with a similar survey, so that both texts might be regarded as, first of all, works on *physica*.

If Constantine left room for doubt, the preface to the *Premnon physicon* made it quite clear that the study of the human body belongs directly and preeminently to *physica*. This work was based upon Nemesius's *De natura hominis* (*On the Nature of Man*), a survey of anatomy and physiology drawn from medical sources, together with a philosophical account of the soul. The translator, Alfanus (ca. 1020–1085), was a poet, medical scholar, and archbishop of Salerno. The work circulated anonymously during the twelfth century, but a distinct persona emerges from Alfanus's preface—that of a high ecclesiastic, a Latin fluent in Greek, a master of the liberal arts and of *physica*, and the adviser and teacher of a prince.[69]

Nemesius (late fourth century), also a bishop, had written for the general

[67] Hugh's allusion to the *Isagoge* is identified by Jerome Taylor in his edition of the *Didascalicon* (cit. n. 61), pp. 78–79 and p. 206nn75–78.

[68] Constantine the African, *L'arte universale della medicina; Pantegni*, ed. and trans. Marco T. Malato and Umberto de Martini, Part 1, Book 1 (Rome: Istituto di Storia della Medicina, Univ. Rome, 1961), p. 39.

[69] *Nemesii episcopi Premnon physicon . . . a N. Alfano . . . in Latinum translatus*, ed. C. J. Burkhard (Leipzig: Teubner, 1917), pp. 1–4 (preface) and pp. 62–72 of Nemesius's text (on the elements). On this work and its influence see J. R. Williams, "The *Microcosmographia* of Trier MS. 1041," *Isis*, 1934–1935, *22*:106–135, esp. pp. 116–121; McKeon, "Medicine and Philosophy" (cit. n. 5), pp. 221–223; and Theodore Silverstein, "Guillaume de Conches and Nemesius of Emesa: On the Sources of the 'New Science' of the Twelfth Century," *Harry Austryn Wolfson Jubilee Volume II* (Jerusalem: American Academy for Jewish Research, 1965), pp. 720–734.

reader and had begun by invoking the broad framework of the macrocosm and the microcosm. Accordingly, Alfanus explained that he had prepared the following treatise in order to begin his prince's education in natural science with the study of man. Hence his naming (actually, renaming) the work: "Its title will be *Premnon physicōn,* that is, 'the stem of natural things [*stipes naturalium*],' because as from one stem many branches arise, so from the font of this doctrine many streams of the science of natural things spring forth." Thus a treatise on human nature was rhetorically transformed into a primer on general natural science. What Alfanus had in mind was a pluralistic natural science that "teaches the natures of all things, one by one, down to their individual parts." His treatise deals with "the universal principles of natures," which presumably refers to the account of the four elements. By then treating in detail the nature of one particular thing, his book will serve as an exemplary model for dealing with other things, "as those who have become clear-sighted from [the study of] very few things then proceed to examine many." But as to the choice of an example, "this little work will not improperly take its beginning from man, as from that which is better known to us [*a notiori*], whom indeed the philosophers claim to bear the image of the whole, and on this ground call him the microcosm."

Alfanus also sought to show that this knowledge had broad relations to many other branches of higher learning:

> The lessons of the liberal arts needed for the construction of this book are to be gathered as effective instruments, and afterward these very arts will also profit not a little [from the result]; and not only these arts, but also the doctrines of medicine and theology. And in a word, although [the subject] is to be ascribed primarily to *physica,* it has nevertheless been sucked from all the breasts of philosophy, and it will in turn offer to the mother herself sustenance that will not be altogether useless.[70]

Medicine is mentioned respectfully, but to show that *medicina* is dependent upon *physica* for knowledge of human nature.

Similar assumptions based upon the doctrine of the microcosm help to explain why twelfth-century scholars were so firm in their conviction that the study of the human body is not only a part of natural science but an indispensable part that cannot be conceded to any other discipline. In fact, a significant goal for twelfth-century cosmologists was precisely to fill the gap in the Platonic program that Calcidius had created by largely omitting the account of the human body from the *Timaeus.* This intention is already apparent in the *Quaestiones naturales* of Adelard of Bath (ca. 1070 to after 1142–1146), one of the pioneering works in twelfth-century science. Whereas Isidore's *De rerum natura* focused on the macrocosm with only passing reference to the microcosm, and the *Premnon physicon* did the reverse, Adelard gave approximately equal attention to both subjects. But his references to *physici* were notably more frequent in his discussion of the human body than in that of the cosmos.[71]

During the twelfth century such an emphasis was able to solidify into a virtual identity of *physica* with the microcosm, and hence with medicine, because of

[70] Alfanus, preface to *Premnon physicon,* p. 3.

[71] *Die Quaestiones naturales des Adelardus von Bath,* ed. Martin Müller (Beiträge zur Geschichte der Philosophie und Theologie des Mittelalters, 31.2) (Münster: Aschendorff, 1934), citing *physici:* p. 12 (on animals); pp. 21, 30, 31, 37, 42 (on man); p. 58 (on meteorology).

prolonged uncertainty as to whether the macrocosm properly belonged to *physica* at all.[72] As we have seen, earlier traditions made it possible to link cosmology with mathematics (by way of the quadrivium) or with theology (by way of Genesis), and both traditions were still influential during the twelfth century. The cosmologists also looked to the *Timaeus* as a principal model, but the dialogue itself does not clearly identify the branch of philosophy to which it belongs, while in the opening paragraphs of his commentary Calcidius links the work with a variety of disciplines, notably the quadrivium. On the other hand, Erigena provided a definite link between cosmology and *physica,* but one with theological pitfalls. Honorius Augustodunensis (ca. 1080–ca. 1140) wrote an abstract of *On the Division of Nature,* which he entitled *Clavis physicae* (*The Key to Natural Philosophy*), but by thus equating *physica* with the most comprehensive of Erigena's views of *natura,* Honorius placed God himself under this discipline.[73] More cautiously, Thierry of Chartres (d. after 1156) placed God above natural science when interpreting Genesis "according to *physica,*" but even so his attempt to attribute the events of creation to the powers of the four elements was not exactly unobjectionable.

William of Conches (ca. 1085–after 1154) chose a different rubric for such speculations in *De philosophia mundi,* an account of the macrocosm and the microcosm written during the 1120s. The title "On the Philosophy of the World" reflects William's view that the study of the cosmos as a whole pertains to philosophy per se, and this permits him to recognize *physica* as a more limited branch of philosophy that applies the four elements and the related four qualities to the explanation of individual natures.[74] Bound up with this distinction is William's proposal that the term *elementa* should refer only to the invisible constituents of such natures, and not to the visible but impure regions that make up the whole cosmos.[75] This conception of *physica* was potentially broad in itself, but William's Platonic frame of reference readily narrowed it to the one truly important such entity, the human body. He followed much the same rationale as Alfanus did in the preface to the *Premnon physicon,* except that he applied it to the *res naturales* as set forth in Constantine's *Pantegni.* Indeed, William saluted

[72] Cf. Winthrop Wetherbee, "Philosophy, Cosmology, and the Twelfth-Century Renaissance," and Tullio Gregory, "The Platonic Inheritance," with Charles Burnett's account of "physical speculation," all in *Twelfth-Century Philosophy,* ed. Dronke (cit. n. 1), pp. 21–53, 54–80, and 166–176, respectively.

[73] Honorius Augustodunensis, *Clavis physicae,* ed. Paolo Lucentini (Rome: Edizioni di Storia e Letteratura, 1974), pp. 4–5. On Thierry of Chartres see Wetherbee, "Philosophy, Cosmology," pp. 26–32.

[74] William of Conches, *De philosophia mundi* 1.21, in *Patrologia Latina,* Vol. 172, cols. 39–102, on col. 50. On the relationship of *De philosophia mundi* to the *Timaeus* see Edouard Jeauneau in *Guillaume de Conches: Glosae super Platonem,* ed. Jeauneau (Textes Philosophiques du Moyen Age, 13) (Paris: J. Vrin, 1965), intro. pp. 9–16, and text p. 58. Compare William's definition of *philosophia* (*De philosophia mundi* 1.1) as "eorum quae sunt et non videntur et eorum quae sunt et videntur vera comprehensio" with his *Glosae super Platonem,* p. 61, "physica vero est de naturis et complexionibus corporum." For more on William see Tullio Gregory, *Anima mundi: La filosofia di Guglielmo di Conches e la Scuola di Chartres* (Pubblicazioni dell'Istituto di Filosofia dell'Università di Roma, 3) (Florence: Sansoni, 1955); and Dorothy Elford, "William of Conches," in *Twelfth-Century Philosophy,* ed. Dronke (cit. n. 1), pp. 308–327.

[75] Theodore Silverstein, "*Elementatum:* Its Appearance among the Twelfth-Century Cosmogonists," *Mediaeval Studies,* 1954, *16*:156–162; Silverstein, "Guillaume de Conches and the Elements: *Homiomeria and Organica,*" *Mediaeval Stud.*, 1964, *26*:363–367; Silverstein, "Guillaume de Conches and Nemesius of Emesa" (cit. n. 69); McKeon, "Medicine and Philosophy" (cit. n. 5), pp. 231–243; and Elford, "William of Conches" (cit. n. 74), pp. 310–317.

Constantine himself as *physicus,* much as he recognized Plato as *philosophus* in a definitive sense.[76]

In the introduction to his commentary on the *Timaeus,* William sought to show that the subject of the dialogue was indeed philosophy in the broadest sense.[77] He invoked the more complex Aristotelian taxonomy of philosophy, so that he had to stretch the contents of the *Timaeus* to cover a variety of disciplines. In particular, he attributed the causal explanations of the cosmos and the human soul to theology, which meant that his own theologically controversial views on these matters would not fall under *physica.*[78] The latter then turns out to include "the natures and temperaments of bodies" or, more specifically, the theory of the four elements and its applications to animals.

Although it was primarily in view of these topics that William acknowledged Constantine as *physicus,* the endorsement was unqualified, and in *De philosophia mundi* he himself delved into the causes and treatments of diseases. He also invoked the notion of the "prudent *physicus*" who takes health-related action based on his own theoretical insights.[79] In his commentary on the *Timaeus* he even offered a medical explanation of the evil eye: the visual ray carries with it the temperament of the emitter, so that when it reaches the face of the one seen it damages *his* temperament, "because contraries are damaged by contraries." He then added, "Hence it is that old women, although they are ignorant of *physica,* nevertheless having derived the principle from some *physicus,* lick the face and then spit, and so, as it were, they extract the poison with their tongues."[80]

After having taught as a grammarian for many years, possibly at the school of Chartres, William joined the court of the duke of Normandy.[81] Here, during the 1140s, he revised *De philosophia mundi* under the new title *Dragmaticon,* formally retracting some opinions that had been deemed heretical and recasting the work as a dialogue between himself and the duke. He wished it to be known that his position at the court was that of *philosophus,* although it is quite apparent that he could also deploy the explanatory tools of a *physicus.* Thus William explained that an infant cries at childbirth because of the sudden change from a warm and moist environment to a cold and dry one; nurses are able to soothe the infant by putting it into warm water, "though I do not know whether the nurses were taught this by nature or by some *physicus.*"[82] One suspects that William himself would have been ready to offer such helpful advice, and indeed the last book of *Dragmaticon* portrays him in an extended conversation with the duke

[76] See William, *De philosophia mundi* 4.8 (cit. n. 74; cols. 87–88), where he makes short shrift of plants and animals; on his familiarity with the *Premnon physicon* see *ibid.,* 1.21 (cols. 48–51), and Silverstein, "Guillaume de Conches and Nemesius of Emesa"; for Constantine and Plato see *De philosophia mundi* 1.21 (col. 50): "Constantinus ergo ut physicus de naturis corporum tractans, simplices illorum et minimas particulas, elementa, quasi prima principia vocavit. Philosophi vero de creatione mundi agentes, non de naturis singulorum corporum, ista quatuor quae videntur, elementa mundi dixerunt." Cf. *ibid.,* 4.32 (col. 98), "Plato, omnium philosophorum doctissimus."

[77] William, *Glosae super Platonem* (cit. n. 74), pp. 60–62.

[78] See Gregory, *Anima mundi* (cit. n. 74), esp. pp. 4–10, 115–121, 133–167, on William and theological controversy.

[79] William, *De philosophia mundi* 4.7, 19, 21 (cit. n. 74; cols. 87, 93), and 2.27 (col. 70).

[80] William, *Glosae super Platonem* (cit. n. 74), p. 238.

[81] Jeauneau, intro. to *Glosae super Platonem,* pp. 9–10.

[82] William of Conches, *Dragmaticon,* published as *Dialogus de substantiis physicis: Ante annos ducentos confectus, a Vuilhelmo Aneponymo philosopho* (Strasbourg, 1567; facs. rpt. Frankfurt am Main: Minerva, 1967), p. 249; see also pp. 216, 233, 259, 260, 274, 278–279, 288, 295, 304. See Elford, "William of Conches" (cit. n. 74), on the relationship of the *Dragmaticon* to *De philosophia mundi.*

about the human body in both health and disease. In this context he declares, "*Physica* considers the qualities of other things [such as herbs] for the sake of man."[83]

Both *De philosophia mundi* and the *Dragmaticon* survive in nearly seventy manuscripts each,[84] indicating an unusually wide influence during the period after the revival of interest in natural science but preceding and accompanying the full assimilation of Aristotle's *libri naturales*. And far from opposing or even just tacitly accepting a medical conception of *physica,* William defined the latter in such a way as to make it virtually indistinguishable from rational medicine except for being potentially broader in scope.

This implication becomes quite apparent in the commentaries of Bartholomaeus of Salerno (fl. ca. 1150?), whom Kristeller has identified as a pioneer in expounding the new theoretical medicine. In his commentary on the *Isagoge,* Bartholomaeus accepted the equation of *physica* with the theory of the four elements and its applications.[85] He proposed that the latter fell into just two broad categories: meteorology, which treats of the cosmic elements, and theoretical medicine, which treats of the human body and all other bodies composed of the four elements. Thus meteorology was the only topic that permitted Bartholomaeus to make a distinction between *medicina* and *physica*. Even so, he had to make the extravagant claim that medicine treats of *all* mixed bodies, which he justified on the grounds that any mixed body can in principle affect the health of the human body.

But nonmedical contemporaries might find that *physica* could just as well lay direct claim to such medical applications, as we see in the *Cosmographia* of Bernard Silvestris (ca. 1100–ca. 1160). This was an allegory written during the late 1140s in which a variety of characters take part in the creation of the macrocosm and the microcosm. But the *Cosmographia* is also an encyclopedia, in that Bernard periodically summarizes various disciplines. One such body of knowledge is associated with the character Physis, who plays the last crucial role of creating the human body.[86] She initially fails at this task because she ignores William of Conches's distinction between the visible cosmic elements and the invisible constituents of things. But then she realizes that she must first resort to the four qualities to create the four elements in a pure form, after which she can in turn fashion the humors, homogeneous parts, and organic parts, whose functions Bernard surveys at some length.[87] Thus much like William of Conches's *physica,* Bernard's Physis represents the *res naturales,* the first part of medicine, in which the human body is built up from its constituent principles.

[83] William, *Dragmaticon,* pp. 234–304, quotation on p. 234. See Elford, "William of Conches," pp. 318–327, on the medical strands in William's cosmological thought.

[84] Elford, "William of Conches," p. 308.

[85] Paul Oskar Kristeller, "Bartholomaeus, Musandinus and Maurus of Salerno and Other Early Commentators of the 'Articella,' with a Tentative List of Texts and Manuscripts," *Italia Medioevale e Umanistica,* 1976, *19*:57–87, on pp. 85–86, quoting excerpts from Bartholomaeus's *Glossa in Johannitium* (Winchester, MS 24, fol. 22v ff.). Kristeller (pp. 86–87) also reports similar views in Bartholomaeus's *Glossa in artem Galeni* (Winchester, MS 24, fol. 52 ff.).

[86] Brian Stock, *Myth and Science in the Twelfth Century: A Study of Bernard Silvester* (Princeton, N.J.: Princeton Univ. Press, 1972), pp. 19–23, 14–17, 220–226. It is especially striking that this character Physis is quite distinct from the transcendent figure of Natura.

[87] *The Cosmographia of Bernardus Silvestris,* ed. and trans. Winthrop Wetherbee (New York/London: Columbia Univ. Press, 1973), pp. 117–120, 120–127.

But Bernard also gave Physis both more general and more specifically medical dimensions. When the other goddesses of creation first enter her presence they find her accompanied by her two daughters, Theoria and Praxis, which seems to allude to the opening words of the *Isagoge*. Bernard goes on to show that Physis is able to explain anything in sublunary nature by resorting to the elements, the qualities, or the humors, that is, the first parts of the *res naturales*. However, she is also moved by compassion to direct these powers toward the compounding of remedies that will counteract imbalances in these same factors.[88] Thus Physis can both represent the universal study of natures and take on a medical identity through a moral decision to serve human needs.

Whereas Bernard relied on personification and allegory, John of Salisbury expressed similar themes in more literal terms. In his *Metalogicon,* completed in 1159, he declared: "In *physica,* above all foretell the cause of the disease, then cure it and remove it. Subsequently restore and build up the health of your patient by restorative and preservative means, until he has recovered to the fullest." *Physica* thus entails both the practices characteristic of rational medicine and an exhortation to carry out those practices faithfully. Indeed, John was here comparing this form of the medical art with that idealization of the military art later known as "chivalry."[89] In his view, both arts include a strong moral component, in which an ethic of competence and knowledge is inseparable from an ethic of good intention.

John's identification of this medical ideal with *physica* presupposed the traditional tripartite division of philosophy. The subject of the *Metalogicon* was logic, but in the introduction John acknowledged the preeminence of ethics: "I have purposely incorporated into this treatise some observations concerning morals, since I am convinced that all things read or written are useless except so far as they have a good influence on one's manner of life. Any pretext of philosophy that does not bear fruit in the cultivation of virtue and the guidance of one's conduct is futile and false."[90] Subsequently, John incorporated *physica* into this ideology, when discussing the origins of philosophy among the Greeks:

> They esteemed knowledge of the truth as the greatest good in human life. They accordingly made careful investigations into the natures of all things [*omnium rerum naturas*], so as to determine which should be avoided as evil, which spurned as not good, which things should be sought as the simple good, which should be preferred as the greater good, and finally which things are counted as good or bad according to circumstance. Hence there were born the [first] two parts of philosophy, natural and moral, which by other names are called *ethica* and *physica*.[91]

[88] *Ibid.*, p. 111, cf. Wetherbee, intro., p. 43, and Stock, *Myth and Science* (cit. n. 86), pp. 221–226; and *Cosmographia,* p. 112.

[89] John of Salisbury, *Metalogicon* 863d (cit. n. 63). See also John, *Policraticus* 417c, 475c–476b (cit. n. 17), for related comments; and Maurice Keen, *Chivalry* (New Haven, Conn./London: Yale Univ. Press, 1984), pp. 1–17, esp. p. 5 on John's role in helping to shape this ideology. See also Wetherbee, intro. to his trans. of Bernard's *Cosmographia* (cit. n. 87), pp. 57–59.

[90] John, *Metalogicon* 1, prologue, 825b, in *The* Metalogicon *of John of Salisbury: A Twelfth-Century Defense of the Verbal and Logical Arts of the Trivium,* ed. and trans. Daniel D. McGarry (Berkeley/Los Angeles: Univ. California Press, 1955), p. 6. See also Edouard Jeauneau, "Jean de Salisbury et la lecture des philosophes," in *The World of John of Salisbury,* ed. Michael Wilks (Oxford: Basil Blackwell for the Ecclesiastical History Society, 1984), pp. 77–88, esp. p. 82.

[91] John, *Metalogicon* 2.2, 858b–c, trans. McGarry, p. 76, but modified. John refers these developments specifically to "the Peripatetics," but it seems clear that he had in mind the original beginnings of philosophy in Greece.

Thus *physica* and *ethica* both represented knowledge of the truth, which permitted the making of well-informed choices; *logica* was later added to regulate such judgments.

John knew that some authorities would limit the matter of choice and avoidance to "ethical questions [*ethicas questiones*]," but he simply disagreed: "I think that they pertain to natural ones [*physicas*] as well." He continued: "In the former, materials for selection or rejection are provided by virtue, vice, and the like. In the latter, such are provided by the health, disease, causes, signs, and circumstances of individuals."[92] John acknowledged that *physica* was concerned with many things beyond such medical choices, but in *Policraticus* he noted that speculations on ultimate questions such as creation and the soul were apt to lead to heresy. He thus preferred that the *physici* direct their attention to "the animal temperament" and "the natures and causes of individual things"; this permitted them to serve as highly competent advisers regarding "health, disease, and neutrality," the hallmarks of rational medicine.[93]

John then went on to differentiate such *physici* from the *medici,* to whom one turned when more active therapeutics were needed, and who were not always honest in their dealings with patients. He acknowledged that the title *physicus* was being extended to such practitioners, and this reminds us of his attack upon Salerno and Montpellier as the centers of a sham *physica*.[94] However, it is well to recall that John's own loyalties were to northern schools, notably Paris and, possibly, Chartres. Moreover, he was also a former student and admirer of William of Conches, who had been among the first scholars in Europe to cultivate the new medicine as part of general philosophy, who had also been publicly accused of heresy for his nonmedical views, and who had ended his career at a princely court.[95]

Thus if one assumes that William was the sort of *physicus* John had in mind, then his distinction can be understood as that between the true master of natural philosophy, who provides medical advice as an ethical consequence of his ability to do so, and the false *physicus,* who (like other Cornificians) acquires a certain

[92] *Ibid.*, 2.15, 872c–d, trans. McGarry, p. 108, modified. McGarry identifies John's reference as to Aristotle, *Topics* 1.11, 104b1 ff.

[93] John, *Metalogicon* 2.15, 872d–873a; and John, *Policraticus* 2.29, 475d, 476a (quotations); cf. 2.2, 417c. In *Metalogicon* 2.20, 880c, John cites Galen's *Ars medica* regarding this definition of medicine, which is also found near the beginning of the *Isagoge* (cit. n. 66), fol. 2r. On possible historical links between this definition of medicine and Stoic ethics see von Staden, *Herophilus* (cit. n. 13), pp. 89–114.

[94] John, *Policraticus* 2.29, 476a–d. In the title of *Policraticus* 2.29 John uses the phrase *physici practici* to designate the same group whom he calls *medici practici* in the body of the chapter. In the opening words he cites *experimentorum indiciis* as the basis on which these persons practice prognosis, as distinct from the *physici theoretici,* who rely on *physica*.

[95] I do not suggest that John had *only* William in mind, but that William is a concrete example of the sort of cosmologist–medical expert whom John associates with the term *physici theoretici*. On John's close but ambivalent relations with the cosmologists see Wetherbee, "Philosophy, Cosmology" (cit. n. 72), pp. 21–22, 34–35, 41–42, 48–49; on John and William see Elford, "William of Conches" (cit. n. 74), p. 308. On John's affiliations with northern schools, see David Luscombe, "John of Salisbury in Recent Scholarship," Olga Weijers, "The Chronology of John of Salisbury's Studies in France," and Charles Burnett, "The Content and Affiliation of the Scientific Manuscripts Written at, or Brought to, Chartres in the Time of John of Salisbury," all in *World of John of Salisbury,* ed. Wilks (cit. n. 90), pp. 21–37 (esp. pp. 23–25), pp. 109–116; and pp. 127–160 (esp. pp. 128–130), respectively. For a more positive estimate of natural philosophy at Salerno see Danielle Jacquart, "Aristotelian Thought in Salerno," in *Twelfth-Century Philosophy,* ed. Dronke (cit. n. 1), pp. 407–428; and Mark Jordan's article in this volume.

smattering of learning for the immediate purpose of a career in medical practice. True *physica* might then be an ideal that few could realize, but as an ideal it would also be highly immune to destruction by the many who abused it.

IV. EPILOGUE

With John of Salisbury we find a well-integrated notion of *physica* as a scientific, medical, and ethical ideal, although his views also reveal some tensions. First, as noted above, he was aware that some authorities (probably meaning Aristotle) would hold *physica* aloof from daily problems. Second, he was familiar with the classical association of *physica* with cosmology. And third, he equated *physica* primarily with a body of knowledge whose leading authorities instead called it *medicina*.[96] The latter two of these tensions were undergoing at least partial resolution even as John wrote. With the arrival of new astrological knowledge, medical *physica* was itself acquiring a fully cosmological dimension, though without losing its strongly microcosmic focus.[97] Furthermore, the dichotomy between *physica* and *medicina* was being transformed into a less ambiguous one between *physica* and *chirurgia* (literally, "natural philosophy" versus "hand work").[98]

However, insofar as the first tension was rooted in specifically Aristotelian notions, it would only become progressively greater.[99] John was writing near the end of the era during which Aristotle was known chiefly as an authority on logic, even by a scholar as accomplished as John. John could therefore feel free to define the scope and purpose of natural philosophy in his own way, but as his successors gradually became familiar with the *libri naturales,* they found that their contents were not in fact predominantly medical. Furthermore, from other newly available authors such as Avicenna they learned that this body of knowledge does have an important relationship to rational medicine, but a relationship whose very articulation demands that the two disciplines be kept clearly distinct.

This new emphasis on the distinction between natural science and medicine would take firm hold in the medical faculties of the late medieval universities, but it could not so readily eliminate the indigenous hybrid, medical *physica,* from the broader mainstream of European culture. We catch a clear glimpse of this older ideology in a work written in 1256 by Aldobrandino or Alebrant (d. 1287), a Sienese physician who had settled in France. The text was a vernacular guide to health entitled *Le régime du corps,* but in the preface Alebrant explained that his subject was actually called "physique."[100] He acknowledged that his information

[96] At *Metalogicon* 2.15, 872d, John gives as a typical nonmedical question in *physica,* "Utrum mundus eternus uel non"; the instance is from Aristotle, *Topics* 104b7, 105b25. Conversely, at *Metalogicon* 2.20, 880c he states: "Galienus in Tegni medicinam dicit scientiam sanorum, egrorum, et neutrorum."

[97] Charles W. Clark, "The Zodiac Man in Medieval Medical Astrology" (Ph.D. diss., Univ. Colorado, 1979; Ann Arbor, Mich.: University Microfilms International, 1987), pp. 178–282; and Charles Burnett, "Hermann of Carinthia," in *Twelfth-Century Philosophy,* ed. Dronke (cit. n. 1), pp. 387–404.

[98] See *The Commentary on Martianus Capella's De nuptiis Philologiae et Mercurii Attributed to Bernardus Silvestris,* ed. Haijo Jan Westra (Toronto: Pontifical Institute of Mediaeval Studies, 1986), p. 119: "Quedam medicine species est quam 'cirurgiam' Greci dicunt, id est opus manuum. Ipsa enim in stimmatibus et scissuris consistit. . . . Haec scientia, cum tota sit in actu, absque phisica que causas perpendit, parum potest. Ideoque gaudet eius consortio." But cf. p. 131. See also Amundsen, "Medicine and Surgery" (cit. n. 1), esp. pp. 53–57.

[99] See nn. 7 and 8, above.

[100] *Le régime du corps de maître Aldebrandin de Sienne,* ed. Louis Landouzy and Roger Pépin (Paris: Champion, 1911; Geneva: Slatkine Reprints, 1978), pp. 3–7.

was derived from works on astronomy, natural science, and medicine, but he underscored the special status of *physique* by way of a cosmogony whose Christian-Platonic roots are quite apparent. Thus in creating the world, God first made the heavens and next the four elements; from the latter he created all other things, including plants, trees, birds, fish, and all other animals. Last of all he created man, the most noble creature, in his own image, and gave him lordship over all the other things that he had previously made to serve human needs.

Like all of these other creatures, however, the human body is composed of the four elements and therefore inevitably undergoes corruption, old age, and finally death. For this reason, God gave to man a special gift, namely,

> a science that is called *physique,* by which he protects the health which was originally given to him, and by which he can also remove disease, although *physique* was created especially to protect health; and so it could not be understood better than that *physique* is a science for making man live all of his [allotted] years, and so has been created for conducting man to his natural death.

Subsequently, Alebrant offered a more formal definition: "*Physique* is a science by which one understands all the characteristics of the body of man, and by which one protects the health of the body, and removes its diseases; and we can say that this science has two parts, of which one is called theory, and the other practice." This definition makes two explicit references to the human body, but none to either natural science or medicine. There were good reasons for both omissions, because this is a translation of the opening passage of the *Canon* of Avicenna, where the word being defined is *medicina.*[101]

Had Alebrant been writing an academic commentary, he could not similarly have ignored the word *medicina,* nor Avicenna's establishment of a distinction between this body of knowledge and *scientia naturalis.*[102] But absent such constraints, we are permitted to see medical *physica* in all its peculiarity, for as to its detailed substance it was, in large measure, theoretical medicine by another name, and yet it was also an ideology so powerful as to have induced Alebrant to write that other name in place of medicine. The two principal motifs of this ideology were a close and optimistic association between the word *physica* and the human body in its "natural" or healthy state, and the assumption that natural science in a broader sense should serve human needs. The first may have been inconsistent with classical linguistic norms, and the second with the values of Aristotelian philosophy. But their intrinsic appeal is quite apparent.

[101] *Ibid.,* pp. 5, 6; and introduction, p. lxii.

[102] Avicenna, *Canon* 1.1.1, intro. and Ch. 1. See also, e.g., Siraisi, *Taddeo Alderotti* (cit. n. 7), pp. 118–127.

The Construction of a Philosophical Medicine

Exegesis and Argument in Salernitan Teaching on the Soul

*By Mark D. Jordan**

THE INFLUX OF NEW SCIENCE into the Latin West was an influx of new texts. These demanded new ways of reading. It was not enough that the texts be put into a passable Latin. They had to be glossed, interpreted, reconciled, and corrected in circumstances very different from the circumstances of their Arabic or Byzantine transmission. The tasks were undertaken most clearly in medical teaching. It is in teaching that new terminologies are appropriated and fixed, that new texts are regularly construed, that new bodies of knowledge can be situated within the prevailing hierarchies of arts and sciences. Some of the best evidence for changes in the twelfth century's teaching of medicine comes from the "school" at Salerno, and especially from expositions of its canon of medical textbooks known as the *articella*. Because this group of texts served as something like an introduction to the study of medical theory, it attracted the attention of several generations of Salernitan masters. Their commentaries on the *articella* illustrate with particular clarity the changes in Salernitan teaching and its attempts to construct a properly comprehensive curriculum for the new medicine.[1]

I. THE *ISAGOGE* AND THE *ARTICELLA*

The earliest evidence for the canon later called the *articella* is found in manuscripts from just after 1100.[2] At first the canon comprises five texts: the *Isagoge* of "Johannitius," Hippocrates' *Aphorisms* and *Prognostics,* Theophilus on urines, and Philaretus on pulses. "Johannitius" is Ḥunain ibn Isḥāq and the *Isagoge* is an anonymous Latin abridgment of one of the versions of his *Ques-*

* Medieval Institute, University of Notre Dame, 715 Hesburgh Library, Notre Dame, Indiana 46556.

[1] No study of the *articella* commentaries can begin without acknowledging a fundamental debt to the discoveries and explanations published since 1945 by Paul Oskar Kristeller. The most important of these studies have now been anthologized in Italian translation as *Studi sulla Scuola medica salernitana* (Naples: Istituto Italiano per gli Studi Filosofici, 1986).

[2] In this and subsequent arguments about chronology, I summarize conclusions argued more fully in "Medicine as Science in the Early Commentaries on 'Johannitius,' " *Traditio,* 1987, *43*:121–145, on pp. 130–137. In the present essay I have also left aside most complications of the manuscript tradition (e.g., the plurality of commentaries bearing family resemblance to the Digby texts, the two versions of Bartholomaeus), and so I confine myself in most instances to citing a single manuscript.

tions on Medicine.[3] The other texts of the canon seem to be of mixed Arabic and Byzantine derivation. There is as yet no satisfying account for their combination.[4] What is clear, and certainly more pertinent, is that the texts of the *articella* quickly became one focus for Salernitan advances in the teaching of theoretical medicine. This is evident in the canon itself, which underwent repeated modifications, but is best seen in the commentaries on it.

The chronology of these commentaries can be fixed only by a combination of inferences and conjectures, which I will just summarize here. Shortly after the original *articella* of five elements became available at Salerno, the texts occasioned a set of elementary commentaries by an author or authors who relied exclusively on sources translated and taught there, and who may be identified with the author of a famous Salernitan anatomical demonstration. These were named by Paul Oskar Kristeller the "Digby" commentaries, after the Oxford manuscript in which he found them together. They served in turn as the basis for a second set of commentaries, even simpler, that were attested in a twelfth-century manuscript of the Chartres chapter library—hence the "Chartres" commentaries. Just after mid century, another group of commentators appeared at Salerno, all of them identifiable masters and even colleagues. The first of these commentaries is that of "Archimatthaeus" on the *Isagoge* alone. There follow the complete commentaries of Bartholomaeus in at least two versions, one of them reported or redacted by Peter Musandinus. In both versions Bartholomaeus comments upon the *articella* of six elements, and he may have been responsible for adding Galen's *Tegni* to the canon. The last of the twelfth-century Salernitan commentaries are those written by Maurus, a student of the two previous commentators and of their commentaries.

These five sets of commentaries are much too large to be studied in detail by even a dozen essays. But the study of evolution in teaching can be seen only in details, and so some selection of particular passages must be made. The most important single work of the canon for theoretical medicine is the *Isagoge* of Johannitius. The *Isagoge* is also the only piece for which there survives a commentary by Archimatthaeus, and so it provides the fullest scope for comparisons. In what follows, then, I will concentrate on the teaching of the *Isagoge*—indeed, on the teaching of some of its preliminary chapters. In the chapters from "On Powers" (*De virtutibus*) (Ch. 9) through "On Spirits" (*De spiritibus*) (Ch. 13), Johannitius introduces the powers and operations of the soul.[5] The doctrine of this section is of obvious importance to medical theory but also to philosophical and theological anthropology. The topics raised here figure prominently in

[3] The best treatment of the *Isagoge*'s relations to its possible Arabic originals is now Danielle Jacquart, "A l'aube de la renaissance médicale des XIe–XIIe siècles: L'*Isagoge Johannitii* et son traducteur," *Bibliothèque de l'Ecole des Chartes,* 1986, *144*:209–240.

[4] One standard supposition is that Constantine translated the *Isagoge* and that he or his immediate disciples assembled the *articella*. See, e.g., Tiziana Pesenti, "Editora medica tra Quattro e Cinquecento: L'*Articella* e il *Fasciculus medicine,*" in *Trattati scientifici nel Veneto fra il XV e XVI secolo* (Venice: Neri Pozza Editore, 1985), pp. 1–28, on p. 1. But Jacquart notes a number of differences between the glossary of the *Isagoge* and that of Constantine's mature works: "A l'aube de la renaissance," pp. 234–236.

[5] Medical terms present some of the greatest difficulties for any reflective translation, since they are notoriously laden with theory. In what follows, I want to keep at a minimum assumptions about the observed phenomena being named, and so I offer English terms "under erasure" and with frequent reminders of the underlying Latin.

ancient, patristic, and medieval works of very different genres, including basic accounts of biology, moralizing depictions of the passions, analyses of sensation, and hexaemeral commentaries on Genesis. They offer a particularly clear case for studying adjustments or dislocations in the hierarchy of learning.

Johannitius's doctrine in these sections can be sketched quickly enough. His treatment of the soul is preceded by a prologue on the divisions and aims of medicine, a chapter enumerating the "natural things" (Ch. 1), two chapters on elements and their mixtures (Chs. 2–3), four chapters on the humors (Chs. 4–7), and a chapter on the kinds of members (Ch. 8).[6] The section on the soul comes next, comprising five chapters. In Chapter 9 Johannitius teaches that there are three powers (*virtutes*): the natural, the spiritual (i.e., connected with *spiritus*), and the animal (connected with *anima*). The natural power is divided into powers that are served (the generative, nutritive, and augmentative) and powers that serve (the appetitive, retentive, digestive, and expulsive). From the spiritual power there proceed, according to Chapter 10, an operative power (which dilates and contracts the heart) and then an operated power (which produces the passions). The animal power, the subject of Chapter 11, is divided into an ordering, composing, and discerning power; then a voluntary power; then a sensible power. The ordering power produces imagination (in the ancient sense), cogitation or thinking, and memory, each of which is seated in a separate ventricle of the brain. The voluntary power moves muscles; the sensible power spreads out into the five senses. Chapter 12 divides the operations (*operationes*) into simple and compound. Simple operations, such as appetite or digestion, produce their effects just by themselves, in virtue of elementary properties. Compound operations, such as desire, require the combination of two simple operations; in desire, the combination would be of appetite and sensation. Finally, in Chapter 13, Johannitius distinguishes three "spirits" (*spiritus*)—the natural, governed by the liver; the vital, by the heart; and the animal, by the brain. Each spirit spreads out to the whole body from its seat—the natural through the pulseless veins (*venas*), the vital through the arteries (*arterias*), and the animal through the "nerves" (*nervos*).[7]

Even so bald a summary will show something of the character of Johannitius's teaching. It is extremely schematic, being more concerned to inculcate terms and classifications than to offer arguments. The telegraphic style may be due in part to the editing of the Latin version of an original that proceeds catechetically by question and answer. But the schematic character is also explained by Johannitius's intention, which is to provide the briefest possible introduction to the central terms and relations of medicine as a prelude to Galen's *Tegni*. A Latin text comparable in tone, though surely not in doctrine, would be Isidore's *Etymolo-*

[6] I follow the chapter divisions in the early Beneventan manuscript, London, Wellcome Institute MS 801A (once Bury St. Edmunds M.27). These conform for the most part to the divisions in the Padua edition of N. Petri, ca. 1476.

[7] There are textual difficulties in the manuscript tradition and in the Arabic original. See Jacquart, "A l'aube de la renaissance" (cit. n. 3), pp. 217–218, 228–229. I simply follow the text of Wellcome MS 801A. Note that Johannitius preserves the three kinds of *pneuma* in Galenic doctrine, whatever the confusion of terminology. Other accounts, notably that in Qusṭa ibn Lūqā's *De differentia spiritus et animae*, assert only two kinds. For a partial history of these teachings in the Latin West see James J. Bono, "Medical Spirits and the Medieval Language of Life," *Traditio*, 1984, *40*:91–130.

gies, Book 4. Indeed, Johannitius's text can sometimes seem so straightforward that it becomes puzzling what there is to be commented upon in it.

II. THE PEDAGOGY OF THE EARLIEST COMMENTARIES

In a study of development the easiest supposition is always that the simplest text comes first. Unfortunately, but perhaps interestingly for easy theories of development, the simplest commentary on Johannitius does not appear to be the first. The justification of this claim will require some attention to textual details. The attention will be repaid in two ways. First, there is the issue of simple chronology, on which theories of development stand or fall. But second, there is the more important question whether the development of the medical curriculum ought properly to be called Salernitan at all. Perhaps it began outside of medical circles, say among teachers of the liberal arts at Chartres.

The simplest set of commentaries on the *articella* of five texts once appeared together in a twelfth-century manuscript of the cathedral chapter at Chartres (MS 171). There are brief descriptions of the manuscript's contents in several sources,[8] but the codex itself was destroyed during World War II. Kristeller identified scattered copies in other codices for all of its commentaries except that on Johannitius. A copy of the first part of that commentary was later identified in a marginal gloss of a manuscript once at the abbey of Bury St. Edmunds.[9] This gloss is both incomplete and somewhat scrambled, but there is no reason to suppose that it does not reproduce some of the missing Johannitius commentary from Chartres.

What is the relation of these passages to the Digby commentary on Johannitius? There can be no doubt of a strong textual relation. Almost everything that appears in the Chartres commentary can be found in Digby, though the converse is not true. Comparisons of remarks on the textual order of Johannitius are particularly telling. For both Chartres and Digby, Constantine's *Pantegni* and Alfanus's version of Nemesius, the *Premnon physicon,* supply most of the matter for the commentary, but the remarks on the textual or logical order of the *Isagoge* clearly cannot be derived from Constantine or Nemesius. Just as clearly, the Chartres and Digby commentaries explain the order in the same sense and with many of the same words.[10]

[8] *Catalogue générale des manuscrits des bibliothèques de France,* 8vo series, *Départements,* Vol. XI (Paris: Plon, 1890), p. 90; and Loren C. MacKinney, *Early Medieval Medicine, with Special Reference to France and Chartres* (Baltimore: Johns Hopkins Press, 1937), pp. 210–211, n. 282. Most important, MacKinney had acquired photostats of the opening page of each commentary in the manuscript, together with two further pages from the commentary on Philaretus.

[9] Formerly Helmingham Hall MS 58, now in the private London collection of Philip Robinson.

[10] Cf. Helmingham MS 58, fol. 77ra marg (the Chartres commentary), with Cambridge, Peterhouse MS 251, fol. 55ra (the Digby commentary), on treating the members before the powers. The *Pantegni* is Constantine the African's partial and periphrastic rendering of the *al-Maleki* of ᶜAli ibn al-ᶜAbbas al-Majūsī (d. 994); the *Premnon physicon* is Alfanus of Salerno's partial and defective translation of Nemesius's *De natura hominis.* The *al-Maleki* comprises twenty books divided into two parts, the first concerned more with explanation and the second more with clinical practice. Constantine translated the whole of the first part (the *Theorica*), but only the opening two books and half of the ninth from the second (the *Practica*). On Nemesius see Helen Brown Wicher, "Nemesius of Emesa," *Catalogus translationum et commentariorum: Medieval and Renaissance Latin Translations and Commentaries,* Vol. VI, ed. F. Edward Cranz (Washington, D.C.: Catholic Univ. America Press, 1986), pp. 46–50.

There is circumstantial evidence that seems to suggest an early use of the *articella* at Chartres. The Chartres chapter library contained another twelfth-century medical manuscript comprising the *articella* of five texts and Constantine's *Pantegni*. Just this cluster of sources appears in the *De philosophia mundi* of William of Conches, well known for his association with Chartres. Indeed, William there mocks those who refuse to read the works of Constantine and the *physici*.[11] Other citations to Constantine and Johannitius occur in William's *Glosae super Platonem*.[12] In fact, the six explicit citations of Constantine make him the most frequently cited author in the *Glosae* after Boethius.

The claim for the Chartrian origin of the simplest set of commentaries could be further supported by an argument from the *accessus* pattern with which the commentaries so prominently begin. It shows strong affinities with introductory patterns in other authors usually associated with Chartres.[13] The correct form of the *accessus* in different fields seems also to have been a matter of concern at Chartres. We are told by the anonymous author of an *accessus* to William's glosses on Juvenal that there was a dispute between William and Bernard of Chartres about the universal appropriateness of one common form. Bernard held that it was inappropriate to ask of literary works under which branch of philosophy they were to be placed, since they were neither parts of philosophy nor works about philosophy. William of Conches insisted, however, that this particular heading should appear in the *accessus* to all authors.[14] So the anonymous narrator includes it in his own *accessus* to William on Juvenal. The particular heading also appears in the *accessus* patterns of authors familiar with Chartrian works, such as Dominicus Gundissalinus.[15] Of course, the reference to philosophy appears as well in the opening lines of the Chartres commentary on Johannitius: "Six things are required at the beginning of this work, namely, the matter, the manner of proceeding, the intention [of the author], the usefulness [to the hearer], what part of philosophy it falls under, the cause and title of the work."[16]

[11] William of Conches, *De philosophia mundi,* ed. Gregor Maurach with Heidemarie Telle (Pretoria: Univ. South Africa, 1980), references to the *Pantegni* at 1.20 (pp. 26–27), 1.23 (pp. 28–29), 1.28 (p. 30), 1.41 (p. 37), 4.37 (p. 106); to Johannitius at 1.7 (p. 28), 4.18 (p. 104); to Theophilus at 4.17 (p. 106). Mockery: *ibid.*, 1.23 (p. 28): "Sunt quidam, qui neque Constantini scripta neque alterius physici umquam legerunt, ex superbia ab aliquo discere dedignantes."

[12] *Guillaume de Conches: Glosae super Platonem,* ed. Edouard Jeauneau (Textes Philosophiques du Moyen Age, 13) (Paris: J. Vrin, 1965), references to Constantine at sects. 51 (p. 120), 53 (p. 128), 59 (p. 130), 157 (p. 263), 164 (pp. 272–273); to Johannitius at sect. 137 (p. 238).

[13] For Thierry of Chartres, *Commentarius in Ciceronis De inventione,* as in N. M. Häring, "Thierry of Chartres and Dominicus Gundissalinus," *Mediaeval Studies,* 1964, *26*:271–296, on p. 281; Thierry, on *De inventione,* as in Karin M. Fredborg, "The Commentary of Thierry of Chartres on Cicero's *De inventione,*" *Cahiers de l'Institut du Moyen Age Grec et Latin* (Univ. Copenhagen), 1971, *7*:1–36, on p. 9; Thierry on Boethius and *Tractatus de sex dierum operibus,* as in Häring, *Commentaries on Boethius by Thierry of Chartres and His School* (Toronto: Pontifical Institute of Mediaeval Studies, 1971), pp. 57, 555, respectively; for Clarembald of Arras, *Tractatus super librum Boetii De Trinitate,* as in Häring, *Life and Works of Clarembald of Arras . . .* (Toronto: Pontifical Institute of Mediaeval Studies, 1965), p. 74; and for William of Conches, *Accessus ad Timaeum* 2, ed. Edouard Jeauneau (Paris: J. Vrin, 1965), p. 58.

[14] In *Guillaume de Conches: Glosae in Iuvenalem,* ed. Bradford Wilson (Paris: J. Vrin, 1980), pp. 89–90.

[15] *Dominicus Gundissalinus De divisione philosophiae,* ed. Ludwig Baur (Beiträge zur Geschichte der Philosophie des Mittelalters, 4.2–3) (Münster: Aschendorff, 1903), in the discussion appended to the treatment of the parts of practical philosophy, p. 140, lines 13–16.

[16] The Chartres commentary, Chartres MS 171, fol. 1ra; Helmingham MS 58, fol. 76ra marg.: "Sex requiruntur in principio huius operis, scilicet materia, modus tractandi, intentio [auctoris *add.*

Might this be another clue to a Chartrian or at least northern origin for the practice of writing short commentaries on the *articella?*

The answer, I think, is that none of these arguments demonstrate either the priority of the Chartres commentary or a Chartrian origin for the practice of writing commentaries on the *articella*. The commentary's simplicity is not that of a first commentary so much as that of an abridgment. In parallel passages in Chartres and Digby for which no outside source is known, the Chartres commentary frequently makes less sense and proceeds less smoothly than the Digby.

Nor are the circumstantial arguments conclusive. Constantine found readers at Chartres early in the twelfth century, but he also found them in other places. There is an abundant early use of the *Pantegni* in William of St. Thierry's *De natura corporis et animae,* a work dated around 1140. Moreover, the sequence of topics in William's first book follows, not the *Pantegni,* but the *Isagoge*.[17] William is perfectly candid about his dependence on the *philosophi* and *physici,* and he cites Constantine's *Liber de gradibus* once by name. Related material, though less extensive and less technical, appears in Adelard of Bath's *Quaestiones naturales* and even in Hugh of St. Victor's *De unione corporis et spiritus*. Again, Constantine's influence reaches out to writers as different as Bernard Silvestris and Daniel of Morley.[18] Moreover, even a casual survey of extant manuscripts of the *Pantegni* will disclose enough twelfth-century copies to suggest a fairly wide reception.[19] No special conclusions can be drawn, then, from the reading of Constantine at Chartres.

The argument from the *accessus* patterns is similarly inconclusive. The Chartres commentary uses an *accessus* pattern going back ultimately to Boethius and found in enormous variety throughout twelfth-century texts.[20] Constantine himself uses a six-part *accessus:* intention, usefulness, title, part of teaching to

---

Helm.], utilitas [audiendi *add. Helm.*], cui parti philosophie subponatur, causa operis et titulus."

[17] William of St. Thierry, *De natura corporis et animae,* in *Patrologiae cursus completus, series Latinae,* ed. J.-P. Migne (Paris, 1845–), Vol. 180, cols. 695–708, noting particularly the main sequence: elements (cols. 695–696), humors (cols. 696–700), spirits (cols. 700–702), cerebral divisions and operations (cols. 702–704), the senses, with much emphasis on sight (cols. 704–707), and the ages (col. 707). For the dating and general remarks on sources see Bernard McGinn, *Three Treatises on Man: A Cistercian Anthropology* (Kalamazoo, Mich.: Cistercian Publications, 1977), pp. 22–24, 28.

[18] *Die Quaestiones naturales des Adelardus von Bath,* ed. Martin Müller (Beiträge zur Geschichte der Philosophie und Theologie des Mittelalters, 31.2) (Münster: Aschendorff, 1934), 17–18 (pp. 22–23), 20 (pp. 24–25), 23 (pp. 26–31), 31 (p. 36); Hugh of St. Victor, *De unione corporis et spiritus,* in *Patrologia Latina,* Vol. 177, cols. 286B–288B; Bernard Silvestris, *Cosmographia,* ed. Peter Dronke (Leiden: E. J. Brill, 1978), 9.6 (p. 139), 13.11 (p. 149), 13.12 (p. 149), and 14 throughout (cf. Brian Stock, *Myth and Science in the Twelfth Century: A Study of Bernard Silvester* [Princeton, N.J.: Princeton Univ. Press, 1972], pp. 26, 212–213, 221–222); and, e.g., Daniel of Morley, *De naturis inferiorum et superiorum* 1, ed. Karl Sudhoff, *Archiv für Geschichte der Naturwissenschaften und der Technik,* 1917, *8*:1–40, on pp. 18–22, the doctrine of the elements.

[19] E.g., Cambridge, Trinity College MS R.14.33 (James 906), from Bury St. Edmunds; Erfurt, Wissenschaftliche Allgemeinsbibliothek MS Amplon. Q.184, dated 1147; Florence, Biblioteca Nazionale Centrale MS Conv. soppr. F.7.1115; Hildesheim, Dombibliothek MS 748; The Hague, Koninklijke Bibliotheek MS 73.J-6, possibly late eleventh century; London, British Library Additional MS 22.719, from Priory of St. Nicholas, Exeter; Vatican, Bibliotheca Apostolica, MS Pal. lat. 1160, possibly early thirteenth century; and *ibid.*, MS lat. 2455.

[20] For a summary of work on *accessus* patterns and a proposed narrative of their development see A. J. Minnis, *Medieval Theory of Authorship* (2nd ed., Philadelphia: Univ. Pennsylvania Press, 1988), pp. 9–72, though Minnis says relatively little about medical texts. For some of these, and the confection of an *accessus* pattern for teaching parts of bodies, see Roger French, "A Note on the Anatomical *Accessus* of the Middle Ages," *Medical History,* 1979, *23*:461–468.

which the work belongs, name of the author, and division.[21] In doing so, he joins a long medical tradition reaching back through the Arabic transmission of the *Summa Alexandrinorum* to the sixth-century teaching practices at Alexandria.[22] The relations of medicine to the rest of learning are particularly stressed by Constantine. He begins the *Pantegni* by announcing that all secular and divine letters must be classified under the three principal parts of science: logic, ethics, and *physica*. Medicine, he argues, falls under all three.

Furthermore, there is in fact evidence to suggest that the practice of commenting on the *articella* originated elsewhere than at Chartres. It comes in a series of textual connections between the Digby commentary and a well-known Salernitan anatomy discovered by August Henschel in the Breslau codex, printed by Salvatore De Renzi, and translated into English by George Corner. What is most interesting is that this *Anatomia* contains references by the author to his own commentaries on Johannitius, the *Aphorisms,* and Philaretus.[23] These are, of course, three of the elements of the *articella* of five, and so we may take the author as referring to his own commentaries on the *articella.* Now the most telling of the references is that to the commentary on Johannitius, because it refers to a doctrine about the processes of vision. As will become clear below, the treatment of vision draws extended attention from the commentators, who disagree with each other in judging among the received accounts. The author of the anatomical demonstration refers to an explanation of vision that is the same as the doctrine of the Digby commentary on Johannitius—the union of the *spiritus* and the clear air, which serves as its instrument.[24] There is no discussion of this particular passage in the corresponding portion of the Chartres commentary. Moreover, the other references to the *Aphorisms* and Philaretus also find matching passages in the Digby commentaries.[25]

A more striking congruence even than these, I think, is found in the introduc-

[21] Constantine, *Pantegni* 1.3. When citing the *Pantegni,* I mean the *Pantegni theorica.* For present purposes I omit discussion of the manuscript tradition. My references to it are general enough to be verified in both printed editions (Lyons, 1515; Basel, 1539), neither of which reproduces the versions of the text available in the twelfth century.

[22] See Max Meyerhof, "Von Alexandrien nach Bagdad," *Sitzungsberichte der Preussischen Akademie der Wissenschaften, Philosophisch-historische Klasse,* 1930, pp. 389–429; L. G. Westerink, "Philosophy and Medicine in Late Antiquity," *Janus,* 1964, *51*:169–177, on pp. 170–171; Stephanus the Philosopher, *A Commentary on the Prognosticon of Hippocrates,* ed. John M. Duffy (Corpus Medicorum Graecorum, 11.1.2) (Berlin: Akademie-Verlag, 1983), on pp. 26–34, and the introductory remarks on the eight-part *accessus.*

[23] For the history of the text's discovery and publication see George W. Corner, *Anatomical Texts of the Earlier Middle Ages* (Washington, D.C.: Carnegie Institution, 1927), p. 21. I use the Latin text as published by Salvatore De Renzi, *Collectio Salernitana,* Vol. II (Naples: Filiatre-Sebezio, 1851–1857; rpt. Bologna: Forni, 1967), pp. 391–401, with corrections derived from Corner and Karl Sudhoff, "Die erste Tieranatomie von Salerno und ein neuer salernitanischer Anatomietext," *Archiv für Geschichte der Mathematik, der Naturwissenschaften, und der Technik,* 1927, *10*:141–147. For the author's references see De Renzi, *Collectio Salernitana,* Vol. II, p. 401, line 31 (Johannitius); p. 395, lines 16–17 (*Aphorisms*); p. 395, line 25 (Philaretus); see also the comments by Corner, *Anatomical Texts,* p. 22. O'Neill has recently shown the dependence of the *Anatomia* on traditional Latin sources and rebutted once again its attribution to Copho; see Ynez Violé O'Neill, "Another Look at the 'Anatomia Porci,' " *Viator,* 1970, *1*:115–124.

[24] De Renzi, *Collectio Salernitana,* Vol. II, p. 401, lines 26–31; cf. Corner, *Anatomical Texts,* p. 66. The doctrine and its sources will be discussed below.

[25] The cited lemma of the *Aphorisms* refers to particula 5.68; for the Digby commentary *ad loc.,* see British Library, Royal MS 8.C.IV, fol. 205ra = Bern, Burgerbibliothek MS A52, fol. 58ra, the passage "Si quis posteriora parte capitis patitur flebothometur de vena frontis. . . ." For the reference to Philaretus see the anatomy at Royal MS 8.C.IV, fol. 166va–vb.

tion to the study of the members in both the *Anatomia* and the Digby commentary. As Corner notes, the very beginning of the *Anatomia* corresponds almost verbatim to a passage in the *Pantegni*.[26] Indeed, the whole of *Anatomia* before the beginning of the instructions for dissection depends upon *Pantegni* 2.1. An almost identical text occurs in the Digby commentary's remarks on Johannitius, Chapter 8, "On Members." The *Anatomia* and the Digby commentary both paraphrase, at different lengths, the same passage from Constantine, but some details of their diction show that they are paraphrasing it with the same peculiarities. These shared additions to the underlying text, together with the matches in Digby for the *Anatomia*'s cross-references, suggest that the author of the *Anatomia* may well be the author of the Digby commentaries. If that is correct, then we have a confirmation of the Salernitan origin of the Digby commentaries and some further support for the stylistic judgment that the Chartres commentary is an abridgment of it.

The dependence of the Digby commentary itself on Constantine and Nemesius has been asserted several times but can now be summarized in tabular form. Given below are the chapters from the Digby commentary and their sources. Sources marked with an asterisk appear verbatim in the commentary.

9	*De virtutibus*	*Pantegni* 4.1,* 4.2,* 4.4
10	*De virtute spirituali*	*Pantegni* 4.5
11	*De virtute animali*	*Pantegni* 4.1, 4.11, 4.12,* 4.13,* 4.14* *Premnon physicon*[27] 75.12–17,* 76.1,* 76.20, 77.5–8, 77.15–18*
12	*De operationibus*	None
13	*De spiritibus*	None

The commentators, of course, do not usually attribute the borrowings, even for the verbatim quotations. But the Digby commentator does cite Constantine several times in these chapters. In one particularly interesting passage he seems to argue that although Johannitius is following a different order from Constantine's, he is offering the same doctrine.[28] In fact the order of the Digby commentary on the ninth chapter is just the order of the first two chapters of Constantine's Book 4, and the commentator falls easily into Constantine's order whenever he comments on related topics in Johannitius.

It is not too much to say, then, that Digby comments on Johannitius by attaching the corresponding passages from Constantine and Nemesius. This collation makes at least three pedagogical gains, each of which tells us something about the intended audience. The first gain is exegetical in the strictest sense. It has to do with clarifying the order of the underlying text, with glossing certain terms, with filling out arguments, and with strengthening textual connections. In dealing with textual order, the Digby commentator introduces the most rudimentary

[26] Corner, *Anatomical Texts* (cit. n. 23), p. 23.

[27] *Nemesii episcopi Premnon physicon . . . liber a N. Alfano . . . in Latinum translatus*, ed. C. J. Burkhard (Leipzig: Teubner, 1917).

[28] Digby commentary, Peterhouse MS 251, fol. 53rb, Ch. 9: "Dicit namque Constantinus in *Pantegni* quod . . . hoc totum in libro [iste] est, licet alio ordine repereatur Constantino." Cf. Ch. 9, fol. 52vb: "Sed Constantinus in *Pantegni* . . . ; Ch. 10, fol. 53vb: "Qua de causa sapiens Constantinus in *Pantegni*. . . ." Constantine is also the most frequently cited source in the *Anatomia*.

form of *quaestio*—that is, a simple objection against an apparent inconsistency in the text.[29] There are many instances in which the commentator defends Johannitius against the charge of superfluity or explains apparent discrepancies by suggesting that Johannitius has left something unsaid.[30] With such remarks, with the glosses, with the supplementary steps of argument, the commentator is quite clearly filling the need of readers for an introduction to medicine even more rudimentary than the schematic and laconic Johannitius. If Johannitius seems to have the brevity of an *aide-mémoire* or a summary review, the Digby commentary seems by contrast to be a primer.

The second pedagogical gain is the introduction of the *quaestio* in a more proper sense, that is, a textual question arising from the conflict of *auctoritates*. The rise of the *quaestio disputata* as the preferred teaching form is well known in all academic fields during the twelfth century. If the suggested chronology is correct, the Digby commentary would be one of the earliest uses of something like the *quaestio* in medical commentary.[31] The Digby commentator frequently enough notes opposing positions in order to reply to them.[32] But something much more like a ranking of authorities and a determination of the truth is provided in the explicitly interpolated treatment of vision.

The commentator introduces the longish digression as a supplement to Johannitius, who provides in Chapter 11 nothing but an enumeration of the senses as effects or constituents of the sensible animal power. To this, the Digby commentator adds an account of sensation, in which the longest part concerns sight. The treatment begins with the principle that sensation is to be understood by the analogous disposition of elements in the great world and the microcosm.[33] So sight, with its fiery nature, is subtler than the other senses and ought to come first. But what follows this analogy is a summary statement on the physics of sight of Hipparchus, "the geometers," the Epicureans, and Constantine. The source for the first three views is Nemesius; for the fourth, the *Pantegni*.[34] These

[29] This is to be distinguished from the much more ancient form of the *quaestio* literature studied for Salerno by Brian Lawn.

[30] Digby commentary, Peterhouse MS 251, fol. 35ra, "non pretermittendum quod auctor in libro superfluum dixit"; fol. 35ra, "liber aperte non ponit sed per extremum subinnuit"; fol. 35rb, "Sed hic querendum quare liber prius de spirituali quam de animali, quae dignior est, exsequitur. Et dicamus ideo quia in pluribus est quam animalis, vel ideo quia lege temporis precedit in actione."

[31] See Danielle Jacquart in B. C. Bazàn, J. W. Wippel, G. Fransen, and D. Jacquart, *Les questions disputées et les questions quodlibétiques. . .* (Typologie des Sources du Moyen Age Occidental, 44–45) (Turnhout: Brepols, 1985), pp. 290–291. For good recent summaries of scholarship on the *quaestio* see *ibid.*, esp. pp. 25–31, 243–246, 291–299. There has been some suggestion that the Latin West might have learned disputation from Islamic medical texts; see George Makdisi, "The Scholastic Method in Medieval Education: An Inquiry into Its Origins in Law and Theology," *Speculum*, 1974, *49*:640–661, on p. 659. But Constantine's *Pantegni* and the texts of the *articella* tend to go in an entirely different direction, toward the univocity of the manual.

[32] E.g., Digby commentary, Peterhouse MS 251, fol. 53rb, "Dicamus quod quidam volunt quod. . . . Nobis aliter placet hoc modo. . . ."

[33] *Ibid.*, fol. 53va, "Videndum est igitur quod hii V sensus IV elementorum proprietates secuntur, et secundum eorum ordinem in maiori mundo et hoc etiam in microcosmo et in minore mundo disponuntur." "Microcosmus" does occur in Nemesius (*Premnon physicon*, cit. n. 27, 22.14), but not in this context.

[34] Nemesius, *Premnon physicon* (cit. n. 27), 75.12–76.2; and Constantine, *Pantegni* 4.11, Basel ed. (cit. n. 21), fols. 92–93. Note that the commentator omits from Nemesius's doxography the views of Aristotle, Plato, and Galen. Much of the argument here resembles that in Johannitius's optical work, known to the Latin West as the *De oculis* of Constantine. For a summary of the work and its importance in the development of Arabic optics and ophthalmology see David C. Lindberg, *Theories*

sources show why the commentator feels compelled to expatiate. If the *Isagoge* says nothing here about the process of sight, the controlling *auctoritates* record and investigate lengthy controversies on just this point.

The Digby commentator meets these controversies by laying out both the historical opinions and the logical possibilities. There are only three possibilities: either the object seen directs something to the sight by which it is made manifest, or else the power of sight goes out to the thing and understands it, or else neither goes to the other and the connection is made by something standing between them.[35] Now the first account is impossible because no power could radiate instantly from visible objects to all eyes, nor could such a great power enter by so small a hole as the pupil. The second account is impossible, similarly, because it would require the visible spirit to rush out and embrace simultaneously a great number of even very large things. Hence, following the third account, vision must take place through a medium, which is the clarity of the air. The visible spirit passes from the optic nerve through the eyes and out the pupil, where it is "joined in some way and made continuous" with the circumambient air. The air, if clear and lucid, serves as a supplement or instrument to the sight, just as the body does for touch, since air represents the impressed forms of things joined to it. The air transmits its changes to the visible spirit, which passes them back through the optic nerve to the brain. Now this is a simplification of the accounts in Nemesius and Constantine. Nemesius's historical references and Constantine's account of the crystalline have been omitted here, though the crystalline will figure in the discussion of the eye's anatomy.[36] But more telling than the simplification is the arrangement of the doxography and its formal analysis in terms of logical possibilities. The Digby commentary's section on vision is not just a textual supplement from other sources; it is an inchoate *quaestio* of a sort recognizably like those emerging in theology, the liberal arts, and law.

The third pedagogical gain in the Digby commentary is the pruning of Constantine's clinical references and detailed anatomies. The motive here is certainly that of general simplification for beginners. But the changes also have the effect of supplying more to the reader by way of *physica*. The *Isagoge* was written for students who would have had available to them alternate sources for the study of *physica*—if not direct access to the chief Aristotelian works and their commentaries, then at least handbooks and compendia by which physical doctrine would be taught. The *Pantegni,* so far as it is an adaptation and not merely a translation, already begins to supply a more general treatment of physics—of the elements, say, and of physical causality. The Digby commentary must go further in meeting the same need, not only for the earlier chapters on the elements and nature but also and perhaps more importantly for the chapters on the soul. The commentator meets the need in one way by supplying numerous definitions; he does so in another and more interesting way by inserting simplified accounts of the basic

of Vision from al-Kindi to Kepler (Chicago: Univ. Chicago Press, 1976), pp. 33–43; and Bruce Stansfield Eastwood, *The Elements of Vision: The Micro-Cosmology of Galenic Visual Theory According to Hunayn Ibn Ishaq* (Transactions of the American Philosophical Society, 72.5) (Philadelphia, 1982). While it is possible that the *De oculis* stands behind the commentary's trichotomy of views, the authoritative text seems to me rather to be *Pantegni* 4.11.

[35] Digby commentary, Peterhouse MS 251, fol. 54ra.

[36] *Ibid.,* fol. 55va.

functions. A clear example of the latter can be seen in the treatment of the cerebral locations for the animal powers.[37]

In Chapter 11 Johannitius assigns the three animal powers to locations in the head: "imagination in the front, cogitation or reason in the brain, memory in the occiput."[38] He says nothing more. The Digby commentator is left to supply a more helpful description. So, first, he names the three anatomical sites by a single technical term, *cellula,* and so emphasizes that they are parts of the one brain.[39] Second, the commentary also suggests something of the physical mechanism involved in the operation of these powers. The imagination is said to be linked "through a certain nerve" (*per quendam nervum*) to the reason, which discerns "by distilling" (*decoquendo*). The reason passes its product by mechanical means to the memory, again through a nerve.[40] Third and finally, the commentator connects these various physical operations to the different elemental qualities in each of the "cells." So the middle cell, hot and wet, is able to digest, that is, to understand, the products of the hot and dry imagination.[41]

The sources for this doctrine are found in Nemesius and Constantine. Nemesius connects imagination with the anterior brain cavities, reason with the middle, and memory with the posterior. But in all cases he is careful to say that the relation is as agent to instrument.[42] He also provides an argument for the cerebral localizations from the effects of particular lesions, but again insists on the instrumental relation in discussing sensation and especially intellection.[43] Constantine, as might be expected, is more forthcoming with anatomical detail, though he nowhere lays out as neatly as Nemesius the schema of the ventricles.[44] His account provides some of the detail on the lines of nerves running from brain to senses and muscles. Although Constantine remarks on the special dignity of human reason, he is not so careful as Nemesius on the relation of soul to brain—nor does he allude to anything like a Platonic account of intellection.

In comparison to these, its sources, the Digby commentary seems innocent of

[37] The topic has a long and dialectical history before the twelfth century, of which only small pieces were known to the Salernitan commentators. For its later development under the influence of Islamic sources see E. R. Harvey, *The Inward Wits* (London: Warburg Institute/Univ. London, 1975).

[38] Digby commentary, BL Royal MS 12.B.XII, fol. 211v: "fantasia in fronte, cogitatio vel ratio in cerebro, memoria in occipitio." The version in this manuscript corresponds most nearly to that in the lemmata of the Peterhouse version of Digby.

[39] Digby commentary, Peterhouse MS 251, fol. 53vb: "Virtus vero ymaginationis operatur in cellula. . . ," and so on for the rest of the passage.

[40] *Ibid.:* "et quod ymaginacio per quendam nervum rationi dirigit. Ipsa vero ratio quod ab ymaginatione suscipit et decoquendo discernit, et post quoddam nervo mediante memorie immittit."

[41] *Ibid.:* "Prima cellula calida est et sicca, nullum aeris et parum cerebri continens. In quo aere anima imaginatur et haec phantasia non potest fieri nisi quinque sensuum precedente officio de rebus iam anima sensibus perceptis, et ideo minus subtilis probatur. Media cellula calida et humida, ut ymaginatorum et appetitu quoddam per caliditatem et siccitatem acceptorum quoddammodo fiat digestio, i.e., discretio. Tertia cellula frigida et sicca, quae quod sibi a ratione comititur fideliter retinet."

[42] For the imagination (*phantastica*): "Instrumenta vero eius sunt anteriores cerebri ventres et animalis spiritus, qui in ipsis est, et nervi, qui sunt ex ipsis rorantes animalem spiritum et compositio sensuum": Nemesius, *Premnon physicon* (cit. n. 27), 73.15–18. For the reason: "Huius vero instrumentum est medius venter cerebri et animalis spiritus, qui est in ipso" (87.16–17). For the memory: "Huius vero instrumenta sunt posterior cerebri venter, quem parencephalida et parencranida vocant et spiritus animalis, qui est in eo" (89.7–9).

[43] *Ibid.,* 89.7–90.21, with reference to Galen on madness; 74.25–27, 87.21–88.5, with the distinction between *intellectus* (*epistēmē*) and *dinoscentia* (*dianoia*).

[44] Constantine, *Pantegni* (cit. n. 21), 4.9. There are other mentions of the ventricles in *Pantegni* 4.10–11, 4.13—and, much earlier, in 1.8.

the dangers in a too-mechanical account of the brain. In comparison with Johannitius, the commentary must be seen as wanting to explain by means of physical causes the operations that Johannitius merely enumerates. The commentary not only supplements the physical doctrine in Johannitius but also adds—here and in other passages, such as that on the origin of humors—such causal explanations as can be gathered from other sources. The need for these explanations is a consequence of the lack of a presumed *physica*. The lack of any other access to the basic doctrines of *physica* requires that the Digby commentator incorporate the elements of explanation into his reading of Johannitius.

By contrast, the primary intention of the Chartres commentary is grammatical rather than explanatory. It is almost exclusively concerned to make sense of the text. Hence, it offers only one definition or example where Digby will offer three or four. Again, any supplementary technical matter is presented by Chartres, so far as we have it, in the manner of recitation rather than argument. It is not difficult to imagine, then, that the Chartres commentaries were made as pedagogical *abbreviationes* of the Digby texts.

The difference between them allows a clearer differentiation of pedagogical tasks. There is the task of exegesis, both for the single text and for a group of authorities. A controversy among authorities, an incoherence in one of them, or a silence in one or more leads to the second task, which is more akin to what we would call explanation. To differentiate the tasks is not to distinguish them. Exegesis is undertaken for the sake of understanding, and so is explanation. But the differentiation does help to make clear the pedagogical situation that elicited these first commentaries and their successors.

III. THE PEDAGOGY OF ARCHIMATTHAEUS AND MAURUS

Archimatthaeus remains a nebulous figure, and much caution is required in approaching him. In what follows, I will mean by "Archimatthaeus" only the author of a commentary on Johannitius preserved at Trier under that name. Internal evidence puts the composition of the commentary after the mid-twelfth century, thus after the Digby and Chartres commentaries. Indeed, the commentary seems an extension of Digby. It may be an extension in the most obvious sense if I am right in hearing verbal echoes. The echoes are quite pronounced in the beginning, where the same images and exactly the same schema for the *accessus* appear. There are also striking similarities in the chapters on the soul. Here the borrowings are not so much of verbal formulas as of solutions to certain textual problems. So Archimatthaeus follows Digby's explanation on the relation of the treatment of powers to that of other members.[45] The same is also true of certain features in the account of vision, as will become clear below.

Still, our understanding of the verbal or doctrinal borrowings must be qualified by the recognition that Archimatthaeus is much more sophisticated than the Digby commentator in his procedures for textual exegesis. He is more explicit, for example, in the division of ampliative commentary from textual exposition.

[45] Archimatthaeus, commentary, Trier, Priesterseminars MS 76, fol. 6ra: "Post tractatum membrorum Io. tractat de virtutibus, licet de virtutibus primo deberet egisse per quarum actiones membrorum fiunt et que membra precedunt. Sed quia virtutes numquam adeo bene operantur sicut iam membris constitutis, et membra sunt quasi fundamenta virtutum, Io. primo tractat de membris."

The remarks on each chapter are divided into two parts. The first is a development or defense of the doctrine presented in Johannitius; the second is an explicitly announced exposition of the letter of Johannitius, which takes the form of short lemmata interspersed with glosses and paraphrases. The textual exposition is often announced with some such formula as "the letter reads as follows" (*littera sic legitur*).[46] The care for clear division is strong enough to lead Archimatthaeus to divide Chapter 11 almost into two chapters, so as to have a separate *tractatus* within which both to explain and to gloss the text on the senses.

The division of each *tractatus* into development and exposition also affords a regular occasion for introducing additional material and argument. Archimatthaeus does not feel the need to apologize for his dialectical digressions. He remains concerned with the textual order in Johannitius but draws on a much fuller range of authorities to supplement Johannitius's definitions. In the early chapters Archimatthaeus cites Boethius's commentary on the *Topics,* Galen on the *Aphorisms,* Isaac Israeli's *De dietis universalibus,* Macrobius, and Aristotle's *Physics,* not to speak of Constantine and Nemesius. Some of this erudition is unhelpful, as when he explains the term *zodiaca* (for *animalis*) as deriving either from the Arabic or from an allusion to the descent of souls through the zodiac, a doctrine perhaps learned from Macrobius.[47] But there are more important pieces of learning, among which I would count the Aristotelian definition of nature.[48]

The multiplication of authorities does not force a rapid development of the *quaestio* form. Indeed, there is less use of the *quaestio* about textual order in Archimatthaeus than in Digby. On the other hand, Archimatthaeus seems able to combine matter from different sources in order to yield more systematic explanations of topics raised by the text. This is evident in the treatment of sight, where he simplifies the logical possibilities in Digby and, indeed, seems to correct the views of that earlier commentary.

Archimatthaeus treats sight and the other senses systematically under three headings: efficient cause, correct instrument (*instrumentum idoneum*), and the cooperation of external things.[49] For sight, the efficient cause is the animal spirit, the instrument is the eyes, and the cooperating externality is clear air. After reviewing the arrangement of optic nerves, Archimatthaeus begins to refute theories about what happens when the spirit comes into an eye. Some think, he reports, that the visible spirit passes out of the eyes to the thing, where it is changed before returning to the eyes. But this view is not credible, because no amount of spirit would be able to reach all the way to the sun and the moon. In this rebuttal we recognize Digby's refutation of the same position (his second

[46] *Ibid.,* fol. 6vb. This practice has many medieval antecedents, but it is also interestingly reminiscent of the textual divisions in late Alexandrian medical commentary. See Marcel Richard, *"Apo phonēs," Byzantion,* 1950, *20*:191–222, on p. 199.

[47] Archimatthaeus, Trier MS 76, fol. 7va: "Dicta est zodiaca virtus a zoa, quod arabice dicitur anima, vel a zodiaco, quia anime secundum sententiam philosophorum cum infundebantur corporibus descendebant per zodiacum circulum, et de Letheo fonte potabant ut eorum qui sciverant obliviscerentur." Cf. Macrobius, *Somnium Scipionis* 1.12.11–12, ed. Luigi Scarpa (Padua: Liviana, 1981), pp. 159–161.

[48] Archimatthaeus, Trier MS 76, fol. 6rb: "Natura est principium motus et quietis rei per se mobilis"; cf. Aristotle, *Physics* 2.1, 192b14–15, esp. the Greco-Latin translation of the *Physics.* See Danielle Jacquart, "Aristotelian Thought in Salerno," in *A History of Twelfth-Century Western Philosophy,* ed. Peter Dronke (Cambridge: Cambridge Univ. Press, 1988), pp. 407–428, on p. 416 and n. 35.

[49] Archimatthaeus, Trier MS 76, fol. 7vb.

logical possibility). Archimatthaeus continues: Others say that the visible spirit leaves the eyes and mixes with the circumambient air, taking on its properties. Now this view—which seems to be exactly that of the Digby commentary[50]—is judged by Archimatthaeus to be somewhat nearer the truth. Still it entails that the air pass into the eye and also that it be simultaneously black and white when we see black and white simultaneously. Both entailments are improbable. So Archimatthaeus judges it better to say that the visible spirit goes to the boundary of the eye, where it mixes with the air and becomes one with it. The spirit then sends its rays out to the object. The rays are changed in the air and sent reverberating back to the visible spirit, which is itself then changed according to the properties of external things. Here one sees an attempt to make the basic account in Digby more precise, as well as a desire to leave aside the letter of the Nemesian texts that were paraphrased so meticulously in Digby. The confluence of sources is the occasion not so much of disputative exegesis as of a fuller and more systematic presentation of supplements to the underlying text.

The same is true at the level of expanded physical explanation. Here the remarks on cerebral divisions can again serve as a case in point. Archimatthaeus begins by distinguishing two grades of animal power, the principal one of which is related to the brain. He justifies the brain's physical appearance, then divides it into the traditional three cells. He adds brief, Boethian definitions of each power.[51] He also distinguishes clearly between *ratio* and *intellectus,* with an allusion to Plato.[52] The greatest development occurs, however, in the explanation of the elemental qualities of the three cells. There Archimatthaeus likens the effects of understanding to those of digestion because of the operation of heat, and explains loss of memory by analogy to a seal being pressed into something too soft to keep its impression. He explains the interrelations of the various cells in sequence and in detail—for example, when speaking of contractions that send spirit through the tubes (*vermiculi*) into the opening (*foramen*) of the next cell.[53] Constantine certainly figures as an authority, but his text is not being paraphrased slavishly and some new matter is being added.

The most remarkable use of authority in this section, however, comes when Archimatthaeus makes up the silence in Digby on the ultimate status of the soul's physical operations. Archimatthaeus's *tractatus* on the animal powers begins with praise of the soul's eminence anchored by a quotation from Augustine. "The principal [animal power] is that which the power of the sole principle operates also in the sole principle, and which emits the principal operations from itself [*a se*]. The principle is in the brain, and also in the soul, as Augustine asserts. In the soul, since it is in single members, though not less in the lesser, nor more in the

[50] Digby commentary, Peterhouse MS 251, fol. 54rb: "Aer exterior corporibus propinquis secundum ipsorum corporum colorem inmutatur. . . . Quod cum mutetur ipse aer visibili spiritui et aeri infert istam mutationem visibilis autem spiritus infert menti." Cf. Archimatthaeus, Trier MS 76, fol. 6vb: "Unde alii dicunt quia spiritus visibilis si exit oculos, sed aer que est iuxta rem videndam inmutatur a proprietatibus eius, et alius inmutatur ab illo aere, et sic de ceteris donec ventum sit ad aerem proxime oculos circumdantem, a quo visibilis spiritus inmutatur."

[51] One source, e.g., would be Boethius, *De consolatione philosophiae* 4.4.

[52] Cf. Nemesius, *Premnon physicon* (cit. n. 27), 87.24–88.3, on Plato's distinction between *dinoscentia* and *intellectus*. Cf. also Macrobius, *Somnium Scipionis* (cit. n. 47), 1.14.6–7, but note that *intellectus* for him means an effect of prudence (1.8.7) and so is not equivalent to *nous*.

[53] Archimatthaeus, Trier MS 76, fol. 7vb.

greater. It principally inhabits [*inhabitat*] the brain because of the principal operations that it there emits from itself."[54] This is by no means a "solution" to the "problem" of the soul's connection with a physical body, but it is at least a recognition that the issues touched upon in the medical account of cerebral locations are necessarily connected to questions addressed in philosophical authors and in theological *auctoritates*. The pedagogy is not only raising issues of physical explanation but also connecting them with other kinds of explanation and description.

The commentary of Maurus is best understood as a reworking of Archimatthaeus. This does not mean that he is abridging Archimatthaeus, as Chartres seems to abridge Digby. On the contrary, Maurus is rendering the exegetical patterns clearer while adding carefully demarcated doctrinal supplements. The commentary on the ninth chapter is typical. The text and the arguments begin by following Archimatthaeus very closely, with only minor adjustments for the sake of precision. So Maurus notes in regard to the various definitions of *virtus* that the word "is taken in a loose and in a strict sense." At the end of the exposition, just where Archimatthaeus is about to turn to the running textual commentary, Maurus interposes an explanation of how the powers are derived from elemental qualities. The explanation is clearly a digression, and so Maurus ends it with the phrase "let us now return to the letter [*ad litteram*]."[55] He then proceeds to a running textual gloss in the manner of Archimatthaeus. It is also interesting that the digression treats what Maurus regards as a traditional question about such matters: "It is rightly usual to doubt [*digne dubitari solet*] about the informing power." This is not only a disputative question, rather than a simply textual one, but also a question typically raised. Here and elsewhere Maurus is clearly working within a tradition of questions provoked by Johannitius.

There are other adjustments throughout Maurus's commentary that improve on Archimatthaeus. The number of citations is trimmed, sharply delineated exegetical sections are kept focused on the text, and Johannitius's textual units—both chapters and larger treatises—are regularly noted. Again, the causal pattern used by Archimatthaeus in describing the senses is expanded for sight and hearing to include the heading of *attentio animae*.[56] Maurus does not add the heading for smell, taste, or touch because these do not require attention for their operation.

With Maurus, the commentary form seems to have reached its maturity at Salerno. The text itself is clarified and methodically explicated. Doctrinal supplements are added in response to typical questions or doubts, but the supplements are clearly marked and are never allowed to overwhelm the exposition. In short, the *lectio* still predominates in Maurus, and with it the centrality of the text as a teaching device. The final commentary to be considered, which may actually be a few years earlier than that of Maurus, takes an entirely different direction. Instead of pursuing the line of exegesis, it turns back to the line of explanation.

[54] *Ibid.*, fols. 6vb–7ra (the original is also awkward). Cf. Augustine, *De Genesi ad literam* 7.18, ed. J. Zycha (Corpus Scriptorum Ecclesiasticorum Latinorum, 28) (Vienna, 1884), p. 215; and Augustine, *De Trinitate* 10.14–15, ed. W. J. Mountain (Corpus Christianorum Latinorum, 50) (Turnhout: Brepols, 1968), p. 3.

[55] Maurus, commentary, Bruges MS 474, fols. 122va, 123rb.

[56] *Ibid.*, fols. 124va, 125ra.

IV. THE PEDAGOGY OF BARTHOLOMAEUS

With Bartholomaeus of Salerno, we reach, as Kristeller long ago suspected, something like a doctrinal leap in the commentaries. The text of Johannitius becomes an occasion for expounding crucial arguments or problems with the aid of material quite different from the text, especially material from the Aristotelian tradition. To the *auctoritates* for the five chapters on the soul, Bartholomaeus adds such texts as Aristotle's *On Generation* and Qusṭā ibn Lūqā's *Liber de differentia spiritus et animae*.[57]

The use of Aristotle is particularly important. Bartholomaeus's first citation of him comes in the prologue as part of a discussion of the place of medicine in the hierarchy of the sciences. Bartholomaeus offers a definition of theory as the study of nature and then supplies the Aristotelian definition of nature as "the principle of motion and rest in the thing that is per se movable." The same phrase had appeared in Archimatthaeus, but with an attribution to Aristotle "in libro generationis"; Bartholomaeus corrects this to "in phisicis," which he seems to read in the Greco-Latin translation. He immediately adds a classification of the species of motion from the *Categories,* which he reads in the "editio composita" or "vulgata."[58] The next citation to Aristotle comes a little later in the prologue, in the discussion of the causes of motion. Bartholomaeus explains that the philosophers divided the cosmos into the superlunary and the sublunary: "Whence Aristotle *On Generation and Corruption,* 'The sun in approaching generates, in receding corrupts.' " Now the *translatio vetus* renders this remark as "in the sun's approaching there is generation, but in its receding diminution." Gerard of Cremona gives it as "there is generation when the sun draws near and there is diminution when it distances itself." The version cited by Bartholomaeus conforms to neither of these, but it matches a citation of the same text in Daniel of Morley's *De naturis inferiorum et superiorum*.[59] Daniel's work is dated to 1175–1185 and may be posterior to Bartholomaeus, though it is equally plausible that they share a common source, which need not be anything like a whole work by Aristotle.[60]

[57] The text of Bartholomaeus reads, "Plato in libro de differentia spiritus et anime" (Winchester, Winchester College MS 24, fol. 31va, though not in the corresponding passage of the copy in Oxford, Corpus Christi College MS 293B, fol. 86ra). I take it as a reference to the discussion of Plato's account in Qusṭa ibn Lūqā's *Liber de differentia spiritus et animae* 2, ed. Carl Sigmund Barach, in *Excerpta e libro Alfredi Anglici De motu cordis . . .* (Innsbruck, 1878; rpt. Frankfurt: Minerva, 1968), pp. 133–140, rather than as a mistaken attribution of authorship.

[58] Bartholomaeus, Winchester MS 24, fol. 22vb = Corpus Christi MS 293B, fol. 75rb; citing Aristotle, *Categories* 14, as in *Aristoteles Latinus,* Vol. I, Pts. 1–5, ed. L. Minio-Paluello (Bruges/Paris: Desclée de Brouwer, 1961), p. 77, lines 19–20.

[59] Bartholomaeus, Winchester MS 24, fol. 23ra: "Unde Aristoteles in libro de generatione et corruptione, Sol adveniendo generat, recedendo corrumpit"; citing Aristotle, *De generatione et corruptione* 2.10, 336b17–18, as in *Aristoteles Latinus,* Vol. IX, Pt. 1, ed. Joanna Judycka (Leiden: Brill, 1986), p. 75, lines 17–18: "adveniente quidem sole generatio est, recedente autem diminutio." For Gerard of Cremona see Judycka's editorial remarks, *ibid.,* p. 36: "sol quando appropinquat est generatio et quando elongatur est diminutio"; see also *ibid.,* on Daniel of Morley, *De naturis inferiorum et superiorum* (cit. n. 18), p. 32.

[60] The earliest extant copies of the *translatio vetus* are both from Normandy and date from the last quarter of the twelfth century (Judycka, in *Aristoteles Latinus,* Vol. IX, pp. xxxviii–xxxix). It would not be difficult to conjecture their circulation in and around the Norman kingdom of Sicily and in England. But given that the citations in Bartholomaeus and Daniel match each other exactly and differ from both full versions, it seems highly unlikely that they occurred by separate corruption of the *vetus*.

More important than the individual citations to Aristotle are the regular uses of Aristotelian doctrine. In explicating the nature of *spiritus,* Bartholomaeus cites from Aristotle the saying that a small quantity (*pugillo*) of water makes many such quantities of air, then goes on to apply this rule to physical explanations of the structure of the *castanea cortex*. Aristotle is cited again as the source of an opinion that the heart is the source of the passions.[61] This leads Bartholomaeus to a general conclusion about the causation of the passions by extrinsic causes, which he analyzes one by one. These imported materials predominate, so that even when he adverts to the text, he most often immediately leaves it to continue a line of argument. There are still brief sections of exegesis, but they are out of proportion to the sections of doctrinal development and are also frequently interrupted by digressions.

Consider, for example, Bartholomaeus's general treatment of sensation. He begins the section on animal power by supplying the correct Greek etymology for *zodiaca* and by noting that it is a textual variant for *animalis*.[62] He then rehearses the familiar topics: the kinds of animal power, their assignment to the three cells, and the elemental properties of those cells. He lavishes more attention on the less familiar topic of how reason can move from sensation to grasp what is beyond sensation, including both incorporeal substances and hidden forms or natures.[63]

Reason leads theological contemplation directly up the hierarchy from corporeal to incorporeal. In *physica* a more complicated comparison of creatures' dispositions and motions is required to infer the creator as their cause and author. Reason is able to "penetrate" the flavors, colors, and odors processed in the imagination; by investigation of causes, it grasps "the hidden natures of things." Bartholomaeus gives a surprisingly specific example. By sense and imagination we perceive the taste of mustard and pepper as sharp and pungent. Now sharpness and pungency derive from heat and subtlety; heat and subtlety have the effect of cutting through humors and dissolving gaseousness (*ventositas*). Thus reason understands from pepper's taste both that and why it has the power to cut through humors and to dissolve gaseousness. Note that the form of reasoning here is to discover a hidden causality linking apparently unconnected properties. Moreover, Bartholomaeus has seen the importance of attempting what may loosely be called an account of cognition. Of course, he makes no more than the first motions of such an attempt. In this limitation we may see what is still lacking in his Aristotelianism. A more striking confirmation of its limits will be seen in the discussion of the brain and in the arguments over sight.

Bartholomaeus returns from this example of causal reasoning to rehearse the distinction between *intellectus* and *ratio,* with Boethian conclusions about the

[61] Bartholomaeus, Winchester MS 24, fol. 30va = Corpus Christi MS 293B, fol. 85rb, to which cf. Aristotle, *De generatione et corruptione* 2.6, 333a23 (see Jacquart, "Aristotelian Thought" [cit. n. 48], p. 422 and n. 60, for the reappearance of this citation in Urso of Salerno's *Aphorisms*); and Winchester MS 24, fol. 30vb = Corpus Christi MS 293B, fol. 85rb–va, to which cf. *De anima* 1.1, 403a–b.

[62] Bartholomaeus, Winchester MS 24, fol. 31ra = Corpus Christi MS 293B, fol. 85va: "Animalis autem virtus a Grecis zodiaca dicitur [*CC:* apellatur]. Unde in quibusdam codicis [*CC:* libris] zodiaca pro animali reperitur." The textual variation may be that recorded in the Monte Cassino manuscript of the *Isagoge* or it may be the internal inconsistency in terminology of the main version; see Jacquart, "A l'aube de la renaissance" (cit. n. 3), pp. 218, 228.

[63] Bartholomaeus, Winchester MS 24, fol. 31rb.

different procedures in *physica,* mathematics, and metaphysics. He quotes the famous and untranslatable triplet "in mathematicis doctrinaliter, in divinis intellectualiter, in phisicis rationabiliter versari oportet," but notes that *intellectus* and *ratio* are often used interchangeably and that there is only one *cellula* of the brain corresponding to both. He ends by claiming that reason's power to discern can be explained by the mixture and hydraulic interaction of the different elemental properties of the cells. So too for memory and imagination.[64]

What is lacking in the explanations of mechanism is any clear formulation of the aporias discussed by Aristotle in the *De anima.* It is surprising to find Bartholomaeus moving so quickly through a discussion of the brain's *cellulae* without stating the questions that such a discussion must raise for a reader of Aristotle—not to say a reader of Boethius or Calcidius. On the other hand, Bartholomaeus may be aware of the lacuna. Two passages suggest this. The first occurs in the discussion of sight, the second at the very end of the consideration of the soul.

In the discussion of sight much is familiar, though there are small improvements throughout. For example, the tripartite explanatory schema in Archimatthaeus—efficient cause, instrument, cooperation of exterior things—is now rendered as intrinsic and extrinsic efficient cause together with instrument. Again, Bartholomaeus appropriates Constantine's references to the crystalline, which had been passed over by earlier commentaries on this particular passage. But his greater achievements are explanatory. Thus he remarks on the curious ontology of visual rays, which are neither bodies nor substances, since they must share space with other bodies.[65] He treats at great length the dispute over the relation of visible spirit and object. He concludes, with arguments both old and new, that sight must occur through a medium. But he insists that the air is not a medium in the sense that it takes on forms from objects and brings them back to the eye. The air can neither hold forms nor keep them distinct. The medium is rather the rays emitted from the eye by the visible spirit, which remains within it. These rays move to the objects with "astonishing speed," rebounding against them back through the eye, up the optic nerve, to the cell of imagination. The care of formulation here suggests that Bartholomaeus is correcting views much like those of Archimatthaeus. The speed and character of these rays can be grasped by analogous experience—whether of candles or shining celestial bodies—and by reflecting on the double agility of the medium, which is both airy and luminous. Bartholomaeus ends by reminding the student that the nature of the rays, which are neither substance nor accident, will remain obscure until the nature of the soul is itself known.

This is an explicit admission of a gap in the account, and it points forward to the last passage in the commentary on the powers of the soul. An objector argues that spirits do not seem to be natural substances, since they are daily destroyed and reconstituted in the body. They have no permanence. The objection elicits a final—and perhaps emblematic—clarification of science and the soul. Science is not concerned with individuals, Bartholomaeus responds, but with universals. So science considers the kinds of spirit, not this particular spirit generated from the

[64] *Ibid.*, fol. 31va, quoting Boethius, *De Trinitate* 2.

[65] Bartholomaeus, Winchester MS 24, fol. 31vb. One might be tempted, here again, to suspect some influence from the *De oculis*.

blood's fumes and proportioned to them. Now some philosophers thought that such a spirit was the soul—thus Democritus, on the report of Nemesius.[66] But others saw that the spirit was the soul's instrument, by which it carries out its actions in the body.[67] With these remarks, Bartholomaeus shows again his interest in the character of scientific knowing and in precise physical ontology. But he also discloses the incompleteness of his account of the soul. The questions have become clear, but the answers to them are still stronger on details of efficient causality than on formal explanations. The integral reception of Aristotle still awaits a final generation of Salernitan masters, whose careful study and sustained reflection can best be seen in a work like Urso's *On the Mixture of the Elements* (*De commixtionibus elementorum*).

V. EXEGESIS, ARGUMENT, AND MEDICAL PEDAGOGY

The reading of a few chapters from the Salernitan commentaries on Johannitius was meant to illuminate two major developments. First was the evolution of exegetical forms, of procedures for discovering and presenting the sense of authoritative texts. The difficulty of rediscovering exegetical techniques ought not to be understated. There are difficulties for exegesis in the works themselves: neologisms, strained syntax, unfamiliar references, unaccustomed silences or ironies. There are the further difficulties of combining texts into pedagogical canons—inconsistencies, repetitions. Then there are more particular difficulties for Latin readers in the forms of commentary and the associated techniques of book production—say in moving from exposition by gloss to exposition by running commentary or by disputed question. But there are also enormous conceptual difficulties in basing knowledge on the interpretation of authoritative texts. The Salernitan commentators by no means rose to the hermeneutical sophistication evident in Galen's relation to Hippocrates.[68] Nor could they attain the philological self-consciousness of Ḥunayn's reflections on translating the medical corpus. But the evidence of their commentaries shows that the Salernitans did discover a number of the central problems of teaching by textual interpretation.

Here an important contrast with what was known of Alexandrian teaching suggests itself.[69] As regards literal commentary, the Salernitan masters might seem to be only reinventing the exegetical habits of the Alexandrians. But their increasing use of the forms of disputation shows them to be motivated by something different from the desire for codification that eventually triumphed over the Alexandrian curriculum. If the historical movement at Alexandria was toward a reduced canon stereotypically interpreted, the movement in the *articella* com-

[66] Nemesius, *Premnon physicon* (cit. n. 27), 23.25–24.7.

[67] The authority Bartholomaeus gives for this remark is the "Liber quidem de stabili vita constabilicione" (Winchester MS 24, fol. 33va = Corpus Christi MS 293B, fol. 87vb), but it seems to be a reference to Qusṭa ibn Lūqā's *De differentia spiritus et animae*. See Jacquart, "Aristotelian Thought" (cit. n. 48), p. 426.

[68] See the suggestive remarks by Paola Manuli, "Lo stile del commenta: Galeno e la tradizione ippocratica," in *La scienze ellenistica: Atti delle tre giornate di studio tenutesi a Pavia del 14 al 16 Aprile 1982,* ed. Gabriele Giannantoni and Mario Vegetti (Naples: Bibliopolis, 1984), pp. 375–394.

[69] What is known of the late curriculum at Alexandria comes from such medieval Arabic sources as Ḥunayn's *Epistle,* ᶜAli Ibn Riḍwān's *Useful Book on the Quality of Medical Education,* and Ayyūb al-Ruhāwī's *Practical Ethics of the Physician.* The comparisons that follow are based on such limited and even anecdotal evidence.

mentaries is toward an expanding canon augmented disputatively. This is just the opposite of an unthinking rehearsal of the letter.[70]

My second point is that the exegetical developments at Salerno are paired with a more and more acute awareness of the need for theoretical foundations, for a conceptually justified physical explanation. This is something more than the simple desire, expressed in the *accessus,* to locate medicine within the hierarchy of the speculative and practical sciences. It may even be something more than was anticipated in the Alexandrian curriculum. As it can be reconstructed, that curriculum required preparatory studies in the arts of the trivium and quadrivium but did not require any systematic study of physics in the Aristotelian sense.[71] Certainly the average Greek-speaking student of the sixth century would have had access to a considerable body of ancient *physica* unknown and inaccessible to the twelfth-century Salernitan commentators. Much of the commentators' work, it has been argued, was concerned just to remedy the lack of such a world of presumed or latent learning. But it is also true that the curricular ideal at Salerno seems more ambitious in principle than the Alexandrian—as it is also closer to the ambitious, thoroughly philosophical conception of medicine in Galen. Moreover, the historical tendency of the Salernitan school was precisely toward a medical *physica* of the kind one sees in the works of Urso, especially his *On the Mixture of the Elements* and the *Aphorisms.*

A medical physics, in this sense, is not so much a part of physics applicable to medicine as it is a physics constructed retrospectively from the study of medicine. Its principles are thus sufficiently flexible and sufficiently detailed to bridge the distance between first causes and the very particular phenomena of human bodies. It is here, and here only, that one can speculate about the Salernitan masters' relation to the empirical. The differences between the development of *physica* at Salerno and at other centers—Chartres, Paris, Toledo, Hereford—are due in some way to the requirements placed on the reinvention of physical knowledge within the context of medical pedagogy. These requirements did not make *physica* utilitarian or antispeculative, as the *articella* commentaries have shown at many points. But they may have drawn Salernitan physics away from the cosmological schemes so prevalent in contemporary texts by masters at other schools.

One consequence, I would suggest, is that the Salernitans were thus not in the best position to receive the new Aristotle. On the contrary, the larger aspirations of their teaching seem to have failed under the double blow of the Aristotelian *libri naturales* and Avicenna's *Canon.* The Aristotelian corpus imposed its own pedagogy, in which medicine had a small and ambiguous place. The *Canon,* on the other side, offered a more attractive and more technical treatment of medical issues. With its ambitious wish to maintain medicine's central place in the hierarchy of science, the Salernitan pedagogy was caught between the two.

[70] See the charges made against the *articella* commentaries by Charles Talbot, "Medical Education in the Middle Ages," in *The History of Medical Education,* ed. C. D. O'Malley (Berkeley/Los Angeles: Univ. California Press, 1970), pp. 73–87, on pp. 77–79.

[71] See Albert Z. Iskandar, "An Attempted Reconstruction of the Late Alexandrian Medical Curriculum," *Med. Hist.,* 1976, *20*:235–258, with the summary chart on pp. 257–258; and Iskandar, "Development of Medical Education among the Arabic-Speaking Peoples," in *The Light of Nature: Essays in the History and Philosophy of Science Presented to A. C. Crombie,* ed. J. D. North and J. J. Roche (Dordrecht/Boston: Martinus Nijhoff, 1985), pp. 7–20, on pp. 12–14.

The Nature and Limits of Medical Certitude at Early Fourteenth-Century Montpellier

*By Michael R. McVaugh**

THE THIRTEENTH CENTURY was a decisive period of transition for learned or academic medicine. It was then that medicine integrated itself into the university network that was coming into existence and thus broadened its basis of instruction so that an arts graduate of 1300 could have proceeded on to study medicine at Bologna or Padua, Paris or Montpellier—to name only the most important of the faculties that had developed. At the same time, medicine had to come to grips with a vast body of new scientific writings in translation, including most of the important works of Galen, together with the encyclopedias and commentaries of his Arabic expositors; furthermore, the need to master these works justified the autonomy of medicine within the new studia. As a result, the sense of a medical profession began to evolve among the possessors of this learning and was communicated to the lay public, which now began to legislate academic training for those who wished to practice medicine. The intellectual consequences of the transition emerged in the last quarter of the thirteenth century in a newly self-conscious medical literature that was itself a product of these changes.

I. THE ASSIMILATION OF "THE NEW GALEN"

The critical part played in these developments by the new medical writings is not always well understood. Medicine had possessed, after all, its own definite corpus of technical literature long before the creation of the universities, a corpus rooted—like the one that would supplant it—in the Galeno-Hippocratic tradition. This collection, the *articella* or *ars medicine,* was in use as a basis for teaching by the early twelfth century at Salerno and perhaps elsewhere.[1] It was even more "Hippocratic" than it was "Galenic" in character (especially after the *Regimen acutorum* was incorporated into the collection, at the beginning of the thirteenth century[2]), and it did not present a systematic medicine in detail: two of its constituents (the *Tegni* and the *Isagoge*) did indeed outline the Galenic system, but in so summary a fashion as to leave much to the imagination. The remaining books in the *articella* treated medical practice, diagnosis, prognosis, and thera-

* Department of History, University of North Carolina, Chapel Hill, North Carolina 27514.

[1] See Mark D. Jordan's article in this volume.

[2] Paul Oskar Kristeller, "Bartholomaeus, Musandinus, and Maurus of Salerno and Other Early Commentators of the 'Articella,' with a Tentative List of Texts and Manuscripts," *Italia Medioevale e Umanistica,* 1976, *19*:57–87, esp. pp. 64–70.

peutics. The more theoretical areas of Galenic medicine, physiology (and its concomitant anatomy) and pathology, received scant attention and must have been of little concern to twelfth- and even thirteenth-century medical students (the *articella* was still the core of scholastic medical education at Paris, Bologna, and Montpellier in the last quarter of the thirteenth century).

Consequently, the assimilation of what one historian has called "the new Galen"[3] met with difficulties that went beyond its mere volume—though that was certainly a problem. Much of this material had been translated at Toledo by Gerard of Cremona and his school: there are nine works of Galen in the list of Gerard's translations made at his death in 1187 (as well as the *Canon* of Avicenna and the *Liber Almansoris* of Rhazes), and this list was not exhaustive: among Gerard's other Galenic translations was the *De ingenio sanitatis* (the *Methodus medendi*), which is by itself as long as the whole *articella*. The works available by the late thirteenth century included many of Galen's more important contributions to general physiology (*De complexionibus, De iuvamentis membrorum*) as well as all his most important writings on pathology: *De morbo et accidenti, De interioribus*. When we consider that thirteenth-century readers were also having to confront Galen's repetitive, opinionated, often rambling language (not improved by passage from Greek through Syriac and Arabic to Latin) at the same time that they were attempting to understand this nearly overwhelming quantity of technically complex and unfamiliar material, we can begin to see why the process of assimilation was not only lengthy but also painful for students of medicine.

An indirect sign of these difficulties is that it was Avicenna's *Canon* rather than a Galenic work that was the first item among the new translations to become an accepted authority. From the 1220s on, one finds the *Canon* cited with increasing frequency, until a generation later it is a standard source with which to expound the *articella*.[4] It is easy to understand why what was in effect a systematically arranged encyclopedia of Galenic medicine should have been mastered before any of the works from which ultimately it had been derived—particularly since it had passed through fewer distorting translations. The *Canon* was able to provide an explanatory framework within which the next generation of medical scholars could feel more comfortable with Galen himself.

Avicenna's work may have lightened the difficulties of assimilation, but it could not do away with them. The problem of quantity remained a serious one: the *Canon* itself was so large as to be virtually unmanageable and was eventually studied in sections. Given the amount of medical literature available, it was no

[3] Luis García-Ballester, "Arnau de Vilanova (c. 1240–1311) y la reforma de los estudios médicos en Montpellier (1309): El Hipócrates latino y la introducción del nuevo Galeno," *Dynamis*, 1982, *2*:97–158; and see too Nancy G. Siraisi, *Taddeo Alderotti and His Pupils: Two Generations of Italian Medical Learning* (Princeton, N.J.: Princeton Univ. Press, 1981), pp. 100ff.

[4] Nancy G. Siraisi, *Avicenna in Renaissance Italy* (Princeton, N.J.: Princeton Univ. Press, 1987), Ch. 3; and, with a more restricted focus, Danielle Jacquart, "La réception du *Canon* d'Avicenne: Comparaison entre Montpellier et Paris aux XIIIe et XIVe siècles," in *Actes du 100e Congrès National des Sociétés Savants* (Montpellier, 1985), Section d'Histoire des Sciences et des Techniques, Vol. II, pp. 69–77. On the *Canon*'s role in mid-century exposition of the *articella* see Michael R. McVaugh, "The 'Humidum Radicale' in Thirteenth-Century Medicine," *Traditio*, 1974, *30*:268–270. Konrad Goehl, *Guido d'Arezzo der Jüngere und sein "Liber mitis"* (Würzburger Medizinhistorische Forschungen, 32) (Hannover: Pattensen, 1984), pp. 12, 15–31, shows that the *Canon* was in use at Parma before the end of the twelfth century; however, its widespread dissemination as a teaching text occurred in the thirteenth century.

longer possible to assume that a scholar could easily search out whatever text he might happen to want: there was simply too much to commit to memory. This is variously acknowledged in texts of the period. An original and ambitious technique for rendering the new Galen manageable was the invention of Jean de St. Amand, a canon of Tournai who seems to have studied medicine at Paris in the generation after 1250. His *Revocativum memoriae,* completed sometime after 1285, was designed "so that students who pass sleepless nights looking for information among the works of Galen may be relieved of their struggles and worry and may more quickly find what their thirsting, exhausted intelligence is seeking."[5] Besides a brief introduction, it contains, first, a précis of nine of the most important works in the new Galen—*De morbo et accidenti, Megategni, Liber interiorum, De creticis diebus, De crisi, De simplicibus medicinis, De complexionibus, De malicia complexionis diverse, De iuvamentis membrorum;* then a précis of works of the *articella*—Hippocrates' *Prognostics,* Galen's *Tegni,* and Hippocrates' *Regimen acutorum* with Galen's commentary; and finally a cursory table of contents of Hippocrates' *Aphorisms.* St. Amand explained that the *Aphorisms* were unsuited to précis, "since, not being all on the same subject, they are as it were conclusions based on a variety of opinions and discoveries based on reason and experience."[6] Using these summaries, a student could (and can still) quickly remind himself of the argument and content of all the essential Galenic sources and go immediately to any part desired; and in order to provide a standard system of reference, St. Amand devised his own subdivision of these works into chapters, identifying each chapter by its incipit.

These précis found their complement in the second part of the *Revocativum memoriae,* where St. Amand presented an assemblage of 4,400 Galenic and Aristotelian statements, arranged alphabetically by topic—582 of them, from *abstinentia* to *ydromel*—and capable of being identified by his system of chaptering: "In what book, chapter, and part of a chapter it may be found, so that 'a' means the beginning, 'b' the middle, and 'c' the end of a chapter."[7] Sometimes he quoted a passage verbatim, sometimes he summarized a Galenic discussion in a sentence, and occasionally he provided an analysis and reconciliation of Galenic and Aristotelian views on topics like *digestio* or *sperma.* The immediate result was an index of the terms and concepts of the new Galen that was also itself "a full presentation of the whole of internal pathology, . . . a short partial medical encyclopedia."[8] The encyclopedic coverage provided by the *Revocativum me-*

[5] Jean de St. Amand, introduction to *Revocativum memoriae,* ed. Otto Paderstein, in "Ueber Johannes de Sancto Amando (XIII. Jahrhundert) . . ." (Inaug.-Diss., Berlin, 1892), p. 10.

[6] St. Amand, *Revocativum memoriae,* ed. Friedrich Petzold, in "Ueber die Schrift des Hippocrates *Von der Lebensordnung in akuten Krankheiten* nebst dem Schluss des *Revocativum memoriae* des Johann von St. Amand (13. Jahrhundert)" (Inaug.-Diss., Berlin [1894]), p. 28.

[7] St. Amand, *Revocativum memoriae,* ed. Paderstein, "Johannes de Sancto Amando" (cit. n. 5), p. 10. St. Amand did not in fact make much use of this device.

[8] Julius Leopold Pagel, *Die Concordanciae des Johannes de Sancto Amando* (Berlin, 1894), p. xx. To these two guides to theory St. Amand added two surveys designed to give the *Revocativum memoriae* a practical dimension (see Paderstein, "Johannes de Sancto Amando," p. 11): one on simple medicines (ed. Pagel as *Die Areolae des Johannes de Sancto Amando* [Berlin, 1893]) and another, not his own composition, on surgery.

I find entirely convincing Mark Jordan's suggestion to me that Jean's Galenic concordance adapted recent developments in biblical scholarship to medicine. Its form is in fact very similar to that of the biblical concordance perfected in Paris by the St.-Jacques Dominicans shortly before 1286. See R. H. Rouse and M. A. Rouse, "The Verbal Concordance to the Scriptures," *Archivum Fratrum Praedicatorum,* 1974, *44*:5–30, esp. pp. 17–21.

moriae was extended to writings by other medical authorities (most importantly Avicenna) as well as to lesser Galenic works in a similar collection prepared by Pierre de St. Flour in the 1350s, and the two guides remained important tools in scholastic medical education into the fifteenth century.[9]

At the beginning of the fourteenth century, of course, when the new Galen was still unfamiliar and a little frightening, before it had been assimilated into medical curricula, the *Revocativum memoriae* enjoyed a particular usefulness. The French surgeon Henri de Mondeville depended in part on St. Amand's references to provide him with the precise citations of medical and surgical authorities for which his *Chirurgia* (begun in 1306) is so remarkable. Such citations drew an accusation of *prolixitas* from the "famosi cyrurgici et medici" to whom Mondeville first showed his work, and he felt obliged to justify his method when he revised the *Chirurgia* in 1312:

> I do this for two reasons: first, so that the labor of students in looking for these authorities will be lessened; second, so that once found they may be better and more clearly understood. The reason for the first is that since if anyone wants to inform someone about a statement, he should do this as easily and as succinctly as he can, as Galen says in *De interioribus* 2.7. But it is much easier for the students being taught to find these authorities in specific chapters and places by using concrete references than it is for them to spend their days and nights and even dreams searching for a single reference by going through all their books.[10]

No doubt his description of tired students was indebted to the language of the earlier *Revocativum memoriae,* but it should not be dismissed on that account as mere rhetoric. Beyond any question, Mondeville intended his book for readers with some medical learning, "intelligent, particularly lettered ones, who comprehend at least the general principles of medicine and who understand the language of the art." He emphasized repeatedly that the *scientia medicine* was essential to the surgeon—"it is impossible for someone who does not know the general principles of medicine to be a competent surgeon"—and as though to drive the point home he stressed at the beginning of the *Chirurgia* that he had studied medicine at Montpellier before training in surgery at Montpellier and at Paris. In a famous passage, he spoke savagely of those pretended "surgeons" who were in fact "illiterati, sicut barberii, sortilegi, locatores, insidiatores, falsarii, alchemistae, meretrices, metatrices, obstetrices, vetulae, Judaei conversi, Sarraceni."[11] Mondeville's device of identifying his quotations must be taken seriously as an attempt to meet a pedagogical need, not dismissed as padding, and the shifts to which he was driven in order to ensure precision for his readers—he will occasionally give two chapter numbers for a reference, one deriving from the original translation and the other imposed according to the revised system of the

[9] Julius Leopold Pagel, *Neue litterarische Beitrage zur mittelalterlichen Medicin* (Berlin, 1894); and Eduard Seidler, *Die Heilkunde des ausgehenden Mittelalters in Paris* (*Sudhoffs Archiv,* Beihefte, 8) (Wiesbaden, 1967), pp. 108–109.

[10] Henri de Mondeville, *Chirurgia,* ed. Julius Leopold Pagel, *Archiv für Klinischer Chirurgie,* 1890, *40*:674. Because Pagel's edition in *Die Chirurgie des Heinrich von Mondeville* . . . (Berlin, 1892) is difficult to obtain, I have referred all my citations to the identical text published by Pagel in installments as "Die Chirurgie des Heinrich von Mondeville (Hermondaville)" in *Arch. Klin. Chirurgie,* 1890, *40*:253–311, 653–752, 869–904; 1891, *41*:122–173, 467–504, 705–746, 917–968; *42*:172–228, 426–490, 645–708, 895–924. It may also be helpful to consult the French translation: E. Nicaise, *Chirurgie de Maitre Henri de Mondeville* (Paris, 1893).

[11] Mondeville, *Chirurgia,* ed. Pagel, *40*:263, 675 (cf. p. 657), 263, and 661.

Revocativum memoriae[12]—show how confusing and even daunting academic training in the new medicine must at first have appeared, at Montpellier and Paris or elsewhere.

As far as our evidence goes, the principal agent in Montpellier's assimilation of the new Galen was Arnau de Vilanova, who lectured there during the 1290s and may have returned briefly to teach at the school toward the middle of the next decade.[13] Luis García-Ballester has already shown that Arnau (who as a student at Montpellier in the 1260s would have been introduced to medicine through the *articella*) made a point of lecturing on the new Galenic texts and of citing them rather than the older, more familiar, material, or even Avicenna, in preparing his own monographic writings on points of medical theory.[14] It is quite conceivable that Arnau was one of Mondeville's teachers, and his recently identified abbreviation of Galen's *De interioribus* (completed in 1300) is an attempt to bring the work down to student scale, one reminiscent of Jean de St. Amand's.[15] Finally, and most decisively for eventual assimilation, Arnau was a principal adviser behind Pope Clement V's implementation of a new curriculum for Montpellier in 1309, a curriculum that established the Galenic pathological and physiological treatises as the future basis for all medical education there.[16]

This aspect of Arnau's contribution to the new medicine is well understood, but acquaintance and mastery reflect only a part of the process of assimilation. There remained the problem of integrating the newly mastered medical material with other scientific knowledge so as to create what we may call a system of scholastic (rather than merely Galenic) medicine, a new medical culture. Every medical student who hoped to comprehend Galen would already, as a student of the arts, have assimilated Aristotle, and although Galen's medicine was deeply marked by Aristotelianism, there were still areas of potential incompatibility between the two systems. As is well known, Aristotle and Galen were in obvious disagreement on certain concrete matters—of physiology, for example.[17] Such

[12] "Notandum similiter, quod cyrurgicus quantumcunque sciens et intelligens vix potest de quolibet particulari opere causam reddere efficativam, quare rationes debent aliquando sufficere apparentes maxime in his, quae testantur practicantes famosi per experientiam se vidisse et ideo dicebat Galenus I° de complexionibus ca. 5, secundum Thadeum, 2° tamen capitulo secundum Joh. de Sancto Amando, quod medicus debet confidere in omni re sua secundum experimentum, nec debet quaerere de re quae est, si est; potest tamen bene quaerere causam, quare est et quare sic est, et hoc latius pertractatum est IX° praeambulo praecedenti, in quo etiam condiciones et virtutes rationis et experimenti declarantur." *Ibid., 40*:718.

[13] Unfortunately there is little concrete evidence concerning Arnau's presence at Montpellier in these years; what there is is summarized by Francesco Santi, *Arnau de Vilanova: L'obra espiritual* (Valencia: Diputació Provincial de València, 1987), pp. 106ff.

[14] García-Ballester, "Arnau de Vilanova" (cit. n. 3).

[15] Arnau's *Doctrina Galieni de interioribus,* edited by Richard J. Durling in *Arnaldi de Villanova Opera medica omnia,* ed. Luis García-Ballester, J. A. Paniagua, and Michael R. McVaugh, Vol. XV (Barcelona: Seminarium Historiae Medicae Cantabricense, 1985), pp. 298–351, is a reworking of Books 1–2 of Galen's text in its Arabo-Latin translation. Jean de St. Amand's résumé—edited by Paderstein, in "Johannes de Sancto Amando" (cit. n. 5), pp. 28–45—outlines all six books, much more summarily.

[16] The pope required all bachelors who were to be promoted to master to possess (*habere*) *De complexionibus, De malicia complexionis diverse, De simplici medicina, De morbo et accidenti, De crisi et criticis diebus,* and *De ingenio sanitatis.* The text, which singles out Arnau as one of the pope's advisers in this matter, is published in *Cartulaire de l'Université de Montpellier,* Vol. I (Montpellier, 1890), p. 220.

[17] Margaret Tallmadge May identifies a number of particularly prominent controversies over anatomy and physiology in her translation, Galen, *On the Usefulness of the Parts of the Body,* 2 vols. (Ithaca, N.Y.: Cornell Univ. Press, 1968), Vol. I, pp. 16–17. R. K. French discusses thirteenth-

Presumed portrait of Henri de Mondeville lecturing to his students, from a manuscript of his **Chirurgia** *in the French translation made during his lifetime (1314); Paris, Bibliothèque Nationale, MS fr. 2030, folio 1v. Courtesy Loren C. MacKinney Collection, University of North Carolina.*

disagreements had been apparent to the first purveyors of the new medical texts (not surprisingly, since Galen often pointedly criticized Aristotle for his physiological mistakes), so that occasionally, as we have seen, St. Amand had felt it necessary to expose their disputes frankly and try to settle them. Pietro d'Abano's *Conciliator* (much of which was completed in Paris by 1303[18]) has much the same purpose, but takes up 210 points of medical contention and needs more than 200 printed folios to resolve them, showing how rapidly controversy was bound to develop in the attempt to fuse a medical training with the arts.

Incompatibility of content was one of the more obvious but perhaps less fundamentally important of the difficulties arising from the need to integrate Galenic

---

century awareness of such disagreements in "*De iuvamentis membrorum* and the Reception of Galenic Physiological Anatomy," *Isis,* 1979, *70*:100–101. Still another important point of dispute is examined by Anthony Preus, "Galen's Criticism of Aristotle's Conception Theory," *Journal of the History of Biology,* 1977, *10*:65–85; and Jutta Kollesch, "Galens Auseinandersetzung mit der Aristotelischen Samenlehre," in *Aristoteles Werk und Wirkung,* ed. J. Wiesner, Vol. II (Berlin: Walter de Gruyter, 1987), pp. 17–26.

[18] Lynn Thorndike, *A History of Magic and Experimental Science,* Vol. II (New York: Columbia Univ. Press, 1923), p. 876, concludes that the work was completed in 1303, when Pietro was in Paris, but it may subsequently have been revised; see also Nancy G. Siraisi, *Arts and Sciences at Padua: The Studium of Padua before 1350* (Toronto: Pontifical Institute of Mediaeval Studies, 1973), p. 59 and n. 129.

medicine with academic Aristotelianism. Students who had learned from Aristotle that the highest form of human reasoning was *scientia,* certain truth drawn deductively from established principles, could be tempted to see Galen as a "medical Aristotle" and to organize the vast quantity of his writings in a consistent synthetic system—forgetting what Galen himself had perfectly understood, that medicine was to at least some degree dependent upon particular features of individual cases and as such could not be reduced to a general, all-encompassing theory. Other thoughtful physicians felt differently; to them it seemed clear that the new Galenic medicine could not be fully integrated into scholasticism without first establishing the boundaries between medicine and philosophy.

This issue of what we might call the nature and limits of medical certitude is probably the most pervasive theme in the scholastic medical writings of Arnau de Vilanova, and the position he held was already thoroughly worked out in the earliest surviving work that he produced while a master at Montpellier, the treatise called *De intentione medicorum.* This little treatise was evidently Arnau's first blow in his battle to incorporate Galen, not merely pedagogically but intellectually and philosophically, into academic medical culture. It is an especially interesting text because Arnau here employed a distinction between medical and philosophical truth: he examined four particular areas of apparent factual disagreement between Aristotle and Galen and showed that in each case, properly understood, the two could be recognized as actually in agreement. On this basis Galen's medicine could henceforth be accepted on its own terms, independent of natural philosophy. Taken in conjunction with its later companion work, *De consideratione operis medicine, De intentione medicorum* teaches the reader to aim for the nice balance between reason and experience that permits individual illnesses to be interpreted in the light of more general medical truths.

The crosscurrents and tensions in the medical community of the 1290s come over vividly in the prefaces to these two works, each of which carries an introduction to a particular individual and therefore—rhetorical exaggeration notwithstanding—has a more personal flavor than might otherwise have been the case. The slightly later *De consideratione* was addressed to a friend who had been persuaded by empiric physicians that medicine is directed by chance rather than by reason or *ars,* as Galen and Hippocrates had taught.[19] Arnau's first targets in the medical world, however, had been quite different: academic physicians, those who rated the new medical theory not too low but too high. In his preface to *De intentione,* Arnau sympathized with his friend "who has so often been terrified by the obscurities of Galen's work,"[20] but he was nonetheless scornful of those who imagine they can skip over the foundations of medical study and go directly to theory building. Such men may have the terminology pat, but they have no real understanding of medicine: "Lacking the polish of reason and the balance of judgment, they cannot properly fit to particulars the universals they

[19] García-Ballester, "Arnau de Vilanova" (cit. n. 3), pp. 109–111. It is not impossible that one of the empirics Arnau had in mind was the Dominican Nicholas of Poland, who was in the order's Montpellier *studium* for much of the third quarter of the thirteenth century: see William Eamon and Gundolf Keil, " 'Plebs amat empirica': Nicholas of Poland and His Critique of the Medieval Medical Establishment," *Sudhoffs Archiv,* 1987, *71*:180–196.

[20] "Ea namque obscura que in libris Galieni sepius expavisti." Preface to *De intentione medicorum,* in Arnau de Vilanova, *Opera nuperrime revisa* (Lyons, 1520), fol. 36rb. I have occasionally found it desirable in this study to modify the reading of the printed edition by reference to the manuscript tradition of *De intentione.*

amass from the basic sciences, since they do not have much experience of [individual] things."[21] Arnau intends to show his friend how, guided by experience, a physician is freed from depending upon philosophical truth in pursuing the objectives of his art.

II. ESTABLISHING THE LIMITS OF MEDICAL CERTAINTY

De intentione medicorum is not an obviously polemical work; indeed, it appears to be just the opposite. It is divided into two treatises, of which the much shorter first explains how philosophical and medical truth can be harmonized, while the second illustrates this harmony in four chapters that resolve apparent disagreements between Aristotle and Galen: over the primacy of the heart among the body's organs, over the way in which foods and medicines act on the body, over a supposed state of neutrality between sickness and health, and over the nature of the soul. The tone of the whole is determinedly conciliatory.

It is thus surprising to find that Arnau evidently conceived of this book as an anti-Averroistic treatise. In his *Aphorismi de gradibus,* written toward the end of the 1290s, Arnau departed from an attack on Averroes' pharmaceutical theories in the *Colliget* to generalize that "this author . . . went wrong on every point where he attacked Galen; and hence we have written our treatises *De intentione medicorum, De considerationibus operis medicine,* and the *Epistola de dosi tyriacalium medicinarum* against him in particular, lest the weak fall into error through his opinions."[22] What opinions can Arnau have had in mind? Averroes is an obvious target of the *Aphorismi,* and the same is true of *De dosi,* for this is in part a rebuttal of Averroes' *Tractatus de tyriaca,* which Arnau considered to have been based upon Averroes' misquotation of a statement by Galen. Yet not once does Averroes' name appear in *De intentione* (or, for that matter, in *De consideratione*). Can we really interpret the work as anti-Averroistic in its aims?

Averroes' views on the character and objectives of medical knowledge had become available to the West in the 1280s, in two somewhat contradictory sources. The position he expressed at the beginning of his *Colliget* (translated into Latin in 1285) was a somewhat equivocal one. Here he treated medicine as a mixture of the speculative and the practical. According to Averroes, to some extent medicine depends on the *vera principia* of natural philosophy, and for these principles the physician goes to the *artifex scientie naturalis.* Nevertheless, some matters of practical medical concern (like a knowledge of anatomy or the properties of drugs) can only be mastered by a lengthy and difficult *ars experimentalis.*[23] Though it could be argued from this discussion that medicine is *scientia* as well as *ars,* it was the latter aspect that received more emphasis from Averroes and to some later commentators seemed to represent the *Colliget*'s true position.[24]

[21] "Ea universalia que colligunt in primitivis scientiis, utpote lima rationis et statera carentes iudicii, cum rerum experientiis non habundent, nequeunt decenter particularibus coaptare." *Ibid.*

[22] *Arnaldi de Villanova Opera medica omnia* (cit. n. 15), Vol. II: *Aphorismi de gradibus,* ed. Michael R. McVaugh (Granada/Barcelona: Seminarium Historiae Medicae Granatensis, 1975), p. 201.

[23] Averroes, *Colliget* 1.1, in *Aristotelis opera . . . ,* Vol. X (Venice, 1562; facs. rpt. Frankfurt am Main: Minerva, 1962), fols. 3rb–4rb; and Per-Gunnar Ottosson, *Scholastic Medicine and Philosophy: A Study of Commentaries on Galen's* Tegni (Naples: Bibliopolis, 1984), pp. 73–74.

[24] Nancy Siraisi, "Views on the Certitude of Medical Science among Late Medieval Medical Writers," unpublished typescript, suggests that, e.g., Nicoletto Vernia interpreted the *Colliget* in this way at the end of the fifteenth century.

The other source for Averroes' views on the character of medical knowledge, his commentary on Avicenna's *Cantica* (translated in 1284 by Arnau's nephew Armengaud Blaise of Montpellier, apparently via a Hebrew intermediary), was already being incorporated into Montpellier medicine by the end of the century.[25] This commentary expressed an opinion on the status of medicine that appeared to be quite different from that of the *Colliget*. Avicenna had begun the *Cantica* with the passing assertion that medicine could be divided into theory and practice, and Averroes took advantage of the terse statement so as to reject it; on the contrary, he insisted, *all* of medicine is "theoretical" and scientific, knowable perfectly, and in this respect it is one with the natural sciences.[26]

> Galen said . . . that medicine is the *science* of health and illness and of things relative to them, as well as of the disposition that is neither health nor illness. If this be true, therefore, all its parts will be scientific and theoretical, not just one theoretical and the other practical. We must recognize that among *operationibus rationalibus* there are some . . . , called *artes*, [which] beyond any question are learned like a science—that is, by the demonstrations and definitions through which something is known perfectly and wholly. This is what an *ars* is, properly, and this is the character of the art of medicine. . . . The science of the art of medicine is divided into two sciences or scientific parts. . . . One is the science that considers health, its cause and signs (and illness, with *its* cause and signs); and the second . . . is that with which the physician considers how and with what he may conserve health, and likewise how and with what he may cure illness. . . . The [first] . . . is a science truly and properly, namely such that the end aimed at in it is only to know, not to act [*in ea est solum scire et non opus*]. The other part is peculiar to the consideration of the art of medicine, and it can be called *praxis* or active in that it is related to *operatio*, since its *operatio* and use is had and known mostly by *operando*. . . .[27]

Averroes' commentary on the *Cantica* thus reverses the position of his *Colliget*: it allows the physician some scope for practical experience, but it stresses much more heavily the rational, demonstrable character of medicine in both its theoretical and its operational aspects.[28] Such an attitude was of course Arnau's target

[25] Moritz Steinschneider, *Die hebraeischen Übersetzungen des Mittelalters und die Juden als Dolmetscher* (Berlin, 1893; rpt. Graz: Akademische Druck- u. Verlagsanstalt, 1956), sect. 444 (pp. 697–699); Ern. R[enan], "Armengaud, fils de Blaise," *Histoire Littéraire de la France* 1881, *28*:131–136; and Lynn Thorndike, "Date of the Translation by Ermengaud Blasius of the Work on the Quadrant by Profatius Judaeus," *Isis*, 1937, *26*:309. On the presence of the *Cantica* at Montpellier before 1314 see *Arnaldi de Villanova Opera medica omnia* (cit. n. 15), Vol. III, ed. Michael R. McVaugh (Barcelona: Seminarium Historiae Medicae Cantabricense, 1985), p. 59. The citation of Averroes' commentary on the *Cantica* by Henri de Mondeville in his *Chirurgia* (cit. n. 10) (e.g., ed. Pagel, *40*:666) may also reflect its use at Montpellier.

[26] Avicenna's views on the status of medicine as science or art, theoretical or practical, were in fact subtler than Averroes led his readers to believe. At the beginning of the *Canon* (1.1.1.1) Avicenna explained carefully that, as he used the terms, "theorica" and practica" both pertained to medical science or understanding—a position perhaps not so far removed from that of the *Cantica* commentary—and had no necessary foundation in actual medical practice. See also Michael R. McVaugh, "The Two Faces of a Medical Career: Jordanus de Turre of Montpellier," in *Mathematics and Its Applications to Science and Natural Philosophy in the Middle Ages: Essays in Honor of Marshall Clagett*, ed. Edward Grant and John E. Murdoch (Cambridge: Cambridge Univ. Press, 1987), esp. pp. 309–310.

[27] *Avicennae Cantica . . . cum Averrois Cordubensis Commentariis* 1.1, text 2, in *Aristotelis opera*, Vol. X (cit. n. 23), fols. 221r–v.

[28] An account of Averroes' views that differs slightly in nuance from my discussion of these texts is given by R. K. French, "Gentile da Foligno and the *Via Medicorum*," in *The Light of Nature: Essays in the History and Philosophy of Science Presented to A. C. Crombie*, ed. J. D. North and J. J. Roche (Dordrecht/Boston: Martinus Nijhoff, 1985), pp. 28–29.

in *De intentione medicorum,* and hence he might easily have composed his work in order to counteract the influence of the *Cantica* commentary that his nephew Armengaud had just made available to students.[29] Indeed, the first section of *De intentione* can be read as an attempt specifically to refute Averroes' contention that medicine is wholly *scientia* and is in part aimed at pursuing knowledge for knowledge's sake.

The technical terms Armengaud had selected in translating a Hebrew version of the Arabic original were full of resonances for anyone formed, like Arnau, in an Aristotelian arts tradition—particularly the terms "ars" and "scientia," which Aristotle distinguished perhaps most clearly in the *Ethics.* "Scientific knowledge," says Aristotle, "is judgement about things that are universal and necessary, and the conclusions of demonstration, and all scientific knowledge, follow from first principles." Art, on the other hand, "is concerned [not] with things that are, or come into being, by necessity" but with *making,* bringing something into being, "with contriving and considering how something may come into being which is capable of either being or not being, and whose origin is in the maker and not in the thing made." Science is necessary, art contingent, truth.[30]

The same terms had been used by the Latin translators of standard medical authorities, but inconsistently, reflecting the confusion of intellectual and empirical traditions in early Western medicine.[31] Hippocrates' famous declaration at the opening of the *Aphorisms* that medicine is an *ars* could be matched against the statement made by Avicenna at the beginning of the *Canon:* "Dico quod medicina est scientia."[32] The confusion was heightened by the use of both terms in works ascribed to Galen. The *Tegni* is constructed around the statement that "Medicina est scientia sanorum, egrorum, et neutrorum," but it immediately backs away from the implications of the term and explains that "the word 'scientia' is to be understood *communiter* and not *proprie.*"[33] In the works of the "new Galen," indeed, medicine is regularly referred to as an *ars.* It is easy to understand why attempts to resolve this contradiction were so frequent in the Italian medical faculties from the last quarter of the thirteenth century on, particularly since more than mere terminology was at stake for physicians: a clear consideration of themselves and of their function was at issue. Arnau's *De intentione* suggests that in the 1290s these concerns were surfacing at Montpellier as well.

But Arnau did not address them by confronting science directly with art. He began his work by tacitly assuming the status of medicine as an art: because the physician is therefore an *artifex,* he must start by having in his mind the form of what he wishes to bring about. As his readers would have known (Arnau does not say so), this conclusion is drawn directly from *Metaphysics* 7, where indeed Aristotle actually uses medicine to illustrate how an *artifex* brings something into

[29] That the force of this particular passage had been felt at Montpellier is also implied by Henri de Mondeville's commentary thereupon in his *Chirurgia* (cit. n. 10) (ed. Pagel, *40*:666; ed. Nicaise, pp. 104–105), though of course Mondeville could have come upon it at Paris rather than Montpellier.

[30] Aristotle, *Ethics,* 1140b30–32, 1140a10–14, emphasis added. I use the translation of W. D. Ross, in *Introduction to Aristotle,* ed. Richard McKeon (New York: Random House, 1947), pp. 429, 427. The establishment of these meanings by the late thirteenth century is well shown in the examples given by Hélène Merle, "Ars," *Bulletin de Philosophie Médiévale,* 1986, *28*:95–133, esp. pp. 115–121.

[31] Ottosson, *Scholastic Medicine and Philosophy* (cit. n. 23), pp. 68ff.

[32] And see above, n. 26.

[33] *Microtegni* 1, in *Galieni . . . Opera,* 2 vols. (Venice, 1490), Vol. I, fol. 11r. This is the passage referred to by Averroes in the passage cited above at n. 27.

being: in the case of the physician, this is health, which, Arnau points out, is consistent with Galen's own dictum that "intentio medicorum sola operatio est."[34] Therefore, he goes on, nothing not aimed at *operatio,* at either preserving or restoring health, can properly be called a part of medicine. It is true that *operatio* requires the physician to have a knowledge of general scientific truths (*doctrina cognitiva*), as well as a knowledge of the principles of practice (*doctrina operativa*), but the former can only be called medical insofar as they help direct the latter.[35]

This summary reveals the fundamental disagreement between Arnau's goals in *De intentione* and Averroes' convictions as expressed in his commentary on the *Cantica.* Both would concur that there exist two branches of medical knowledge, theoretical and practical; in this, it may be said, they are in accord with the wider Western medical tradition. But for Arnau both these branches have practice (*operatio*) as their object, while for Averroes that is true only of active or practical knowledge; theoretical knowledge wishes "to know, not to act." Moreover, for Averroes, both seem to be true sciences, "learned . . . by the demonstrations and definitions through which something is known perfectly and wholly"—the very assumption against which Arnau warns students in the preface to his work. Arnau argues instead that because medical practice deals with individuals, medical *doctrina* must concern itself with sense knowledge of these individuals, not with generalizations. This emphatic separation of medicine from science, its restriction to the realm of *ars,* might have seemed undesirable to many contemporaries; for to include medicine among the sciences, as Averroes had done, would ascribe certainty to it and give its practitioners prestige and status.[36]

What then is the relationship in Arnau's mind between the natural philosopher's *scientia* and the physician's *ars?* It is odd that Arnau does not confront this question here in the first, broad, introductory treatise of *De intentione medicorum,* where it would seem natural to do so; but he has an answer to it that can be reconstructed from the second part of the work, where it provides the underpinning for his reconciliation of Aristotle and Galen. The crux is again the issue of purpose or function, of *intentio.* The natural philosopher aims at *perfectam rerum cognitionem,* at a complete understanding of things, of their natures and causes, but the physician needs to understand only what will perfect his treatment of patients (*opus*), and hence he has to concern himself only with manifest sense evidence, while the natural philosopher has to consider the inner, hidden nature of things as well. Even so, properly understood there is no contradiction between the two realms: the "universals that formally constitute the art of medicine" are also those of the natural philosopher, but for the sake of medical practice they must be expressed differently if they are to be "applied to sensible particulars." Medicine will be expressed so as to describe appearances, natural philosophy so as to explain the true nature of things.[37]

[34] See Aristotle, *Metaphysics* 7.7, 1032b. Here (*De intentione* 1 [cit. n. 20], fol. 36va) Arnau appears to be paraphrasing Galen, *De sectis,* text 2, *Opera,* Vol. I, fol. 3rb: "Medicine artis intentio quidem est sanitas, finis vero eius adeptio est."

[35] See also McVaugh, "Two Faces of a Medical Career" (cit. n. 26), esp. pp. 308–311.

[36] Siraisi, *Taddeo Alderotti* (cit. n. 3), pp. 120ff., suggests that Italian masters at the turn of the fourteenth century might have recognized such an advantage in associating medicine with philosophy, though she also points out that too close an association might eventually have been understood to entail risks to the autonomy of medicine.

[37] Arnau, *De intentione* 2.1 (cit. n. 20), fol. 37ra; and 2.2 (fol. 37rb), 2.1 (fol. 36va), and 2.3 (fol. 38va).

Arnau appears to have believed that in making this distinction he was following Galen's own doctrine. He comments elsewhere that

> in several places, especially in *De interioribus* 1, Galen teaches us to distinguish between philosophical and medical knowledge or inquiry, [as also] in his treatise on the definition of medicine and his *De demonstrationibus medicinalibus;* and he has sufficiently explained these matters in the first and third texts of his commentary on the prologue of [Hippocrates'] *Regimen acutorum.* Here it will be enough generally to know that no statement has relevance for medicine unless it assists our practice by teaching something that has a role in that practice, as Galen expressly says in the aforesaid passages.[38]

Actually, not all the Galenic texts to which Arnau here refers put the argument so forcibly. The commentary on the *Regimen acutorum* merely explains that a medical author should refrain from passing on knowledge if "it has no bearing on the object of the art," and that "no one ought to claim that the goal of medicine is simply to understand or to be able to predict diseases," since such knowledge is of value only insofar as it makes it possible to *cure* disease.[39] The passage that Arnau singles out as especially important is evidently *De interioribus* 1.3, where, however, Galen is far from repudiating "philosophical knowledge": he merely complains that physicians and philosophers have confused the meanings that originally attached to the terms *dolor, nocumentum,* and *accidens* and mistakenly equate them with *morbus.*[40] Galen criticizes those who claim that sufferers from headache "have pain and injury in the head, because their suffering really comes about because of a connection to the stomach"; if you want to communicate knowledge (*scientia*) of anything, he insists, you must be able to name it and to understand its nature correctly before you can explain it. Yet while Galen may have been talking here about the importance of founding *scientia* on accurate definitions, Arnau apparently understood his meaning rather differently.

Arnau's idiosyncratic interpretation of Galen's views is made even more apparent in his reworking of Galen's *De interioribus.* This reworking originated, he explains elsewhere, when he found that those first two books contained much that was necessary to the student of medicine but could not be easily understood in the existing translation; consequently, he prepared a revised version of the text.[41] In this revision, in the words of its modern editor, Arnau "has attempted to clarify Galen's somewhat confused and disorderly account. He does not slavishly follow the order of the original but groups the material thematically."[42] As it

38 Arnau de Vilanova, *Repetitio super canonem vita brevis,* in *Opera* (cit. n. 20), fol. 276vb.

39 Galen, Commentary on *Regimen acutorum,* text 3, in *Articella,* 2 vols. bound as 1 (Venice, 1523), [Vol. II], fol. 1rb.

40 "Multi enim medici et philosophi post antiquissimos inventores existentes hec destruxerunt, id est certitudinem horum vocabulorum, nocumenti scilicet doloris et accidentis, que nomina ad significandum actionis habitum sunt inventa. . . . Morbus enim non est aliud nisi motus extra naturam. Quam videlicet mutationem aliquando abusive vocamus nocumentum dolorem et accidens; iccirco qui vult imitari antiquos in nominibus vocet motum extra naturam nocumentum vel dolorem, sed qui sequitur veritatem appellat membrum infirmum in quo motus extra naturam est; quod si abusive eum vocaveris, dices illud hoc nomine quod est nocumentum vel dolor." Galen, *De interioribus* 1.3, *Opera* (cit. n. 33), Vol. II, fol. 115v.

41 Michael R. McVaugh, "The Authorship of the Galenic Compendium *De interioribus,* 'Quoniam diversitas . . . ,' " *Dynamis,* 1981, *1*:228, n. 8. That Arnau completed the revision in 1300 can be inferred from the inventory record published by Luis García-Ballester in *Arnaldi de Villanova Opera medica omnia,* Vol. XV (cit. n. 15), p. 32.

42 Durling, in his edition of Arnau's *Doctrina Galieni de interioribus* (cit. n. 15), pp. 301–302.

happens, Arnau's reworking of *De interioribus* 1.3 illustrates these practices nicely. He has reorganized its contents, moving the middle of the chapter to the beginning, and has simplified Galen's language; in the process, however, he has lost the chapter's original point that some authors confuse *nocumentum* and *dolor* with *morbus*. The revised version merely explains that these terms mean different things, and it is introduced and concluded with sentences that are apparently new to the chapter, added by Arnau to make clear what he thought the chapter meant—a meaning rather different from Galen's. "To investigate whether this or that ought properly to be called *nocumentum* rather than *morbus*, or vice versa, or *passio*, or *accidens*, and so forth, is of no value for a cure. . . . Such investigations are entirely useless to the physician, because he cannot gain a correct understanding of illness through them, nor is it useful to a cure or to knowledge of future things." Consciously or unconsciously, Arnau has turned Galen's chapter into an attack on the pursuit of medical theory for its own sake, whereas the original stressed instead the importance of reasoning from solid foundations.[43]

Arnau was not the only figure in late thirteenth-century academic medicine to have modified Galen's views on the status of medical knowledge. In a difficult passage from *De complexionibus* 1.5, Galen complains about those who attempt to understand medical causation without having mastered the skills of reasoning: either one should base all one's practice on experience, or one should first make oneself expert in logic.[44] Those who do not, those who speculate on natural causes without having prepared themselves to do so, will misunderstand Aristotle on the primary qualities, thinking he meant only one thing by "hot" and "cold" when in fact he used these words in a number of ways. Indeed, Galen goes on, perhaps all the errors committed by men in every *ars*, in every *opus* that they pursue, are due to their failure to understand the meaning of words; his concern for definitions here is in keeping with his similar concern in *De interioribus* 1.3. However, as summarized in Henri de Mondeville's *Chirurgia*, this passage is given a different twist.

> Note that according to Galen, *De complexionibus* 1.5, a doctor should depend on experience [*experimentum*] in all his doings, and should not investigate what a thing is and what it is good for. He may investigate and debate its cause, why it is and why it is as it is, but he should always remember that a surgeon, no matter how knowledgeable and intelligent he may be, cannot easily identify the efficient cause of a particular activity; hence sometimes apparent explanations [*rationes apparentes*] must suffice,

[43] *Ibid.*, 1.4 (in Arnau's enumeration), pp. 313, 315. Cf. this with Galen's own statement, quoted above in n. 40. It may be that the opinion that Arnau here ascribes to Galen—that the philosopher and physician can be distinguished by their different concerns—can be supported by Galenic texts known to Arnau. Galen makes a somewhat similar distinction in, e.g., *De placitis Hippocratis et Platonis* 6.8, in *On the Doctrines of Hippocrates and Plato*, ed. and trans. Phillip de Lacy (Corpus Medicorum Graecorum 5.4.1.2), Vol. II (Berlin: Akademie Verlag, 1981), pp. 418, 419. This particular work was not yet available in Latin translation in Arnau's day, but he might have encountered it in Arabic. And Galen may express the same idea in other works that I have not yet found. The point here is, however, that the Galenic texts actually cited by Arnau do *not* express this opinion; he is reading it into them.

[44] "Est una duarum rerum melior; aut ne medicus nullo modo conetur in huiusmodi ratiocinationibus sed confidat in omni re sua secundum experimentum, aut ut ipse ante fit etiam assuetus in scientia dyaletice et dimittat subtiliter intueri illud quod experimento apparet et perscrutari speculationem in naturis antequam exerceat intelligentia qua inquirit eius intentionem secundum quod oportet." Galen, *De complexionibus* 1.5, *Opera* (cit. n. 33), Vol. II, fol. 9va.

especially in those instances that practitioners attest to on the basis of their own experience.[45]

Once again, Galen's concern for sound reasoning has been transmuted into an argument that medical reasoning is of only limited value; the medical practitioner is concerned with the facts of experience, not with causal explanations.[46] Henri's reinterpretation of Galen is strikingly similar to Arnau's, and this may not be mere coincidence: Henri, it will be remembered, was trained at Montpellier and could even have studied with Arnau in the 1290s.

Taken as a whole, these instances shed further light on the problems associated with the reception of the "new" Galen. His commentary on the *Regimen acutorum* was relatively simple and straightforward and had (perhaps for that reason) long circulated in conjunction with the *articella;* but *De complexionibus* and *De interioribus* were more difficult works, at times verging on incomprehensibility, and could only be understood by an audience intellectually prepared for them. Arnau and Henri had tried to prepare their readers to understand Galen by using summaries, commentaries, and careful references. But how had these first teachers been prepared for *their* reading of Galen? Specifically, what prepared Arnau and shaped *his* understanding?

III. MEDICAL INSTRUMENTALISM AND ITS AVICENNAN ORIGIN

For the first of the controversies to be resolved in the second part of *De intentione,* Arnau selected perhaps the most famous point of disagreement between natural philosophers and physicians in the Middle Ages: is there only one principal member, the heart, as Aristotle said, or was Galen right to identify three or even four such? It was not just its notoriety that made this a natural point of departure for Arnau, for the particular points at issue lent themselves unusually well to the detailed application of his understanding of the objectives peculiar to the medical practitioner.

To understand these issues, we should first outline the foundations of the Galenic physiology in which Arnau had been trained, foundations epitomized in medieval accounts of the seven *res naturales,* the factors indispensable to health: *elementa, complexiones, humores, membra, virtutes, operationes,* and *spiritus.*[47] The four elements are the first constituents of all things. In living organisms, however, they and their qualities are only remote constituents, while the four humors are the immediate constituents of the members. These latter are the

[45] Mondeville, *Chirurgia* (cit. n. 10), ed. Pagel, *40*:673; ed. Nicaise, pp. 114–115. See also the text quoted above, n. 12, which interprets *De complexionibus* 1.5 in the same way. Mondeville applies this precept to practice in *Chirurgia* 2.1.1.2 (ed. Pagel, *40*:881; ed. Nicaise, p. 251).

[46] Christian Probst, "Der Weg des ärztlichen Erkennens bei Heinrich von Mondeville," in *Fachliteratur des Mittelalters,* ed. Gundolf Keil et al. (Stuttgart: J. B. Metzler, 1968), pp. 333–357, seems to interpret the passage from which I have quoted as evidence for both a "rein rational-deduktives Verfahren" (p. 340) and a "reinen Pragmatismus" (p. 341) in Mondeville's method. The discussion of Mondeville's "nominalism" by Marie-Christine Pouchelle, *Corps et chirurgie à l'apogée du moyen âge* (Paris: Flammarion, 1983), esp. pp. 39–61, does not attempt to explore this aspect of the wider medical context in which he wrote.

[47] This summary is based on the expositions provided by Arnau in his *Speculum* of 1308 and Avicenna's *Canon;* the two are quite consistent in their accounts, a fact of some interest given the thesis of this section.

distinguishable functional entities within the body. Their functioning, whether healthy or unhealthy, depends indirectly on the three sets of powers (*virtutes*) that derive ultimately from the soul and act through the members. The members are thus the instruments of the powers in carrying on all the activities of life. The three sets of powers are the animal powers (associated primarily with the brain), which give rise to cognition and locomotion; the vital powers, whose source (*fundamentum*) is the heart, and on which the pulse and respiration as well as the emotions depend; and the natural powers, responsible for generation, growth, and nutrition. Some powers inhere within a member; others may flow into one member from another, carried by the corresponding spirits, which thus serve as their vehicles. It is the physician's task to recognize how the powers bring about the functioning of the members, acting through the members and their qualitative *complexio* of hotness and coldness, or dryness and moistness.

How is this to be done? It is here that Arnau begins his analysis in *De intentione* 2.1 of the relative importance of the principal members in the body, pointing out that a physician's understanding has to be based on the senses. One way that sense experience of the members can be obtained is through dissections, and the information thus gained from the dead can shape the physician's treatment of the living. But of course it is not licit or safe to try to get direct sense knowledge of living organs in this way, by vivisection, and in the living we are therefore limited to observing the functioning (*operationes*) of the members in order to infer their state from the health of their *virtutes*.[48]

Arnau's explanation of how one may gain physiological understanding seems unusual for the importance it gives to anatomical knowledge as more important, more directly sense-based, than clinical information. The European medical faculties are supposed to have been merely initiating the study of anatomy at this time, the end of the thirteenth century, hesitantly and irregularly. Actually Arnau's statement does not necessarily imply the regular practice of anatomical dissection at Montpellier, since it appears to be a conscious echo of the investigative program already set out by Galen in *De interioribus* 1.1:

> It is necessary to learn the essence of every member through anatomy, what it is and then its activity [*actionem*] and its relationship to its neighbors, a relationship that depends on its location. Likewise the function [*iuvamentum*] of every member is a great help in identifying the unhealthy parts, for activity is nothing but the active

[48] García-Ballester has argued ("Arnau de Vilanova" [cit. n. 3], pp. 148–152) that Arnau consciously limited "anatomy" to a book-learned subject, but his argument rests entirely upon the reading of a single word in the following passage: "Quia igitur ad hanc noticiam obtinendam perfecte maxime iuvat subiicere membra sensibus, et hoc de interioribus membris licitum non sit efficere vel securum in vivorum corporibus, ideo docet hoc sensibiliter in mortuis perscrutari; in vivis autem, quia de occulta membris nihil sensui manifestum est nisi sola operatio . . . , ideo docet medicus cognoscere et attendere operationes proprias et immediatas membrorum: per has enim in vivis membrorum occultorum dispositio sensibiliter apprehenditur." *De intentione* 2.1 (cit. n. 20), fol. 36vb, emended by reference to manuscripts. García-Ballester follows the printed text in reading *insensibiliter* for the first *sensibiliter* in the text given above. In the process of preparing an edition of *De intentione,* I have collated nine manuscripts of the text: of the four best, two at this point read *sensibiliter* (including MS Oxford, Merton College 230, which modern editors of Arnau's medical works have identified as a particularly trustworthy witness) and two read *subtiliter;* four other, markedly less valuable, versions read *insensibiliter;* and the ninth reads *insensibiliter al. visibiliter.* My understanding of the overall argument of *De intentione* 2.1 confirms my feeling from textual grounds that *sensibiliter* is the best reading: Arnau is saying "sense evidence is essential, but since it is difficult to obtain this by dissection we must accept the easier clinical information."

motion of moving members, and for every member, even if there is no action, there is still function.[49]

The express purpose of *De interioribus* is to explain how knowledge of the members can be acquired clinically by studying disturbances in their function, a characteristically Galenic theme that underlies his *De ingenio sanitatis* as well; but in his opening statement Galen has made it clear that in his view anatomical knowledge is a prerequisite for the clinician.

Was Arnau then merely parroting Galen when he explained how one gains knowledge of the internal organs? Can we conclude nothing from *De intentione* about his personal experience of dissection as a scientific tool? It is not easy to be sure. Still, it is interesting that when, some years later, Arnau prepared his rewritten condensation of the first two books of *De interioribus,* he kept and even intensified Galen's emphasis on the importance of anatomy.[50] We know, too, that Henri de Mondeville gave an anatomical demonstration at Montpellier in 1304—that is, only a decade or so after Arnau's completion of *De intentione* there—in which he illustrated the structure of the body with visual aids: certainly showing drawings, quite possibly appealing to models and perhaps even to dissected material.[51] So it is not impossible that Arnau's insistence upon the value of the sense-data provided by anatomy sprang from his own exposure to dissections, and not merely from respect for Galen's opinion.

What does a physician need to know about a member, then, to conclude that it is a "principal" one? Such a member must be shown to possess a power (*virtus*) that influences and directs other members in their growth and function. Since the physician gains a knowledge of members in the living only through *operatio,* he will label as "principal" those members in whose activity he sensibly detects the presence of a directive power, and there are four such: heart, liver, brain, and

[49] "Oportet per anothomiam sciri essentiam uniuscuiusque membri qualiter est et postea eorum actionem et vicinorum colligantia que colligantia et positioni pertinet. Item iuvamentum uniuscuiusque membri adiutorium magnum est in investigandis locis infirmis; actio enim non est aliud nisi membrorum movendorum motio activa, verum unumquodque membrum et si nullius sit actionis alicuius tamen est iuvamentum." Galen, *De interioribus* 1.1 (cit. n. 40), fol. 114vb. A translation of the Greek text can be found in Rudolph E. Siegel, *Galen on the Affected Parts* (Basel/New York: Karger, 1976), p. 21.

[50] Cf. Arnau's rendering of the ideas expressed by Galen in the passage quoted in the previous note: "Sapientem enim medicum non oportet hoc racione discutere. Nam debet esse informatus de esse uniuscuiusque interiorum membrorum per anotomiam, an sit cartillaginosa vel carnosa et cetera. Unde visi frustis egredientibus, sensu perpendere potest, quoniam illud membrum patitur, a cuius substancia resolvuntur. Sic ergo patet, quod ad noticiam paciencium membrorum occultorum necesse est habere cognicionem accionis et iuvamenti et anathomie ipsorum. Nam per anathomiam cognoscitur essencia cuiuslibet ac propria composicio." *Doctrina Galieni de interioribus,* ed. Durling (cit. n. 15), p. 306.

[51] For the drawings see Loren C. MacKinney, "The Beginnings of Western Scientific Anatomy: New Evidence and a Revision in Interpretation of Mondeville's Role," *Medical History,* 1962, *6*:233–239. The text of the 1304 demonstration advises the audience, "Quicunque vult anathomiam capitis ostendere . . . , ipse debet habere craneum artificiale, aperibile formatum, per veras commissuras divisum . . . , quod cum anathomiam extrinsecam ostenderit, aperire possit, ut sensibiliter anathomia panniculorum et cerebri videatur." *Die Anatomie des Heinrich von Mondeville,* ed. Julius Leopold Pagel (Berlin, 1889), p. 26. In the anatomical section of the *Chirurgia* he drew up at Paris two years later, Mondeville expanded his advice: "Quicunque vult anatomiam ostendere capitis . . . , si non posset habere verum caput humanum, ipse debet habere craneum artificiale, aperibile, serratum per commissuras." *Chirurgia* (cit. n. 10), ed. Pagel, *40*:276; ed. Nicaise, p. 26. If they do not establish his use of dissection as a technique, these passages prove at the very least that Henri vividly appreciated the sensible advantages that dissection could provide.

testicles. Proper identification is vital. If a physician detects a weakening of sensation in a patient, for example, he must be aware that this indicates a weakness of the brain rather than the heart, so that he can direct his treatment to the proper organ.

How can this understanding of the physician's approach be reconciled with Aristotle's declaration that there is only one principal member, the heart? The answer is that the heart is the *principium primum,* whose powers direct all the others but can only be seen in them, which are therefore as it were the instruments of the heart. From the medical, "instrumentalist" point of view there are indeed four principal members when they are considered as *principia proxima,* whose function and dysfunction are evident to the senses. A perfect analogy can be found in the next functional stage down: while we know that a general *virtus sentiendi* proceeds directly from the brain (thus differentiating it from its own more remote *principium,* the heart), the specific activity of the senses cannot be detected in the brain, only in the individual sense organs—"the brain neither sees nor hears nor tastes nor touches." The physician calls members "principia" only insofar as he can demonstrate the activity of a directive power in them; the heart, as the source of a manifest vital power, is one of four such *principia manifestativa seu proxima.* The physician can ignore the philosophical reality of a *principium primum et originale virtutum* because it has no bearing on the objectives of his art. If a patient's condition reveals damage to the nutritive power it is the organs that act for the heart as nutritive instruments that should be treated, and not the heart itself. Indeed, if in such a case the physician were to think first about the heart it would be harmful, and a distraction from his real concerns, even though at base his understanding is perfectly consistent with the natural philosopher's.[52]

Here, therefore, we find applied to a concrete situation Arnau's interpretation of "Galen's" program for the physician and how it differs from the philosopher's objectives.

> The physician orients his understanding and theorizing toward practice, and hence is not concerned to understand more theory than is sufficient to govern or correct his practice. . . . But the natural philosopher does not orient his thought toward practice but toward a full understanding of the nature of moving body, and the nature of something is not fully grasped by knowledge or rational understanding until its first and proximate causes have [both] been fully understood. . . . Here appears the truth of the doctrine [*sermonis*] that the doctrines of physicians are based on sufficient propositions, whose truth is aimed at achieving a predetermined goal in their art, rather than on necessary ones—that is, they are not based on propositions that are true in every sense or application of the doctrine.[53]

But as we have already learned, Arnau's interpretation diverges significantly from Galen's own defense of careful reasoning. We may ask once more, therefore, what was it that shaped his understanding of Galen's thought?

There can be no doubt that in this particular respect it was Avicenna who was Arnau's teacher, even though he nowhere mentions Avicenna's name in *De intentione,* for the *sermo* from which Arnau quotes is to be found in the *Canon.* The first fen, or treatise, of the first book of the *Canon* is structured around a list

[52] Arnau, *De intentione* 2.1 (cit. n. 20), fols. 36va–37rb.
[53] *Ibid.,* fol. 37ra.

of the *res naturales,* and hence one of the subjects Avicenna treats here is the powers (*virtutes*). His account leads naturally into a discussion of the disagreement between Aristotle and Galen over the question of the primacy of the heart as source of the powers.

> To Aristotle, greatest of all physicians, it seems that the heart is the *principium* of all these functions, but the manifestation of their primary functions appears in these other organs [the Galenic ones], in the same way that physicians say the brain is the principle of sensation but that every sense has an organ through which its functioning takes place. Those who consider as they should will find that things are as they appear to Aristotle, not as they appear to [physicians], and they will find the physicians' treatises based on sufficient rather than necessary propositions, in which they go only by that which appears from things. But qua physician the physician need not consider which is the true one between these two opinions, rather the philosopher must do this. Since it is conceded to the physician that these four members are the [immediate] principles of these powers, he need not worry when considering the nature of treatment whether they derive from still other principles prior to them.[54]

Virtually all of Arnau's chapter can be found, tightly compressed, in this paragraph from Avicenna: the physician's judgment of *principium* based on *manifestatio operationum;* the parallel between brain-heart and heart–sense organ dependence; the contrast between sufficient and necessary propositions; the irrelevance to the physician of the philosopher's prior causal principles. Moreover, these same themes are developed again, in only slightly different form, in the *Canon*'s discussion of the members. Evidently it was Avicenna who taught Arnau what Galen had meant and helped him refute Averroes.[55]

Something that might have helped reinforce Arnau in this Avicennan medical positivism was the parallel presented by astronomy. Since antiquity, physical cosmologists and mathematical astronomers had found themselves at odds: the mathematicians' eccentrics and epicycles gave an accurate description of the planetary motions but were inconsistent with the principles of uniform, circular, earth-centered motion required by the cosmologists. Some scholastic philosophers, however, had suggested a way out, claiming that the mathematical devices were not to be understood as true but merely as hypotheses designed to save the appearances, to meet the needs of calculation.[56] Arnau's argument in *De intentione* is an analogous one: the physician's explanations must meet his need to heal a patient quickly and effectively and need not fulfill the demands of natural philosophy.

What increases the likelihood that the astronomical example was in Arnau's mind is his more explicit extension of that example to medicine a few years later, in his *Aphorismi de gradibus*. Here he argued that any method of calculating a drug's intensity would be acceptable if it gave the right results,

> even though the things measured may not be exactly as described, just as happens with astronomers. For although the planets do not really move in eccentrics and

[54] Avicenna, *Canon* 1.1.6.1 (Venice, 1507; facs. rpt. Hildesheim: Georg Olms, 1964), fol. 23rb.

[55] *Ibid.,* 1.1.4.1, fols. 7va–b. Arnau's attitude toward Avicenna is complicated and not easy to understand. Overtly he professed scorn for his contemporaries because they had been bowled over by the *Canon:* Jacquart, "La réception du *Canon*" (cit. n. 4), p. 6. As *De intentione* shows, however, Arnau himself was not immune to the attractions of the *Canon,* whether he liked to admit it or not.

[56] The material was first discussed extensively by Pierre Duhem, whose account is most conveniently available in *To Save the Phenomena: An Essay on the Idea of Physical Theory from Plato to Galileo* (Chicago/London: Univ. Chicago Press, 1969), pp. 36–45.

> epicycles, still, granted that they do move according to the spaces and measures [*terminos*] by which such circles are defined, it will not matter if these are used to calculate the various motions of the planets; indeed, in this way it is possible to represent them with certainty, as is apparent in solar and lunar eclipses. So too here . . . to meet his objectives, it is enough for the physician to use such measurements as will provide certainty (insofar as possible) in calculation, even though they do not actually correspond to reality.[57]

This willingness to justify the apparent inconsistency between medical theory and natural philosophy by claiming that for the former it was sufficient to save the appearances is not easy to find in Arnau's contemporaries.[58] His use of this device is unlike the astronomer's, however, in that it does not represent an abandonment of the physician's claim to know natural-philosophical truth. His assertion in *De intentione* that philosophical and medical judgments are essentially the same and will lead to the same results in practice[59] is far stronger than anything the astronomers could have claimed, and he will try to justify his assertion by explaining the equivalence of these judgments in each of the four problems he treats in *De intentione*. But Arnau carefully refrains from claiming that medicine is, like natural philosophy, a *scientia*.

Arnau's views are perhaps representative of the general attitude of the Montpellier faculty toward the status of its discipline. The *questiones* on Hippocrates' *Aphorisms* ascribed to Bernard de Angrarra (fl. Montpellier, ca. 1300) reveal much the same outlook. In commenting upon Hippocrates' opening words, "Ars vero longa," which naturally provoke the question "utrum medicina sit scientia?" Bernard quickly decides that medicine is art rather than science, because science is a *habitus* of the intellect considering the causes of things for the sake of speculation, whereas art does so not only for the sake of speculation but for that of *operatio* as well: medicine is "non ut sanitatem corporis humani speculemur sed operemur." He concludes his analysis by making the same distinction in a slightly different way, one rather reminiscent of Arnau's exposition in *De intentione:* "Medicine considers the naturals and non-naturals only insofar as they matter to practice [*opus*], namely, to the conservation or restoration of health; hence its nature is not purely scientific [*speculabilis*] but practically oriented [*factibilis*]."[60] In none of this is there any sense that Bernard thinks the foundations of an art are any more insecure, any less certain, than those of a science. Indeed, when he goes on to gloss Hippocrates' next words, "experimentum vero fallax," he seems to be treating their foundations as epistemologically identical, both drawn from *experimentum*. For *experimentum,* correctly understood, is not uncertain but actually even more to be trusted than *ratio*—though, as Bernard admits, it is not altogether easy to compare *cognitio experimentalis* with *cognitio*

[57] Arnau, *Aphorismi de gradibus* (cit. n. 22), p. 176, lines 19–33.

[58] Siraisi, "Views on the Certitude of Medical Science" (cit. n. 24), suggests that the author of the *Plusquam commentum,* Torrigiano (d. 1319), may have entertained a similar idea, but he certainly does not express it so clearly.

[59] "Sententia philosophi secundum veritatem non deviat a recta operatione, cum eadem sit secundum rem cum sententia medicinalis artificis." Arnau, *De intentione* 2.3 (cit. n. 20), fol. 38va.

[60] Bernard de Angrarra's *questiones* on the *Aphorisms* are unedited, but can be found in MS Erfurt, F. 290, fols. 40–116 (quotation on fol. 40va), as "questiones super omnes amphorismos reportate post magistrum b. de angrara cancellarium." Luke Demaitre has kindly provided me with microfilm of the text and a preliminary transcription. On Bernard see Ernest Wickersheimer, *Dictionnaire biographique des médecins en France au Moyen Age* (Paris: Droz, 1936), Vol. I, pp. 74, 76.

intellectiva. Certainly a single observation by a lone individual is bound to be misleading, and if we are willing to call this act "experimentum" then Hippocrates is right; but properly defined, "experimentum" refers to the collective remembered observations of many. From this collective experience we abstract and establish the "principia cuiuslibet artis sive scientie." Bernard of Angrarra perhaps comes closer than Arnau, therefore, to claiming that the art of medicine is *equivalent* to a science, but they agree that because of its objective it remains *ars* rather than *scientia*.

In this respect Montpellier's attitude was rather different from that of the contemporary northern Italian universities, which were also attempting to determine the boundaries of medical knowledge and were coming to a position closer to that of Averroes. Taddeo Alderotti, the key figure in the introduction of the new Galen at Bologna in the 1280s, argued vigorously that not only medical *theoria* but *practica* (corresponding to Arnau's *doctrina theorica* and *doctrina practica*, and distinct from actual *operatio*) was a *scientia intellectualis*. Pietro d'Abano returned to Padua from Paris in 1306 saying much the same thing: *medicina practica* may be closer to *ars*, but *medicina theorica* can indeed be called a science in the accustomed sense of the word (*communiter*), for it proceeds demonstratively from first principles, such as "cure by contraries." Nancy Siraisi has suggested that by arguing for the scientific status of medicine, medical academics were attempting to set the educated physician above the untrained empiric in the public estimation. This issue of prestige may well have been an important one at Montpellier, too, but *De intentione* is more concerned with the dangers than with the benefits that will arise from imagining that medicine can, as Averroes proposed in the *Cantica* commentary, be understood as a science whose goal is to know rather than an art whose purpose is to heal.

Pietro d'Abano's *Conciliator* also explores the question of the heart's primacy, and his treatment lets us recognize a further distinctive feature of Arnau's thought. The third of Pietro's four attacks upon the problem in his *differentia* 38 resolves it along virtually the same lines as those followed in *De intentione*.[61] Pietro distinguishes between two senses of *principium*—*radicaliter* and *manifestative:* there can be only one principal member in the first sense of the word, but four in the second. He quotes Avicenna to support his contention that the brain and liver are dependent on the heart in the same manner as the senses are dependent on the brain, and decides for Aristotle's opinion as true and dismisses the physicians' as merely a "probable fantasy"—quoting the *Canon* again—based on "sufficient or probable propositions, not true ones." This is as high as Pietro seems willing to go in assessing the merits of the medical tradition, and it is here that the *Conciliator* most sharply diverges from *De intentione*. Pietro did not fully accept the implications of the medical instrumentalism according to which the therapeutic efficacy of a statement is justification enough for holding it, "true" or not.

In this respect Pietro is following closely the position advanced by Averroes in Book 2 of the *Colliget*, much of which is given over to a discussion of the members and their function whose tone recalls the commentary on the *Cantica* in its insistence that medicine must be understood as natural philosophy. For that

[61] Pietro d'Abano, *Conciliator* (Venice, 1565; facs. rpt. Padua: Antenore, 1985), fols. 59rb–60rb.

reason, Averroes explains, Galen and his physician-followers are often wrong, even in their understanding of so fundamental a concept as health:

> If you read the works of Galen, and those of other physicians who follow his teaching, you will recognize that what I say about the classification of health is much better and truer than what *he* said. And you cannot understand this matter well if you do not study the natural sciences; wherefore you must slightly favor the approach of the philosophers over the approach of the physician—this book is not intended for someone who is just beginning his studies.[62]

Occasionally Averroes was willing to concede that the physician might reasonably overlook the truth of recondite philosophical conclusions,[63] but his tolerance did not extend to repudiations of the doctrine of the primacy of the heart and its control of all physiological activity, by Galen or anyone else. In *De interioribus* 3.5 Galen had attacked his predecessor Archigenes for having treated loss of memory by drawing blood from the head, while proclaiming the Aristotelian doctrine that the heart, not the brain, was the seat of memory; sarcastically Galen had asked how an experienced therapeutist like Archigenes could have treated one member if he really believed a different member was the one affected. "All physicians," Galen declared, "accept as basis for the treatment of affections of the reasoning power that the soul is seated in the head."[64] Galen's criticism of Archigenes exemplifies his position that therapeutic effectiveness is the test of truth in medicine, and is an open challenge to the teaching of the natural philosophers. Accepting the challenge, Averroes came to Archigenes' defense in the *Colliget,* and Pietro d'Abano subsequently repeated Averroes' response approvingly in the *Conciliator:* "Est una reprehensio que est magis vituperabilis quam verba Archigenis."[65] Not merely by dogmatically accepting the heart as *principium,* but by rejecting a physician's experience as a sufficient standard of judgment, Pietro displays a markedly less pragmatic view of medical truth than does Arnau. Whereas Arnau had learned from Avicenna and Galen, Pietro was faithful to Averroes and Aristotle.

IV. AVICENNAN INSTRUMENTALISM AT MONTPELLIER

There is some evidence that the *Canon* inspired other Montpellier masters besides Arnau to adopt an instrumentalist interpretation of medical theory. We saw above (at note 45) that Henri de Mondeville interpreted a passage from Galen's *De complexionibus* as saying that a doctor should base his practice on experience, not on causal explanations, an interpretation suggestively close to Avicenna's; and Henri's conclusion, "ideo rationes debent aliquando sufficere apparentes," is almost surely an echo of an Avicennan phrase. But an even more conclusive proof of the influence of the *Canon* on the masters at Montpellier is the similar distinction drawn by Bernard de Gordon in his *Tractatus de marasmo* (composed perhaps 1308–1318), which opens with the question "whether semen materially enters the composition of the fetus." Bernard cites Aristotle, Avicenna, and Averroes in opposition and quotes Galen, Hippocrates, Haly Abbas,

[62] Averroes, *Colliget* 2.5 (cit. n. 23), fol. 15vb.
[63] *Ibid.,* 2.3, 2.7, fols. 14vb, 16vb.
[64] Galen, *De interioribus* 3.5–6, trans. Siegel (cit. n. 49), pp. 75–80.
[65] Averroes, *Colliget* 2.9 (cit. n. 23), fols. 21v–22r; and d'Abano, *Conciliator* (cit. n. 61), fol. 60ra.

and Avicenna again (!) in favor, concluding: "He who wishes to be productive, let him take the way of the physicians [*viam medicorum*]; but let him who wishes to speculate instead take the path of the philosophers [*viam philosophorum*]. Avicenna says in the *Canon* that in this case an ignorance of philosophy does not harm the physician, for, whether this or that be said, generation occurs anyway."[66] The passage Bernard has in mind from Avicenna is not either of the two we have already identified but a third (*Canon* 3.20.1.3: *De causa spermatis*), where Avicenna does indeed once again bring instrumentalism into play to explain why the physician need not concern himself with philosophical issues: "An understanding of which of these two positions is true pertains to natural science, and it does not harm the physician to be ignorant of it."[67]

Bernard's juxtaposition of a "via medicorum" and a "via philosophorum" succinctly captures the contrast worked out so carefully in *De intentione*. The image of a "path and rule" (*viam et regulam*) appropriate to the physician had been offered by Haly Abbas, who saw it embodied in his own *Pantegni;* but Haly did not use the specific phrase "via medicorum," and he did not directly compare physicians and philosophers.[68] *De intentione* follows Haly in using "via" exclusively in this general sense. With his more explicit contrast, Bernard sharpened the differentiation between the two professional approaches, of which Montpellier had just become conscious, and his phrase seems to echo, not Haly, but Averroes, in a passage from the *Colliget* that held the philosopher's path to be the more important.[69]

Still, it is certainly Arnau who most obviously shows the influence of the Avicennan instrumentalism at Montpellier. In the three concluding chapters of *De intentione* Arnau went on to apply it to other areas of disagreement between medicine and philosophy that Avicenna had not touched on. Moreover, in the works that succeeded *De intentione* he continually returned to this theme, citing that early work with remarkable frequency: it was clearly a treatise of which he remained proud, and which argued a point of deep conviction.[70] Sometimes he recognized its Avicennan descent, as when he quoted the chapter *De causa spermatis* to explain why it was unimportant to know "whether sperm is the only material principle of the living body."[71] But he continued to believe that it was

[66] Quoted in Luke E. Demaitre, *Doctor Bernard de Gordon: Professor and Practitioner* (Toronto: Pontifical Institute of Mediaeval Studies, 1980), p. 79.

[67] Avicenna, *Canon* 3.20.1.3 (cit. n. 54), fols. 352rb–va.

[68] "Cum peritum nullo modo deceat medicum hanc transgredi viam et regulam nec ab ea declinare." Haly Abbas, *Liber totius medicine* (Lyons, 1523), fol. 7ra. Haly's views are treated by French, "Gentile da Foligno" (cit. n. 28), pp. 25–27; the discussion perhaps implies, misleadingly, that the actual phrase "via medicorum" is employed by Haly.

[69] Averroes uses the phrase in *Colliget* 2.5 (translated above, at n. 62): "Opportet quod tu teneas de via philosophorum aliquantulum plus quam de via medici." French ("Gentile da Foligno," pp. 28–29), explaining the "professional 'faith' " that Gentile understood the *via medicorum* to entail, shows that the *Colliget* "made available to Gentile a sophisticated discussion about the relationships of science and art which Gentile (but not Averroes) uses to extend the notion of a *via medicorum*"; it is therefore ironic that Averroes should apparently have introduced the phrase.

[70] In addition to the reference in Arnau's *Aphorismi de gradibus* (see n. 57), the following citations may be noted: *De humido radicali* 1.1 (*Opera* [cit. n. 20], fol. 39rb) and 2.3 (fol. 41rb); *De consideratione operis medicinae* 1.3.1.2 (fol. 93rb) and 2.1 (fol. 97vb); and *Speculum medicinae,* Ch. 5 (fol. 2va) and Ch. 88 (fol. 27rb).

[71] Arnau, *De humido radicali,* fol. 39rb. The common interest taken by Arnau and Bernard de Gordon in this problem apparently reflects contemporary discussions at Montpellier: see McVaugh, " 'Humidum Radicale' in Thirteenth-Century Medicine" (cit. n. 4), pp. 259–283. It too, of course, is a topic on which Aristotle and Galen were at odds.

fundamentally a Galenic view: writing of the view ascribed to Galen that bodily organs created out of sperm could not be regenerated, he declared that "we have thoroughly studied Galen's medical judgments [*sermones*] . . . and according to his teaching as found in his works *De fine medicine* and *De demonstrationibus medicine* and in other works on medicine his opinions on medical topics are to be accepted as suited to medical investigation [*medicinali speculationi*] and as sufficient to attaining one's objectives in that art, nothing more."[72] It is not clear that Arnau found these "Galenic" opinions in the books he cites, nor even that he could have encountered these books in Latin,[73] but this scarcely matters, for his Galen is a fusion of the new medical literature: in the broadest sense, a "new Galen" that is independent of any particular text. His "Galenic" understanding of medical truth—carefully worked out, consistently developed, and systematically applied—is one that justifies the physician in pursuing his own peculiar objectives, and that tacitly defines medicine as an art akin to but distinct from the sciences.

This interpretation of Galen as having maintained an instrumentalist view of medical truth seems to have prevailed at Montpellier in the first decade of the fourteenth century, as the examples of Henri and Bernard suggest. It testifies to the school's successful assimilation of the "new Galen" into the framework of Aristotelian learning. How long this interpretation remained current, however, remains to be seen. Arnau died in 1311, and in the decade that followed yet another series of Galenic works began to appear in the European medical world, this time in translations made directly from the Greek (by Niccolò da Reggio); the arrival of these new, purer witnesses to Galenic thought certainly entailed for some writers a new sensitivity to the problems of variability of language and meaning among medical authorities.[74] Whether it also led to a revision of the instrumentalist approach to medical knowledge, which today seems so modern and so right by virtue of its clinical practicality, must await the study of the next generation of European medical writers.

[72] Arnau, *De humido radicali* 2.3, fol. 41rb. Arnau again ascribed this view to Galen ("in tractatu [suo] diffinitionis medicine, et de demonstrationibus medicinalibus") in his *Repetitio super canonem vita brevis,* in *Opera* (cit. n. 20), fol. 276vb.

[73] No Galenic work entitled *De fine medicine* or *De demonstrationibus* is known to have been available in Latin in Arnau's day. *De fine medicine* may refer to Galen's *De constitutione artis medicine,* but this work was not translated into Latin until the fourteenth century. *De demonstrationibus* presumably is a reference to Galen's *De demonstratione,* which Arnau cites in several other works, including *De intentione medicorum* (2.4), but the Greek text of this work was lost early and it was never translated into Latin; only fragments survive: see Iwan von Müller, "Ueber Galens Werk vom wissenschaftlichen Beweis," *Abhandlungen der Philosophisch-Philologischen Klasse der Königlich Bayerischen Akademie der Wissenschaften,* 1897, *20*(2):405–478. Arnau may here be quoting these works at second hand from references in other works by Galen or by Arabic authors, or may perhaps have encountered them in Arabic. I will explore this question more carefully in a fuller study of *De intentione,* now in preparation.

[74] Margaret S. Ogden, "The Galenic Works Cited in Guy de Chauliac's *Chirurgia magna,*" *Journal of the History of Medicine,* 1973, *28*:24–33, illustrates this sensitivity in Guy, who studied at Montpellier probably in the 1320s, and makes plain Guy's conviction that the newer translations were to be preferred as more trustworthy.

Jewish Appreciation of Fourteenth-Century Scholastic Medicine

By Luis García-Ballester, Lola Ferre,** and Eduard Feliu****

FOR MOST OF THE MIDDLE AGES, Jewish physicians maintained a reputation within Latin Christian society that was altogether disproportionate to their numerical presence. The importance of medicine as a calling within the Jewish community, able to draw on the learning of the Greco-Arabic tradition, set Jewish medicine apart from the empirical practice that was nearly universal in the West during the twelfth and thirteenth centuries; it continues to enjoy this high reputation among present-day historians.[1] What has not been fully appreciated is that by the fourteenth century this situation had been to some extent reversed—that the Latin system of medical learning had so developed as to establish itself, in the eyes of at least a few Jewish practitioners, as a model to be imitated and a resource to be exploited.[2]

I. SOURCES OF LEARNING

The intellectual life of the three great Mediterranean cultures of the Middle Ages —Judaic, Christian, and Islamic—was molded by religion; the development of

* Consejo Superior de Investigaciones Científicas, Historia de la Ciencia, Egipciacas 15, 08001 Barcelona, Spain.
** Facultad de Letras, Departamento de Estudios Semíticos, Universidad de Granada, 18071 Granada, Spain.
*** Ferrería 7–9, 08014 Barcelona, Spain.
We would like to thank David Romano, Angel Sáenz-Badillos, Saul Jarcho, Miguel Pérez, Ron Barkai, and Joseph Shatzmiller for criticism and for help with the Hebrew text; and Lola Badia for supplying us with valuable bibliographical references.

[1] Joseph Shatzmiller has studied the social and intellectual world of the Provençal Jewish physicians with particular care: see, e.g., his "Notes sur les médecins Juifs en Provence au Moyen Age," *Revue des Etudes Juives,* 1969, *128*:259–266; "Rationalisme et orthodoxie religieuse chez les Juifs provençaux au commencement du XIVe siècle," *Provence Historique,* 1972, *22*:261–286; "Contacts et échanges entre savants juifs et chrétiens a Montpellier vers 1300," in *Juifs et judaïsme de Languedoc, XIIIe siècle–début XIVe siècle,* ed. Marie-Hembert Vicaire and Bernhard Blumenkranz (Toulouse: Edouard Privat, 1977), pp. 337–344; "Livres médicaux et éducation médicale à propos d'un contrat de Marseille en 1316," *Mediaeval Studies,* 1980, *42*:463–470; "In Search of the 'Book of Figures': Medicine and Astrology in Montpellier at the Turn of the Fourteenth Century," *Association for Jewish Studies Review,* 1982/83, *7/8*:383–407; "On Becoming a Jewish Doctor in the High Middle Ages," *Sefarad,* 1983, *43*:239–250; "Doctors' Fees and Their Medical Responsibility: Evidence from Notarial and Court Records," in *Gli atti privati nel tardo Medioevo: Fonti per la storia sociale,* ed. Paolo Brezzi and Egmont Lee (Rome: Istituto di Studi Romani, 1984), pp. 201–208.

[2] The influence of Latin scholastic medicine on the Jewish medical tradition is only one aspect of a much larger problem that still needs to be studied—the impact of scholastic thought generally on Jewish intellectual history. The first to examine this problem was Shlomo Pines, "Scholasticism after Thomas Aquinas and the Teaching of Hasdai Crescas and His Predecessors," *Proceedings of the Israel Academy of Sciences and Humanities,* 1967, *1*(10):1–101; see also Isadore Twersky, "Joseph ibn Kaspi: Portrait of a Medieval Jewish Intellectual," in *Studies in Medieval Jewish History and Literature,* ed. Twersky (Cambridge, Mass.: Harvard Univ. Press, 1979), pp. 231–257, on pp. 235–236.

natural philosophy (and medicine) in all three was conditioned by issues that were more theological than philosophical.[3] Each claimed authority from a divine revelation, and therefore each encountered difficulty in reconciling revelation with Greek thought, specifically with Aristotelian thought. Aristotelian rationalism encouraged the formulation of a logically consistent theology, which could easily come into sharp conflict with those who considered themselves the only ones capable of interpreting the divine word, or simply with those who defended alternative approaches to understanding the relationship between man and God.

Within Latin Christendom, the university world was able during the thirteenth century (not without difficulty) to harmonize Aristotelian thought and Christian doctrine, somewhat as Maimonides had done for Judaism at the end of the twelfth century and as the *muᶜtazilites* and *mutakallimin* had tried to do for Islam still earlier.[4] Albertus Magnus (ca. 1200–1280) and Thomas Aquinas (1225–1274) symbolize the assimilation of Greek natural philosophy and metaphysics by Christian theology that established Aristotle's teachings as the starting point for all intellectual investigations.[5] In turn, the acceptance of Aristotle facilitated the systematic study of the enormous medical corpus of Galenic writings, which (together with works of the great representatives of Arab Galenism—Haly Abbas, Avicenna, Albucasis, Averroes) began to enter the universities in the last third of the thirteenth century.[6] During the next hundred years, a scholastic Galenism was founded on the texts of the "new Galen," and Latin medicine in late medieval Europe was thereby able to claim for itself a new status as *scientia,* with an associated elevation in social prestige.

The same forces worked upon the Jewish intellectual community to produce three occasionally discernible tendencies: "The Rabbis, whose main interest was in questions of Halakhah and Talmudic exegesis, the Kabbalists, whose main concern was theosophical speculation, and the philosophical group, whose major intellectual interests were secular in nature."[7] The Talmudic and kabbalistic

[3] See David C. Lindberg, "Science and the Early Christian Church," in *God and Nature: Historical Essays on the Encounter between Christianity and Science,* ed. D. C. Lindberg and R. L. Numbers (Berkeley/Los Angeles: Univ. California Press, 1986), pp. 19–48 (revision of an article published in *Isis,* 1983, *74*:509–530); Edward Grant, "Science and Theology in the Middle Ages," in *God and Nature,* pp. 49–75; Joseph Sarachek, *Faith and Reason: The Conflict on the Rationalism of Maïmonides* (1935; rpt. New York: Hermon Press, 1970); and A. I. Sabra, "The Appropriation and Subsequent Naturalization of Greek Science in Medieval Islam: A Preliminary Statement," *History of Science,* 1987, *25*:223–243.

[4] Sarachek, *Faith and Reason;* and Isadore Twersky, "Aspects of the Social and Cultural History of Provençal Jewry," *Journal of World History,* 1968, *2*:185–207. For Islamic culture see S. Munk, *Mélanges de philosophie juive et arabe* (rev. ed., Paris: J. Vrin, 1955), pp. 318–324; and Seyyed Hossein Nasr, *An Introduction to Islamic Cosmological Doctrines: Conceptions of Nature and Methods Used for Its Study by the Ikhwān al-Ṣafāᵓ, al-Bīrūnī, and Ibn Sīnā* (Cambridge, Mass.: Belknap Press of Harvard Univ. Press, 1964).

[5] M.-D. Chenu, *Introduction à l'étude de Saint Thomas d'Aquin* (Montreal: Institut d'Etudes Médiévales, 1954); Albert Zimmermann, ed., *Thomas von Aquin: Werk und Wirkung im Licht neuerer Forschungen* (Berlin/New York: Walter de Gruyter, 1988); James A. Weisheipl, ed., *Albertus Magnus and the Sciences: Commemorative Essays, 1980* (Toronto: Pontifical Institute of Mediaeval Studies, 1980); and G. Meyer and A. Zimmermann, eds., *Albertus Magnus Doctor Universalis, 1280–1980* (Mainz: Matthias-Grünewald Verlag, 1980).

[6] Nancy G. Siraisi, *Taddeo Alderotti and His Pupils: Two Generations of Italian Medical Learning* (Princeton, N.J.: Princeton Univ. Press, 1981); Luis García-Ballester, "Arnau de Vilanova (c. 1240–1311) y la reforma de los estudios médicos en Montpellier (1309): El Hipócrates latino y la introducción del nuevo Galeno," *Dynamis,* 1982, *2*:97–158; and Marie-Thérèse d'Alverny, "Pietro d'Abano traducteur de Galien," *Medioevo,* 1985, *11*:19–64.

[7] Lawrence V. Berman, "Greek into Hebrew: Samuel ben Judah of Marseilles, Fourteenth-Century

commentaries constructed a system of thinking and of understanding of man and the world with its own logic and terminology, different from the Aristotelian way of reasoning: "The historical Kabbalah (in Provence and Catalonia) represented an ongoing effort to systematize existing elements of Jewish theurgy, myth, and mysticism into a full-fledged response to the rationalistic challenge."[8] Even so, antirationalist authors were influenced by the Greek-Islamic natural philosophy embraced by rationalists.[9] In Muslim Spain, where Jewish intellectual life was especially vigorous, the Almohad conquest of the mid-twelfth century added a new element of religious intolerance that weighed heavily on the Jewish minority, particularly on the rationalist group therein, and many families chose to leave the peninsula, either for Muslim lands in the eastern Mediterranean (like Maimonides' family, which went first to the Maghreb and then to Egypt) or for Christian territories in the North (like Judah ibn Tibbon, who moved to southern France).[10]

Jewish intellectual life, wherever it was diffused, had still to deal with the tension between a wholehearted acceptance of divine revelation and the attitude of those we have labeled "rationalists." Critics of the latter group objected that familiarity with the Greco-Arab philosophical tradition (Aristotle, Ptolemy, Galen, Avicenna, Averroes) would lead away from Orthodox Judaism.[11] In particular, various passages from Revelation were in sharp conflict with doctrines established in accordance with logic or at least commonly accepted within Greek natural philosophy as elaborated by Islamic scientists; this further complicated the situation of Jewish rationalists, who had to decide whether to follow a literal or an allegorical interpretation of such passages. The recourse (by men like Maimonides, in the last third of the twelfth century) to the Aristotelian corpus and its commentators (e.g., al-Farabi or Averroes) to clarify the philosophical problems posed by such contradictions led to intellectual achievements that in turn only intensified the antirationalist polemic.[12]

Philosopher and Translator," in *Jewish Medieval and Renaissance Studies,* ed. Alexander Altmann (Cambridge, Mass.: Harvard Univ. Press, 1967), pp. 289–320, on pp. 289–290, 293. On this division of Jewish society into three rival social groups see the Hebrew grammar of Profiat Duran (Perpignan, ca. 1340/45–1414), *Ma^case Efod* (Vienna, 1865), pp. 4–10.

[8] Moshe Idel, *Kabbalah: New Perspectives* (New Haven, Conn.: Yale Univ. Press, 1988), p. 253.

[9] Here we employ the terms *antirationalist* and *rationalist* for the sake of convenience, although they are overly simplistic. See Isadore Twersky, "Rabbi Abraham Ben David of Posquières: His Attitude to and Acquaintance with Secular Learning," *Proceedings of the American Academy for Jewish Research,* 1957, *26*:161–192, on pp. 164, 184–185; and Charles Mopsik, *Lettre sur la sainteté: Le secret de la rélation entre l'homme et la femme dans la cabale* (Paris: Verdier, 1986), pp. 32–34, 48–49.

[10] Judah ibn Tibbon described the conditions of Jewish learning in Provence when he arrived there as an exile from Spain in the middle of the twelfth century: "Also in the lands of the Christians there was a remnant for our people. From the earliest days (of their settlement) there were among them scholars proficient in the knowledge of Torah and Talmud, but they did not occupy themselves with other sciences because their Torah-study was their (sole) profession and because books about other sciences were not available in their region"; cited and trans. by Twersky, "Aspects of the Social and Cultural History" (cit. n. 4), p. 195.

[11] Shatzmiller, "Rationalisme et orthodoxie" (cit. n. 1), p. 281; and Berman, "Greek into Hebrew" (cit. n. 7), p. 303.

[12] The polemic has given rise to a considerable literature: see Sarachek, *Faith and Reason* (cit. n. 3); Georges Vajda, *Introduction à la pensée juive du Moyen Age* (Paris: J. Vrin, 1947), pp. 119ff.; Vajda, *Recherches sur la philosophie et la kabbale dans la pensée juive du Moyen Age* (Paris: Mouton, 1962), pp. 118ff.; Daniel Jeremy Silver, *Maimonidean Criticism and the Maimonidean Controversy, 1180–1240* (Leiden: E. J. Brill, 1965); Charles Touati, "La controverse de 1303–1306 autour des études philosophiques et scientifiques," *Rev. Etud. Juives,* 1968, *127*:21–37; and Joseph Shatz-

The debates over Maimonidean teachings were still very much alive in the heart of the Jewish communities of Provence and the Spanish Mediterranean at the beginning of the fourteenth century, when the events culminating in the expulsion of Jews from France (1306) led to a reinforcement of the rationalists and an intensification of the controversy in cities such as Gerona and Barcelona.[13] The members of this group thought of themselves as faithful to the tradition of Maimonides, as belonging to a "sect of philosophers," "interested in rational speculation, the [true] congregation of believers." Natural philosophy had become for them a way of life—indeed almost a religion, with its own "prophets" (*nebiᵓim*): Greek, Muslim, and Christian natural philosophers and doctors.[14]

Jewish intellectuals had by now something of a philosophical literature available to them in Hebrew, for during the thirteenth century most of Aristotle's works had been translated into that language, along with the commentaries of Averroes, but traditionally it was their knowledge of Arabic that had given them access to the classical texts.[15] It was of course not only Jewish philosophers who recognized the richness of the sources still available only in Arabic; Christian scholars did as well. At Montpellier, Arnau de Vilanova used his knowledge of Arabic to correct passages in the standard Latin translations of medical works;[16] scholars who could not read Arabic asked students who could to translate works of possible interest to them.[17] But it was the Jews who still had the reputation for interest in and mastery of the Arabic-language material, especially those living in the Crown of Aragon or on the frontiers of Islam in southeastern Spain. Guillem de Béziers, who came from Montpellier to teach medicine at the newly created University of Lerida in 1301, asked its founder, King James II of Aragon, to provide him with "quosdam libros arabicos medicinales quos aliqui judei habent," in order to emend the texts that he had found on arriving at the new *studium*.[18] It was men like Samuel ben Judah (1294–ca. 1340) who kept this reputation alive: in 1324 he traveled from Marseilles to Murcia (on the Spanish coast, south of Valencia), recently conquered from Islam by Christians, simply to find a trustworthy Arabic text of Alexander of Aphrodisias's *De anima*.[19]

---

miller, "Towards a Picture of the First Maimonidean Controversy" (in Hebrew), *Zion*, 1969, *34*:126–144. See too the penetrating paper by Twersky, "Joseph ibn Kaspi" (cit. n. 2), esp. pp. 233–234.

[13] Touati, "La controverse de 1303–1306." Yitzhak Baer, *History of the Jews in Christian Spain*, 2 vols. (Philadelphia: Jewish Publication Society of America, 1978), Vol. I, p. 301, reports the *herem* (ban) drawn up by Solomon ben Adret and other religious leaders in Catalonia against scholars interested in Greek philosophy. Rabbi Moses ben Nahman (Nahmanides) of Gerona was a vigorous defender of traditionalism in the debates: see *ibid.*, pp. 102ff.; David Berger, "Nahmanides' Attitude toward Secular Learning and Its Bearing upon His Stance in the Maimonidean Controversy" (M.A. thesis, Columbia Univ., 1965); and Shatzmiller, "Rationalisme et orthodoxie" (cit. n. 1), and literature cited there.

[14] See the various remarks by Samuel ben Judah in the epilogues to his translation from the Arabic of Averroes' *Epitome of Plato's Republic* and of his correction of Jacob ben Makhir's translation of Jabir ibn Aflah's *Epitome of the Almagest*, quoted in Berman, "Greek into Hebrew" (cit. n. 7), pp. 309, 310, 312–313.

[15] Twersky, "Aspects of the Social and Cultural History" (cit. n. 4), pp. 201–202.

[16] Arnau de Vilanova, *Commentum supra tractatum Galieni de malicia complexionis diverse*, ed. Luis García-Ballester and E. Sánchez Salor, in *Arnaldi de Villanova Opera medica omnia*, ed. L. García-Ballester, J. A. Paniagua, and M. R. McVaugh, Vol. XV (Barcelona: Univ. Barcelona, 1985), p. 192, lines 15–18.

[17] Roger Bacon is a famous case in point; see his *Compendium studii*, Ch. 8, in *Fr. Rogeri Bacon Opera quaedam hactenus inedita*, ed. John Serren Brewer (London, 1859), pp. 467–468, 471–472.

[18] Antoni Rubió y Lluch, *Documents per l'història de la cultura catalana mig-eval*, Vol. II (Barcelona: Institut d'Estudis Catalans, 1921), pp. 13–14.

[19] Berman, "Greek into Hebrew" (cit. n. 7), p. 318.

Yet by this time the fame of Jewish scholars as masters of Arabic was no longer fully deserved, as a closer look at that same Samuel ben Judah makes plain. In his prologue to his Hebrew translation of Averroes' *Middle Commentary on the Nicomachean Ethics* Samuel speaks regretfully of his "limited grasp of the Arabic language" and remarks wistfully, "I trust to study it intensively again."[20] After translating Averroes' summary of Plato's *Republic,* he asks any reader who knows Arabic as well as Hebrew to forgive the defects of his translation and to improve it wherever possible, though he admits that such knowledge would be "unique and exceptional."[21] Embedded now in a Christian world, the Jewish community of the western Mediterranean was losing its familiarity with Arabic,[22] and with it its easy access to a rich philosophical and medical heritage. By the second half of the fourteenth century, so distinguished a scholar as Hasday Crescas (ca. 1340–ca. 1411) could no longer read Arabic and was forced to study Aristotle through Hebrew translations of Averroes' commentaries.[23]

A possible alternative for Jewish scholars was natural-philosophical or medical literature in the Romance vernacular. The vitality of medieval Catalan or Provençal is manifested in a variety of medical treatises on practical subjects, surgery and recipes, and such texts did circulate within the Jewish medical world of the fourteenth century: witness the thick marginal glosses in Hebrew that festoon several chapters in the Catalan translation (produced in Mallorca in 1305) of the *Surgery* of Teodorico Borgognoni;[24] or the intercalation of Romance vernacular words in a Hebrew recipe collection from Catalonia, probably also of the fourteenth century.[25] Learned texts from the Christian university world sometimes

[20] *Ibid.,* p. 305.

[21] He acknowledged, "It is not impossible—rather it is possible, nay, it is unavoidable—that whoever examines my translation of the two parts of this science will be perplexed about some passages on account of the badness of the translation, which is a result of my limited grasp of the Arabic language." *Ibid.,* p. 308.

[22] Lawrence V. Berman, "Ibn Rushd's *Middle Commentary on the Nicomachean Ethics* in Medieval Hebrew Literature," in *Multiple Averroès* (Paris: Belles Lettres, 1978), pp. 287–322, on p. 299. An anonymous Hebrew translation of Averroes' *Fasl al-Magal* (a general approach to the role of philosophic speculation in religion) was probably produced by a Provençal Jew who lived in the first years of the fourteenth century and "whose native tongue was not Arabic nor had achieved great prowess in that language": N. Golb, "The Hebrew Translation of Averroes' *Fasl Al-Magal,*" *Proc. Amer. Acad. Jewish Res.,* 1956, *25*:91–113; 1957, *26*:41–64, on pp. 92–95. See also David Romano, "La transmission des sciences arabes par les juifs en Languedoc," in *Juifs et judaïsme de Languedoc,* ed. Vicaire and Blumenkranz (cit. n. 1), pp. 363–386, on p. 379.

[23] Harry A. Wolfson, *Crescas' Critique of Aristotle* (Cambridge, Mass.: Harvard Univ. Press, 1929), p. viii. In the last years of the thirteenth century, Solomon ben Adret had difficulties finding in Barcelona an experienced translator from Arabic into Hebrew to translate Maimonides' *Commentary on the Mishna,* as Rome's Jewish community had asked him to do. He sent a messenger to Aragon (Huesca, Saragossa), where he found Jews skilled in Arabic. See Fred Rosner, *Maimonides' Commentary on the Mishna: Tractate Sanhedrin* (New York: Sepher-Hermon Press, 1981), pp. xi–xii.

[24] The *Surgery* of Teodorico Borgognoni was translated at least twice into Catalan in the first decade of the fourteenth century. One translation was made by Maestre Bernat (fl. 1305), physician of James II of Mallorca (1275–1311) (MS Graz, Universität-Bibliothek 342, fols. 4–282). All chapter titles were translated into Hebrew, with a summary in Hebrew at fol. 25v. The other was made by Guillem Correger, a Valencian surgeon (fl. 1288–1308) (MS Paris, Bibliothèque Nationale, Fons Espagnol 212, fols. 1–89). There is a manuscript fragment in the Municipal Archive in Barcelona (Capsa B-109 [2]) related to the Graz copy; see Antonio Contreras, *La difusión medieval de la* Cyrurgia *de Teodorico Borgognoni (1205–1298) en los paises de habla catalana: La versión catalana de Guillem Correger, Libro I (cirugía general), según el MS Paris Bibliot. Nat., fons espagnol 212, fols. 1–18v* (Tesis de Licenciatura en Medicina) (Santander: Univ. Cantabria, 1986). Contreras and García-Ballester are preparing a critical edition of the Catalan version, to be published in 1992 (Barcelona: Els Nostres Clàsics).

[25] Gregorio del Olmo Lete and José R. Magdalena Nom de Deu, "Documento hebreo-catalán de farmacopea medieval," *Anuario de Filología,* 1980, *6*:159–187. For a nonexhaustive list of medical

were translated from Latin into Romance and then translated from Romance into Hebrew. Three treatises, at least, by the Montpellier master Bernard de Gordon —his *Tractatus de prognosticis* (*de crisi*) (*Haqdamat ha-yediᶜah*); his chef d'oeuvre, the *Lilium medicine* (*Shoshan ha-refuᵓah*); and his *De phlebotomia* (*Ha-maᵓamar ba-haqazah*)— and one by Arnau de Vilanova (his *Regimen sanitatis ad regem Aragonum* [*Hanhagat ha-beriᵓut*]) passed into Hebrew by this route, very much as Arabic texts had earlier been wont to undergo Latinization via an intermediate language.[26] But such Hebrew versions suffered from the compounding of error and obscurity inevitable in two-stage translation, and this made it preferable to turn, where possible, to the Latin original.[27] The ignorance of Latin by the Jewish translator was another reason for the low quality of a translation. The translator of the *De phlebotomia* admitted that his ignorance of Latin had made it impossible to verify quotations from Galen and had rendered his translation less valuable.[28]

In fact, a knowledge of Latin in the medieval Jewish world was increasing as its familiarity with Arabic withered. The intellectual tide was turning; the Latin West was becoming the creative power. Few Jews could read Avicenna's *Canon* in the original Arabic any longer, and the Hebrew translation was defective enough that "reading it keeps us from the truth," as a contemporary critic put it, even though it circulated more widely than any other medical text in Hebrew.[29] The "rediscovery of learning" by Christian Europe and the success of its universities meant that a wide variety of scientific literature was now fairly easily available in Latin, and prohibitions against its sale to Jews were not systematically enforced.[30] Mosse Aventida, a Jewish physician of Monzón in Aragon, had copies of the *Canon* and the *Practica* of Ibn Sarabi (Serapion) in his library (1381), not in Arabic but "in latino xristianico scripti"; in the first half of the next century, Bendich de Borriano (d. 1441), a Jewish physician of Arles, bequeathed

---

works translated into Provençal see Clovis Brunel, *Bibliographie des manuscrits littéraires en ancien provençal* (Paris, 1935; rpt. Geneva: Slatkine, 1973), pp. 38 (no. 121, *Chirurgie* d'Albucasis), 102 (no. 355, surgical treatises). We suspect some medical works belonging to the Jewish physician Astruc de Sestiers (d. 1439) may have been written in Romance; unfortunately the notarial description does not include the incipit. See Danielle Iancu-Agou, "L'inventaire de la bibliothèque et du mobilier d'un médecin juif d'Aix-en-Provence au milieu du XVe siècle," *Rev. Etud. Juives*, 1975, *134*:47–80, on p. 51 (nos. 120, 130, 134, 145, 154).

[26] On these texts see Moritz Steinschneider, *Die hebraeischen Übersetzungen des Mittelalters und die Juden als Dolmetscher* (Berlin, 1893; rpt. Graz: Akademische Druck- u. Verlagsanstalt, 1956), pp. 778–787. To give an example of the dissemination of such works: the library of the Jewish physician Astruc de Sestiers contained Hebrew translations from the Latin (or conceivably Romance) of works by three Montpellier academics of the fourteenth century—the *Regimen sanitatis* (*Hanhagat ha-beriᵓut*) of Arnau de Vilanova, the *Introductorium iuvenum* (*Maishir ha-mathilim*) of Gérard de Solo, and the *Lilium* (*Shoshan ha-refuᵓah*) of Bernard. See Iancu-Agou, "L'inventaire de la bibliothèque," nos. 146, 162, and 176, on pp. 59–61. On the *Lilium*'s translator, Bonsenyor Salamon (d. ca. 1415 in Perpignan), see Richard W. Emery, "Documents Concerning Some Jewish Scholars in Perpignan in the Fourteenth and Early Fifteenth Centuries," in *Michael: On the History of the Jews in the Diaspora*, ed. Shlomo Simonshon and Joseph Shatzmiller (Tel-Aviv: Diaspora Research Institute, 1976), pp. 27–48, on pp. 43–45.

[27] See Appendix D, lines 113–114. Here and subsequently the reader is referred to the Hebrew texts included (with their English translations) as appendixes to this article.

[28] The treatise was translated in 1378 by an anonymous translator in the Catalonian town of Castelló d'Empúries: Steinschneider, *Die hebraeischen Übersetzungen* (cit. n. 26), p. 787.

[29] Benjamin Richler, "Manuscripts of Avicenna's *Kanon* in Hebrew Translation, a Revised and Up-to-Date List," *Koroth*, 1982, *8*:145*–168*; see Appendix D, lines 40–43.

[30] Appendix D, lines 119–121.

to his brother-in-law Salamies Manelli (also a physician) copies of the *Canon* in both Hebrew and Latin.[31] But while most Mediterranean Jews could read or at least speak the Romance vernacular, Latin was known only to a minority, and this gave rise to a remarkable feature of the intellectual reversal of the fourteenth century: a new impulse to translate Latin medical and natural-philosophical texts directly into Hebrew.

This countercurrent has so far been noted only in scattered cases, like that of the Castilian Jew Meir Alguadez ben Solomon, who retranslated the *Nicomachean Ethics* in the 1390s from Latin because the Hebrew version Samuel ben Judah had prepared directly from the Arabic, seventy years before, was incomprehensible; for Meir Alguadez, Latin was more precise and articulate as a language of translation.[32] In this article, therefore, we mean to draw attention to a series of Jewish translators from the western Mediterranean whose careers encompassed the fourteenth century and who shared an admiration for the learned medical tradition at the Christian *studium* of Montpellier: Estori bar Moses ha-Parhi (1306), translator of the *Tabula antidotarii* of Armengaud Blaise (fl. 1290–1310) (see Appendix A); Israel ben Joseph Caslari (1327), translator of the *Regimen sanitatis* of Arnau de Vilanova (d. 1311) (Appendix B); Joseph bar Judah ha-Sefardi, translator of the *Regimen sanitatis* of Arnau de Vilanova (Appendix C); Leon Joseph of Carcassonne (fl. 1394–1402), translator of the *Practica super nono Almansoris* of Gérard de Solo (fl. 1335) (Appendix D); and Abraham ben Meshullam ben Solomon Abigdor (1379), translator of the *Introductorium* of Bernard Alberti (fl. 1340) (Appendix E). Each of these translators prefixed a personal statement as an introduction to his work, explaining (in varying detail) why he had undertaken the task. Difficult to transcribe and translate, and of course partaking of the self-justifying rhetoric characteristic of almost all prefaces, these introductions are nevertheless remarkably revealing, individually and as a whole, for they show that the intellectual reversal was not merely a response to the shifting fortunes of Arabic and Latin. A few members of the Jewish intellectual community, we will see, had with time come to feel that the well-structured scholastic training provided by the Latin *studia* had a power their own schools no longer possessed. As the century advanced, the interest of medical translators shifted from obviously practical works by Christians especially famous for their medical practice and distinguished clientele, to theoretical, typically academic productions by university masters. These fourteenth-

[31] "Item tercius et quintus libri Avicenne in latino xristianico scripti. Item alius liber scriptus in latino xristianico vocatus Sarampio. . . . Item [lego] quartam et quintam particulas sive libros Avicenne in uno volumine in papiro descriptos lictera ebrayca; item primum (et) quartum librum Avicenne in pergameno in uno volumine lingua latina scripto; item tercium librum sive particulam Avicenne in pergameno lingua latina descriptum." Rubió y Lluch, *Documents* (cit. n. 18), Vol. II, p. 247. Cited by Danielle Iancu-Agou, "Une vente de livres hébreux à Arles en 1434: Tableau de l'élite arlésienne au milieu du XVe siècle," *Rev. Etud. Juives,* 1987, *146*:5–62, on p. 55.

[32] "Die Übersetzung aus dem Arabischen war ihm umverständlich, darum schien es ihm zweckmässig, das Buch aus dem Lateinischen zu übersetzen, vielleicht würden sie dadurch verständlicher [*yoter mebu ᵓarim*], 'da wir unter christlichen Weisen wohnen,' welche jene Bücher studirt haben, auch Commentare derselben sich (bei ihnen) finden, welche alles Dunkle erläutern. Dennoch säumte Meïr, sein Vorhaben auszuführen, aus verschiedenen Gründen. Ein Übersetzer müsse beide Sprachen gründlich verstehen, die Gabe besitzen, die Worte des Vf. in einer anderen Sprache wiederzugeben, er müsse ruhig an einem Orte arbeiten können; diese 3 Bedingungen haben ihm gefehlt, und das Lateinische ist weitläufig [*we-lashon nosrí arukhah*]." Steinschneider, *Die hebraeischen Übersetzungen* (cit. n. 26), p. 210.

century translations are, in fact, an attempt to recapture for medieval Jewish medicine not only learned texts but an underlying approach and habit of mind.

II. THE NORMS OF THE HEBREW TRANSLATORS

The dependence of the Jewish minority upon translations of scientific works made it particularly sensitive to the requirements that translations had to meet, and translators routinely commented on the set of norms they felt compelled to observe. Such norms often recalled the instructions laid down in 1199 by Maimonides himself, when he advised Samuel ibn Tibbon of Arles on the translation of the *Guide of the Perplexed* from Maimonides' original Arabic into Hebrew.[33] They can be summarized as follows: the translator should possess a total mastery of the two languages; he should have a thorough understanding of the material he is translating; he must not translate rigidly, word for word; and he must make sure that the constructions of the translation are perfectly understandable. This purely technical advice was still essential for Jewish translators from Latin two hundred years later, but the very different social context in which they lived made them add new requirements to those set out by the master. Samuel ben Judah, the first translator of the *Nicomachean Ethics,* noted the urgency of seeking out at whatever pains a manuscript containing an accurate text, error free, as he himself had done in his trip to Murcia in 1324.[34] He insisted, too, that the translator needed a knowledge not only of the subject he was translating but of all the sciences, for all intellectual disciplines are interrelated. Consequently, the translator could not work in isolation; he should consult other books and other scholars, Jews and Christians.[35]

Fifty years later, Meir Alguadez remarked on a further requirement for the intellectual labor of translation: peace and tranquillity.[36] The increasingly harsh conditions, including expulsion, imposed by Christianity upon its internal Jewish communities during the fourteenth century rendered intellectual development therein particularly difficult.[37] But even forced displacement could have its com-

[33] Maimonides, *Qobetz Teshubot ha-RaMBaM we-iggerotaw* (Leipzig, 1859; rpt. Farnborough, 1969), pp. 26–29; M. Sonne, "Maimonides' Letter to Semu'el b. Tibbon according to an Unknown Text in the Archives of the Jewish Community of Verona" (in Hebrew), *Tarbiz,* 1939, *10*:135–154, 309–332. There are English versions by H. Adler, "Translation of an Epistle Addressed by R. Moses Maimonides to R. Semu'el ibn Tibbon," *Miscellania of Hebrew Literature,* 1975, *1*:219–227; L. D. Stitskin, *Letters of Maimonides, Translated and Edited with Introductions and Notes* (New York: Yeshiva Univ. Press, 1977), pp. 130–136, 197–198; and Franz Kobler, *Letters of Jews through the Ages,* 2 vols. (New York: East & West Library, 1978), pp. 208–213. These two last versions are incomplete. A Spanish translation is given by Maria J. Cano and Dolores Ferre, *Cinco epístolas de Maimónides* (Barcelona: Riopiedras Ediciones, 1988), pp. 111–124.

[34] Berman, "Greek into Hebrew" (cit. n. 7), p. 318.

[35] "I thought and conceived the desire to correct this translation again together with the Christian philosophers, especially the first part of it because they have the separate treatises of the philosopher on that part together with their explanation by Abu Nasr al-Farabi. . . . He [the translator] must be well educated, not only in the science or art which he is translating but also in all sciences conventionally recognized as such, or most of them, because of the interconnectedness of all of the sciences and arts one with the other, since one always uses analogies taken from other sciences." *Ibid.,* pp. 308, 309.

[36] See the passage from Steinschneider, *Die hebraeischen Übersetzungen* (cit. n. 26), quoted above, n. 32.

[37] See the remarks in Appendixes A (line 16) and D (line 79). Samuel ben Judah alludes to the expulsion of the Jews from France, Burgundy, and Languedoc in 1322 in the following passage: "I promised to correct the translation of this science with the help of Christian philosophers but was unable [to do so] on account of the magnitude of the annoyances and persecutions which have

pensations by bringing exiles into contact with new and intellectually stimulating circles of scholars. Estori bar Moses ha-Parhi was among those forced out of France in the expulsion of 1306, only to find a welcome in the Jewish community of Barcelona, as well as, apparently, from some Christians. There he must have encountered Armengaud Blaise, formerly a master at Montpellier but from 1301 to 1306 in the service of James II of Aragon in the Catalan capital, and also Armengaud's *Tabula antidotarii,* which he found "more precious than fine gold, mother-of-pearl, and onyx" and translated in the following year.[38]

At the very end of the century, two new requirements for the Jewish translator and scholar were identified by Leon Joseph of Carcassonne, who was expelled from France in 1394 and settled at Perpignan (in the Crown of Aragon), where he is found practicing medicine in 1414. Sometime thereafter he converted to Christianity, adopting the name "Leonardus Benedictus"; he died before February 1418.[39] Leon's introduction to his translation of Gérard de Solo's *Practica super nono Almansoris* is a particularly personal statement, undoubtedly the most reflective and most illuminating of the texts we are here considering; it will be worthwhile concentrating most of our attention henceforth on it. The first of the new conditions identified by Leon is the necessity of appreciating the scholastic method, based on the *questio* and the *disputatio* and using dialectical reasoning to guarantee the necessary truth of conclusions.[40] The other requirement, on which Leon places even more stress, is sociocultural rather than technical: the necessity of freedom of discourse within the Jewish communities themselves, freedom from the very real tyranny of "those learned in the Torah, who . . . from the strength of their hands and from their many ruses make the mass of the common people believe . . . that these sciences [natural philosophy and theoretical medicine] and those who are concerned with them are sundered from the community of those who possess the Torah." Lacking such freedom, the rationalist minority has been forced to carry on its studies "in secret . . . 'in the clefts of the rocks and in the secret places' " (Song of Sol. 2:14).[41] With these observations, Leon Joseph brought into the open an issue that he found of great concern: Why, he asked, do the Jews no longer display intellectual creativity in the sciences (medicine), and what can be done to change this?

III. ADMIRATION OF THE SCHOLASTIC TEACHING METHOD

Part of the answer lay, for Leon, in the system of intellectual training. Within the Jewish community, medical education had always followed what we might call

overtaken us on the part of this nation which exiles us." Berman, "Greek into Hebrew" (cit. n. 7), p. 310. See also Yom Tov Assis, "Juifs de France réfugiés en Aragon (XIIIe–XIVe siècles)," *Rev. Etud. Juives,* 1983, *142*:285–322, on p. 315.

[38] Appendix A, line 13. Michael R. McVaugh and Lola Ferre are preparing an edition of the Latin original together with the Hebrew translation. Samuel ben Judah's journey to Murcia and his contact with new Arabic manuscripts may also be connected with the expulsion of the Jews in 1322; see Assis, "Juifs de France," p. 317.

[39] Emery, "Documents Concerning Jewish Scholars" (cit. n. 26), pp. 40–43. Leon's son, Astruc Leo, was also a physician at Perpignan.

[40] Appendix D, lines 70–71. On the intellectual impact of Latin scholasticism on Jewish philosophy (as distinct from medicine) see A. Altmann, "Judaism and World Philosophy," in *The Jews: Their Role in Civilization,* ed. L. Finkelstein (New York: Schocken Books, 1971), pp. 65–115; and Pines, "Scholasticism after Aquinas" (cit. n. 2).

[41] Appendix D, lines 29–31, 24–25.

an "open" model. Under this model, anyone might pass on medical knowledge who had it to impart, faithful to personal standards and outside any institutional framework. In such an open system, teaching activity reflects the interests of an individual teacher, not the requirements of a set curriculum; and the success of education, a student's qualification for practice, can be tested only by his success at curing his patients—it is ultimately the patient who will judge who is a physician and who is not. Where books for technical instruction are relatively scarce, as in medieval Judaism (and Islam),[42] the open system built around the master-disciple arrangement further encourages the oral transmission of knowledge and consequently a tradition of empirical and familiar medical practice.

The model that stood opposed to the open system, institutionalized education, was represented for Leon Joseph by the new Latin university, the *studium generale;* the system of certification it offered was closed to non-Christians. At the beginning of the fourteenth century, the medieval university held the monopoly on systematic medical education, and within a few decades it was attempting to add to this a monopoly on medical practice: to claim, that is, that only those individuals who had been trained at a university should be permitted to practice medicine. But while Christian society did indeed begin in the early fourteenth century to increase its efforts to supervise anyone claiming a medical occupation, these measures were never more than a dim ideal; the European university system was still inadequate to turn out the number of professionals necessary to meet the demands of late-medieval society.[43]

For the Jewish (and Muslim) communities of the western Mediterranean, the open model prevailed throughout the fourteenth century, though we have very little documentation to help us understand how it actually functioned then.[44] But this did not prevent them—or at least their rationalist minority—from admiring the very different institutional model of the Christians. Partly this was a consequence of power relationships. Both in Provence and in the Crown of Aragon, a would-be Jewish (or Christian) physician had by law to pass an examination; a Jew would confront a mixed tribunal of Christian and Jewish physicians in a scholastic ritual, with *questiones et responsiones, disputationes, rationes et argumentationes* requiring a mastery of medical authorities. Thus the Jewish physician who wanted to practice among Christians had to be acquainted with the methods and contents of scholastic medicine. Indeed, the ritual of the examination played a role of acculturation for those Jewish physicians who aspired to be respected by Christian authorities.[45]

[42] At the end of his *Middle Commentary on the Ethics,* Averroes "states that only the first four books of the *Ethics* were available to him in Andalusia until Abuʾ Amr ibn Martin brought the integral text of the *Ethics* from Egypt." Berman, "Ibn Rushd's *Middle Commentary*" (cit. n. 22), p. 291. See also Berman, "Revised Hebrew Translation of Averroes' *Middle Commentary on the Nicomachean Ethics,*" in *Seventy-Fifth Anniversary Volume of the Jewish Quarterly Review* (Philadelphia: Dropsie College, 1967), p. 106, n. 4.

[43] Luis García-Ballester, Michael R. McVaugh, and Agustín Rubio Vela, *Medical Licensing and Learning in Fourteenth-Century Valencia* (Transactions of the American Philosophical Society, 79.6) (Philadelphia, 1989).

[44] Examples of the scattered documentation upon which we must depend are L. Barthélemy, *Les médecins à Marseille avant et pendant le Moyen-Age* (Marseilles, 1883), pp. 31 (28 Aug. 1326), 32 (7 Mar. 1431); P. Pansier, "Les médecins juifs à Avignon aux XIIIe, XIVe et XVe siècles," *Janus,* 1910, *15*:421–451, on pp. 442, 443; and Antonio de la Torre and Jordi Rubió i Balaguer, *Documentos para la historia de la Universidad de Barcelona,* Vol. I (Barcelona: Univ. Barcelona, 1971), document 46.

[45] P. Hildenfinger, "Documents relatifs aux juifs d'Arles," *Rev. Etud. Juives,* 1900, *41*:62–97, on p.

Further, power relationships aside, the Latin faculties of medicine seemed to be particularly effective centers of learning, to judge both from the quality and quantity of knowledge they generated and from the achievements of particular individuals trained there. The individual Christian physician as well as the institutional model was becoming an object of admiration: famous practitioners like Armengaud or Arnau (who "surpassed all his predecessors"[46]) early in the century, eminent teachers in the second half. It is again Leon Joseph who provides the telling vignette.

> In his time [Jean de Tournemire] [fl. 1380] was at the head of all the scholars of Montpellier. . . . I saw him with my own eyes and I spoke to him. He was a man who was pleasant to talk to. His behavior was not like that of the other scholars of his generation, who scorned those Jews who practiced the art of medicine. . . . His book speaks of the limbs and of the faculties, and of the illnesses that occur in them. It asks and answers, it explains and clarifies. He teaches like a father to a son, true and proper words, which are pleasing and agreeable to whoever listens.[47]

The existence of such teachers seemed to explain to Leon why, throughout the fourteenth century, progress in the science of medicine had obviously been restricted to men with an academic formation.

The learned medicine of the universities was of course in its content not at all foreign to Jewish scholars and certainly not distinguishably Christian. Its Galenic outline, articulated by the Arabs, was the common intellectual property of all three Mediterranean cultures.[48] It was not so much the content of medicine as the manner in which it was pursued at the universities, its systematic development out of Aristotelian methodology and philosophy, that excited admiration and a desire to study. Leon Joseph confessed, "I directed my attention toward the study of and research into the profane sciences, which are . . . as many as the days of the week [i.e., the seven liberal arts], and each one has its own subject matter. . . . In my eyes, the merits of these sciences were above all praise. This explains my zeal . . . to master them."[49] But to do so it had been necessary to seek the setting of a Christian *studium*. Likewise, when, after years of medical practice, Abraham Abigdor wanted to be able to answer the fundamental questions of medicine, he found no other solution than to direct himself to Montpellier in order "to hear the science of medicine [*hokhmat ha-refu*ᵓ*ah*]"[50]—that is,

---

67, n. 4; English trans. by Cecil Roth, "The Qualifications of Jewish Physicians in the Middle Ages," *Speculum,* 1958, *28*:834–843, on pp. 839–840. Unfortunately, the translation has not preserved the scholastic terminology; see García-Ballester, McVaugh, and Rubío, *Medical Licensing* (cit. n. 43).

[46] Appendix B, lines 47–48.

[47] Appendix D, lines 132–138.

[48] While the diffusion of Galenism through the Latin West is relatively well understood (see n. 6), its spread within the Jewish world is not. It is one thing to identify Hebrew translations of Galen's writings, quite another to speak confidently about their circulation. It seems that the Alexandrine canon of Galen's work was translated into Hebrew by Simson ben Salomo (fl. 1322) in the early fourteenth century (Steinschneider, *Die hebraeischen Übersetzungen* [cit. n. 26], pp. 653–656), but we find very few copies of the Hebrew Galen in the libraries of Jewish physicians. Everything seems to suggest that Jewish Galenism was based on the Arabic encyclopedists, particularly Avicenna. See Danielle Iancu-Agou, "Préoccupations intellectuelles des médecins juifs au Moyen Age: Inventaires de bibliothèques," *Provence Hist.,* 1976, *26*:21–44, on pp. 36–38; Vajda, *Recherches sur la philosophie* (cit. n. 12), pp. 120–137; and Richler, "Manuscripts of Avicenna's *Kanon*" (cit. n. 29).

[49] Appendix D, lines 2–7.

[50] Appendix E, lines 10–11.

scholastic medicine founded upon natural philosophy. For to penetrate this scientific medical world, the liberal arts were a necessary tool; without them it was impossible to begin to learn, since scholastic learning required a preliminary mastery of definitions and distinctions, of the nature of the syllogism, and so forth.[51] These intellectual materials formed the core of medical instruction in the medieval university, which turned around the reading (*lectio*) of medical writings of Greek and Arabic authorities.

The content of the *lectio* in scholastic medicine during the period we are discussing was quite broad. It could include anything from the simple exposition of words or sentences, explanation of meaning, to the most intricate exegesis, but it always involved a detailed dissection of the passage in question.[52] In the course of this analysis, particular problems (*questiones*) might emerge that needed to be resolved. Subsequently, during the fourteenth century, these *questiones* became a new literary form independent of the *lectio;* and this in turn gave rise to academic medical *disputationes,* in which students and auditors were invited to participate. In medicine, as in scholastic education generally, the *questio* became a flexible, effective tool permitting university masters to identify problems, treat them in depth, and develop and defend new replies. In fact, we can follow the changing intellectual concerns of fourteenth-century scholastic medicine by tracing a chain of *questiones* through contemporary commentaries, where they may be found noted down in the margins of manuscripts or extracted into separate lists. *Questiones* reveal that apparently abstruse speculative issues in pathology —for example, the nature of fever and the mode of its production—were not merely subjects of debate in narrow professional circles but topics of lively interest to which students of all levels as well as masters were exposed and might contribute. Thus this institutionalized system of medical education combined the tools of analysis with a method of scientific communication in such a way as to encourage cohesion within a broad academic community.

This allows us to understand still better the enthusiasm that the system aroused in a few Jewish intellectuals like Abraham Abigdor and Leon Joseph, to understand why they chose to travel to Montpellier "to hear the science of medicine from the mouths of Christian doctors and scholars."[53] One of the reasons why the latter physician chose to translate into Hebrew particular works by Gérard de Solo and Jean de Tournemire was precisely that they made, in his opinion, perfect use of the *questio* technique. Indeed, Leon was so taken by this literary form that he enhanced his translation of Gérard's *Practica super nono Almansoris* by adding

> in some chapters questions and answers following the disputes [*questiones, disputationes*] that I have found recorded, scattered here and there, some attributed to this

[51] Chenu, *Introduction à l'étude de Thomas d'Aquin* (cit. n. 5); P. Glorieux, "L'enseignement au Moyen Age: Techniques et méthodes en usage à la Faculté de Théologie de Paris, au XIIIe siècle," *Archives d'Histoire Doctrinale et Littéraire du Moyen Age,* 1968, *43*:65–186; Brian Lawn, *The Salernitan Questions: An Introduction to the History of Medieval and Renaissance Problem Literature* (Oxford: Clarendon Press, 1963); and John Marenbon, *Later Medieval Philosophy (1150–1350): An Introduction* (London/New York: Routledge & Kegan Paul, 1987).

[52] Roger French, "A Note on the Anatomical Accessus of the Middle Ages," *Medical History,* 1979, *23*:461–468, on p. 465.

[53] Appendix E, line 11.

same learned author, others to another eminent and authoritative scholar; others I drew up as well as I was able to when I could find nothing suitable that discussed the problem of that chapter. . . . In the case of each question that is not from the book you will find at the beginning ALYH [Amar Leon Yosef Ha-maᶜtiq = Here speaks Leon Joseph, the translator], which are the initials of my name, so that you may know it is an addition.[54]

To Leon we owe a particularly eloquent eulogy of the scholastic method:

Consequently, I found great benefits in this [the study of the liberal arts], because in general their discussions on these sciences [natural philosophy, law, medicine] do not stray from the subject matter; [the Christians] leave out nothing when it is a question of debating the truth or falsehood of a proposition; they are very rigorous concerning the questions and answers of a debate, which are linked together in such a way as eventually to bring out the truth by means of an analysis of opposing points of view, "like a lily among thorns" [Song of Sol. 2:2].[55]

IV. RESISTANCE TO CLASSING MEDICINE AS LEARNING

These translators' enthusiasm for scholastic medicine was not, however, universally shared by their contemporaries. Inevitably, the intellectual divisions within the Jewish community, the debates between rationalists and antirationalists or traditionalists, had found some reflection in its attitudes toward medicine. Its religious leaders contended that the medical knowledge necessary for the practitioner who tries to cure his patients is, first and foremost, a skill (*tahbulah*), not a science; neither the study of natural philosophy nor a grounding in medical theory is needed in order to heal. Social stimuli encouraged a utilitarian approach to medicine, an instrumentalist view of medical knowledge.[56] Indeed, if Abraham Abigdor and Leon Joseph are to be believed, an empirical medicine was the norm within the community. Medicine was simply not generally understood to be a way of obtaining knowledge of nature and of man, while the open system of education further encouraged a pragmatic approach to the art of healing.

The historian looking back at the fourteenth century finds occasional instances of what these men bemoaned, the decreasing cultivation of the sciences, broadly considered, by the Jewish communities of the western Mediterranean in this period, from Provence around the coast to the frontier with the kingdom of Granada. Some Jewish intellectuals were already beginning to fear an inferiority at the beginning of the fourteenth century. Writing from Montpellier to Solomon ben Adret just before his death, the philosopher-translator Jacob ben Makhir ibn Tibbon (Profatius) advised him that

we need to convince the Gentiles of our knowledge and understanding [of natural philosophy], so that they cannot say that we lack all science and wisdom. We must

[54] Appendix D, lines 155–161.

[55] *Ibid.*, lines 68–72.

[56] This utilitarian approach, which did not reject the use of rational medicine, was shared by, e.g., Solomon ben Adret, who recommended the use of "experimental [empirical] medicine [*refuʾah nisyonit*]," since the utility of remedies—even charms—was demonstrated by experience. See Ron Barkai, "L'ús dels salms en la màgia jueva de l'édat mitjana i el renaixement: El llibre *Shimush Tehilim*," in *La Càbala* (Barcelona: Fundació Caixa de Pensions, 1989), pp. 17–57, esp. pp. 28–30. For a similar movement in medieval Islam see the appealing article by Sabra, "Appropriation of Greek Science" (cit. n. 3), p. 241.

> follow the lead of the Gentiles—at least, of the most learned ones. They have translated scientific works into their various languages, even when the content of these works is at odds with their faith. They respect scholarship and scholars, no matter what creed they may profess.[57]

It was not only the Jews who were conscious of this difference; it was commented upon also by Christian scholars as part of their defense of Christianity. Some put it down to a deficient education. The absence of the liberal arts from the training of those in the Jewish community who wanted to pursue an intellectual activity meant that the Jews (through their most conspicuous intellectual representatives) were prevented from discussing certain topics (be they natural philosophy, medicine, or theology and law), as well as deprived of an analytical tool of proven value. Ramon Llull had this in mind when he asserted in 1313 that "the Jews are a people intellectually coarse, for they have no knowledge of the liberal arts. Therefore, when anyone wishes to speak with them or to discuss scientifically [*subtiliter*] the faith, the seven sacraments, or the ten commandments, they cannot understand what is said to them. . . . Without [the liberal arts] it is impossible to understand anything, even to follow an argument."[58]

But their silence could also be seen by contemporary Christian critics as an indication that Jews were incapable of achievement in natural philosophy or medical theory. Leon Joseph attacked this second interpretation vigorously. "If these and other sciences should remain beyond our reach, it is not because our intelligence is inferior to that of the Christians, for we possess the same capacity for understanding; indeed, it is the circumstances that have kept us apart . . . exile and oppression. . . . For this reason the Christians have continued to advance in the profane sciences while we have continued to lose ground as a consequence of distress and oppression."[59] Thus Leon argued that the failure of Jewish natural philosophy should be explained by a combination of social circumstances that had produced intellectual sterility: in part their exile and their

[57] Given in French translation by Shatzmiller, "Contacts et échanges" (cit. n. 1), p. 337. The letter was published by Abba Mari ben Moïse, *Minhat genaot* (Pressburg, 1838), letter 39, p. 85; we have checked our English translation with the Hebrew text. Shatzmiller, who is stressing the positive exchanges between Jewish and Christian scholars in Montpellier around 1300, expresses surprise at ibn Tibbon's attitude. "Mais une autre idée, qui peut paraître singulière, se dégage encore de ce texte: les connaissances diffusées dans le monde chrétien doivent aussi être acquises par les Juifs. Et l'effort scientifique des Gentils est présenté comme un modèle" (p. 338). As we see, ibn Tibbon follows a model of scientific progress still based on translations from the Arabic. Among his reasons, he does not include Jewish ignorance of the tools of scholasticism—techniques that Jewish rationalists would discover later.

[58] "Iudaei sunt homines grossi intellectus, quia non uescuntur artibus liberalibus, nec eas sciunt. Et propter hoc, quando quis cum eis loquitur, subtiliter disputando de fide, de septem sacramentis et de decem preceptis, id, quod eis dicitur non intelligunt . . . per quam subtilitatem queant intelligere et percipere rationes." Ramon Llull, *Summa sermonum in civitate Maioricensi . . . composita,* in *Raimundi Lulli Opera latina,* ed. Fernando Rodriguez Reboiras and Abraham Soria Flores, Vol. XV (Turnhout: Brepols, 1987), p. 122. Three years before (1310), in Montpellier, Llull had written, "Judaei sunt homines sine scientia et quando catholicus disputat cum ipsis rationabiliter, non intelligunt rationes"; see P. E. Longpré, "Le *Liber de acquisitione terrae sanctae* du Bienheureux Raymond Lulle," *Criterion,* 1927, *3*:264–278, on p. 274. For Llull, of course, words like "subtiliter" and "rationabiliter" in this context entail the Aristotelian way of reasoning *secundum artes liberales;* see Eusebio Colomer, "Las *Artes liberales* en la concepción científica y pedagógica de Ramón Llull," in *Arts libéraux et philosophie au Moyen Age* (Montreal: Institut d'Etudes Médiévales; Paris: J. Vrin, 1969), pp. 683–690.

[59] Appendix D, lines 77–81.

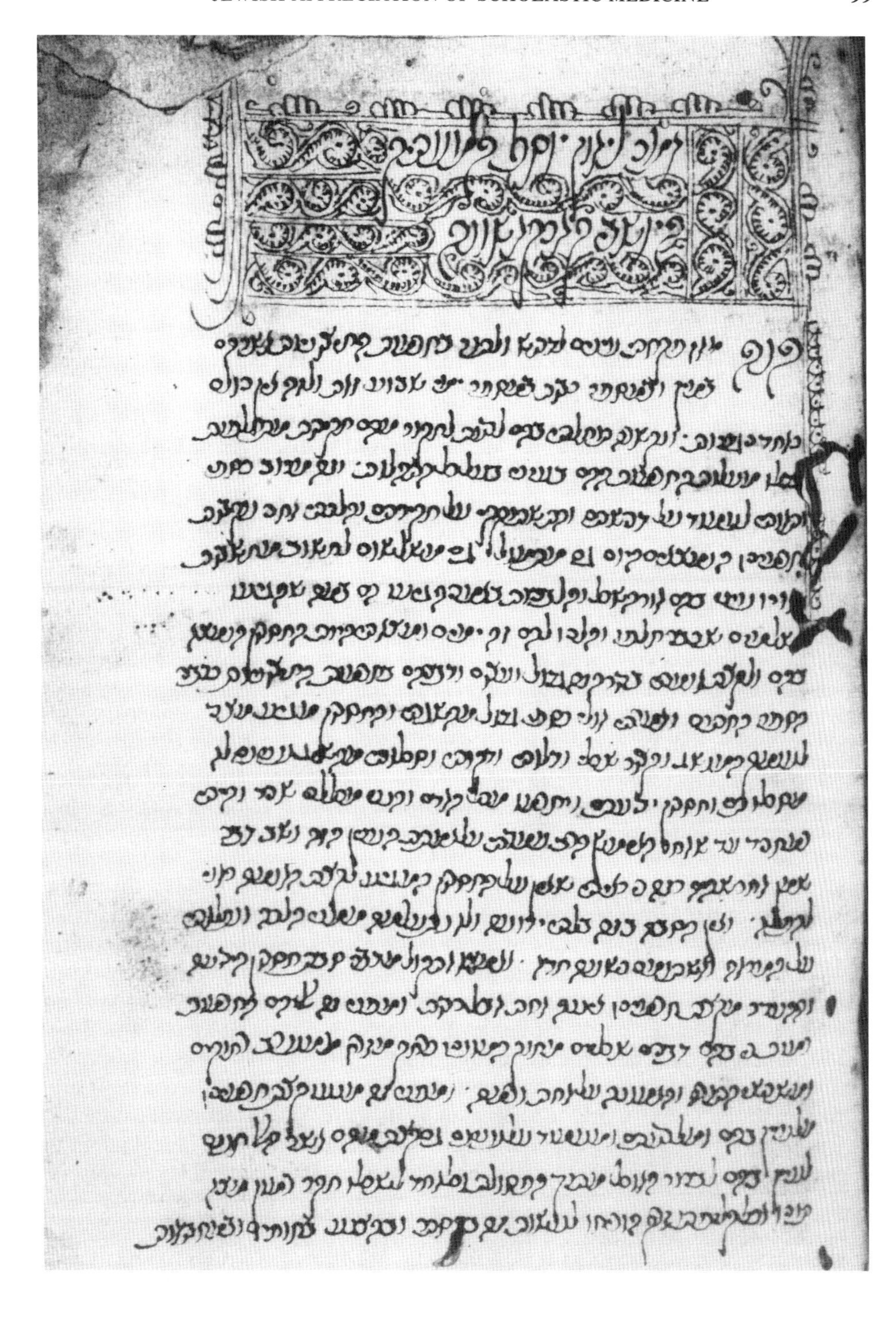

The opening of Leon Joseph's preface to his Hebrew translation of Gerard de Solo's* Practica super nono Almansoris, *Paris, Bibliothèque Nationale, MS héb. 1123. The manuscript is in Sephardic writing. Courtesy of the Bibliothèque Nationale.

oppression by Christians, in part too the inadequacy of the existing Hebrew translations of medical-scientific works, but, most important, the absence of intellectual freedom within the Jewish community. For him, the proven success of the Christians' scholastic medicine was the best reason for condemning the rigidity of the traditionalists within the Jewish community, those "learned in the Torah," whose "words concerning the profane sciences were like 'the words of a sealed book' [Isa. 29:11] . . . and [the sciences] are as far from them as east is from west." Besides, properly understood, scholasticism is really a legacy of earlier Judaism: dialectic was "employed by our own scholars [*hakhamim*] in earlier times in their study of the science of the Torah," and thus Jewish intellectual history leads to and is in keeping with the scholastic method of the Latins.[60] Leon clearly hoped to reinforce his argument by providing the new scholasticism with roots in Jewish tradition.

Nevertheless, those who felt as Leon did remained noticeably on the defensive. The hostility they sometimes had to confront is displayed, for example, in the words of the Talmudic scholar and rabbi Isaac bar Sheshet Perfet, of Barcelona, who lived in Saragossa and Valencia in the second half of the fourteenth century. In one of his letters—written after the upheavals of 1391 that so deeply affected that generation of Spanish Jews—Isaac bar Sheshet extolled the advantages of Talmudic instruction over the study of medicine or natural philosophy.

> We should not judge the laws and commandments of our Torah according to the views of the scientists and physicians, who, if they are to be believed—God forbid—assert that the Torah was not divinely revealed, the which they assume as a result of their false demonstrations. One who interpreted the laws relating to forbidden food according to the rules of the physicians would be richly rewarded by the butchers. . . . But we, for our part, should be loyal to our sages of blessed memory, even if they tell us that right is left, because they received the truth . . . and we should not believe the Greek and Arab scholars who speak only in accordance with their own assumptions and certain experiments.[61]

Isaac thus brings home the potential division in medieval Judaism over the proper role of the physician within the community. Evidently, for the rationalist minority this role was to practice in accordance with a medical science (*hokhmat ha-refu⊃ah*) built up on an understanding of man gained through natural philosophy, but just as evidently some of the religious leaders of the community repudiated such a conception of medicine. For them, a physician's role was pragmatic, simply to treat and to cure, nothing more; certainly not to understand.

We have already suggested that the reflux of medical thought from the Latin Christian to the Jewish world was assisted by its increasingly technical character, the absence of any ideological component. Medical practice was independent of a practitioner's religion, whether at the theoretical or the practical level—despite

[60] *Ibid.*, lines 19–21, 25–32; 12, 21; 72–73.

[61] Quoted by Baer, *History of Jews in Spain* (cit. n. 13), Vol. II, pp. 75–76. A similar assertion was made by Jacob ben Sheshet Gerundi in his polemic against Maimonideanism: "In their madness [the rationalists] follow the science of the Greeks. . . . It is illicit to learn something from the Gentiles." Quoted by Vajda, *Recherches sur la philosophie* (cit. n. 12), pp. 357–358. For illustrations of the complexity of the traditionalists' intellectual world see Twersky, "Rabbi Abraham Ben David" (cit. n. 9); Twersky, "Aspects of the Social and Cultural History" (cit. n. 4); and Shatzmiller, "Rationalisme et orthodoxie" (cit. n. 1), pp. 265–266.

the attempts of the medieval Church to christianize medical activity by, for example, requiring physicians to see to it that dying patients made a last confession, or by simply prohibiting Christians from seeing Jewish physicians.[62] Theoretical medicine as discussed and developed within the universities maintained its autonomy, with a curricular-technical dependence upon the arts faculty but none at all on theology. Doubtless Jews were formally prohibited from enrolling in the *studia,* but they could still find a certain acceptance there, as Leon Joseph did from Jean de Tournemire. This ideological neutrality could have enabled the transfer of a mature Latin Galenism into Judaic society, normally so jealous of its religious identity. Indeed, Leon saw no objection even to translating that portion of Gérard de Solo's commentary in which that author set out his belief in the Holy Trinity; he could not be blamed for translating this, Leon insisted, for those few words at the outset of Gérard de Solo's proemium were merely a personal statement by the Christian author that in no way affected the really interesting part of the *Practica,* its strictly medical content.[63]

It was therefore not the nominally "Christian" character of Latin medical theory that impeded its transmission; its lack of doctrinal character made it all the easier to be accepted by that group of Jews wishing to assimilate the achievements of scholasticism. After 1400, in fact, some Jewish scholars would find those achievements increasingly attractive.[64] How large, how widespread that group was is not something that can be answered here; we are far from wishing to generalize from such a limited number of texts. For the moment, all that the prefaces edited here can prove is that at least a few Jewish physicians in the western Mediterranean found that Latin scholastic medicine opened a new perspective to them during the fourteenth century.

[62] For papal letters and conciliar decrees against the practice of medicine within the Christian community by Jewish physicians see Solomon Grayzel, *The Church and the Jews in the Thirteenth Century* (rev. ed., New York: Hermon Press, 1966), pp. 74–75, 332–337.

[63] Appendix D, lines 168–173. Leon Joseph alludes to the following words from Gérard de Solo's work: "Invoco igitur Deum Patrem qui scit omnia qui sua potentia poterit meam indigentiam relevare cum ab eo dependeat tota celi natura. . . . Invoco Deum Filium qui sua sapientia et excellentia me gubernet quia ipse patris sapientia et a quo omnis sapientia et excellentia me gubernet et a quo omnis sapientia derivatur. Tertio invoco Spiritum Sanctum qui est amicitia et charitas patris." Gérard de Solo, *Nonus liber Almansoris cum expositione Gerardi de Solo* (Venice, 1505), proemium, fols. 13va–b.

[64] B. and C. Lagumina, *Codice diplomatico dei Giudei di Sicilia,* 3 vols. (Salerno, 1884–1895), no. 491, Vol. I, pp. 28–29 (17 Jan. 1466); cited by H. Bresc, *Livre et société en Sicile (1299–1499)* (Palermo: Centro di Studi Filologici e Linguistici Siciliani, 1971), p. 67, n. 1. See also Cecil Roth, *The History of the Jews of Italy* (Philadelphia: Jewish Publication Society of America, 1946), p. 241.

Appendixes follow

APPENDIX A

Prologue to the translation by Estori bar Moses ha-Parhi of Armengaud Blaise's *Antidotarium*[a1]

1 The translator, Estori bar Moses ha-Parhi, says:

13 . . . Then it was I came upon a book written in the language of the Christians, more precious than fine gold, mother-of-pearl, and onyx, which treats of the art of medicine. The usefulness of this book is no trifling matter for one who is wracked 15 in pain, it is as much a treasure to the humble as to the great, / a treasure of which the sage Christian named Armengaud Blaise from Montpellier is the author. It was handed to me here in Barcelona, in the year of my slavery, at the beginning of my new exile, and there I translated it from their language to our own blessed language, thanks to the benevolent care of God for us, His people. After having translated it, I did not wish to publish it for some time, but on the arrival of the Nasi[a2] in the community of wise and judicious men, he asked for a book, and I mentioned this trifle to him, this treatise, which I presented and expounded unto him. /

20 The theme of this book is self-explanatory. An introduction is unnecessary. That which goes before is the first and the last.

APPENDIX B

Prologue to the translation by Israel ben Joseph Caslari of Arnau de Vilanova's *Regimen sanitatis ad regem Aragonum*[b]

The translator says: It is known to him who is interested in the various branches of the art of medicine, and more so to him who reaches the natural roots, that is to say, the principles and foundations of this art, that there is nothing within the sections relating to the regulation and conservation of health that does not be- 5 long to the field of / analogical speculation in the art of medicine; the principles and the roots of his knowledge are taken, in general, from propositions that are to be found in natural science. There is no doubt of their certainty. For most of the nutritive substances that conserve our health have been known by way of analogy, selecting—by means of flavor or smell, for dryness or wetness—those fit from those unfit; all these are attained by one who possesses medical science, he ob- 10 tains them by analogy, / in accordance with natural principles, as we have said.

On the other hand, most of the things that are found in the sections relating to the restoration of health come from the field of scientific experiment, the object of which is not verified, in any case, by way of analogy. For the character of most medicines and remedies aimed at restoring health is known by way of experimental speculation, especially those products that cure diseases by their tertiary powers 15 or / specific qualities, that is to say, the faculty inherent in the form of the medication or the remedy.

Thus the reason for the large quantity of books and for the repetition of texts that are found concerning the conservation and regulation of health and [concerning] the nutritive substances, is the same reason that explains the multiplicity of books and the repetition of texts that exist concerning the restoration of health and the details of medical treatments. The reason for the first part and for the repeti- 20 tion of texts that can be found in it is that temperaments vary in accordance with / the climates in which the books are written.

[a1] MS Parma, R. 347 (13 pages); see Steinschneider, *Die hebraeischen Übersetzungen* (cit. n. 26), p. 778. We have translated only line 1, which contains the translator's name, and lines 13–21, which are directly connected with the theme of our article.

APPENDIX A

1 הקדמה לאשתורי בר משה הפרחי

13 בה אלי ספר מלשון אחרת, יקר מזהב ומפז דר וסוחרת על מלאכת
הרפואה, הלא הוא כלל קצר יועיל לכאב לב, אוצר לקטן יעשה ולגדול,
15 אוצר מיחוס לחכם נוצרי שמו מא" ארמנגב בלדי דמונפשליר, מסרו בידי פה
ברצלונה שנת שעבודי תחלת גלותי החרשה ושם הוצאתיו מלשונם ללשוננו
המאושר כיד השם הטובה עלינו עמו. היה אחר העתקתו ימים אחדים לא
רציתי לפרסמו. ובבא הנשיא לכלל אנשים חכמים ונבונים, דרוש דרש ספר
וספור וספור הקומץ הזה אליו הקרבתיו ולפניו הצגתיו.
20 ענין הספר מבואר בעצמו לא יצטרך אל הקדמה. הקודם הראשון
והוא אחרון. ישביענו מתורתו ישתבח שמו!

APPENDIX B

הקדמה [לישראל בן יוסף כשלרי]

אמר המעתיק כבר נודע לכל מי שיעיין בסדורי מלאכת הרפואה
יותר למי שהשיג השרשים הטבעיים שהם התחלות ויסדות למלאכה ההיא, כי
אין אחד מהדברים הנכנסים בחלקי שמיר[ת] הבריאות והנהגתו יותר מגדר
5 העיון ההקשי אשר במלאכת הרפואה, וההתחלות והשרשים בידיעתו, בכללות
לקוחות מהקדמות מצויות בחכמה הטבעית, אין ספק באמתתה.
וזה כי רוב המזונות שהם שומרים הבריאות נודעו על דרך ההקש,
ויובדלו הנאותים מהבלתי נאותים או מטעמם או מריחם או מיבשם או
מספוגיותם; וכל אלו הדברים נמשכים ונותנים בהקש לבעל חכמת הרפואה
10 עם התחלות טבעיות כמו שבארנו.
ואמנם הדברים הנמשכים בחלקי השבת הבריאות הם רובם מגדר
החכמה הנסיונית, אשר לא יתאמת ענינה בכלל על דרך ההקש. וזה כי רוב
הסמים והתרופות המיוחדים להשיב הבריאות נודע ענינם על דרך העיון
הנסיוני ובפרט אותם הדברים אשר ירפאו החלאים בכחותם השלישיות או
15 בסגולה רוצה לאמר הכח הנמשך אחר צורת הסם או התרופה, ועל כן סבת
רוב הסדורים והשנות המאמרים הנמצאים בחלק שמירת הבריאות והנהגתו
ובמסעדים המזוניים, הסבה העצמה ברבוי החבורים והשנות המאמרים אשר
חוברו בחלקי השבת הבריאות ופרטי הרפואות, לפי שסבת החלק הראשון
והשנות המאמרים המחברים בה היא לפי שסבת המזגים אם לפי טבע
20 האיקלימים אשר חוברו המאמרים בהם.

[a2] President of the rabbinic tribunal.

[b] MS Florence, Mediceo-Laurentiana Library Pl. 88, C. 36. The name of the translator into Hebrew does not appear in the prologue. Steinschneider (*Die hebraeischen Übersetzungen,* p. 779) attributes it to Israel Caslari, to whom the translation of the *Regimen sanitatis ad regem Aragonum* contained in this manuscript belongs.

In fact, remedies are repeatedly talked about, because they need to be tried continually. Each one of the scholars in medicine tries to verify on his own account, not to accept verification by others, the actions of the medicines [and] the limit of their form, called their occult nature; and each one writes a book on what he has proved, as happens in the science of the stars with those aspects that have
25 already been described, / as explained in the *Almagest*.

I say, then, that texts written on the theme of medicines and remedies will be useful to us, and will be preferable to those that discuss the nutritive substances that conserve the health, for these change, sometimes a lot and sometimes a little. In effect, most of the things that conserve health bring about [different] changes in
30 different men; many do not have any purpose. / On the other hand, the efficacy of the medicines and remedies that act as we have described, in accordance with their image, change, in any case, only a little. In this sense, we already know that scammony purges yellow bile and that turpeth purges phlegm; this was determined by the ancients, and it does not appear that it has ever happened that someone should have taken scammony without something similar to the nature of saffron having them come out of his body; and no one has ever taken turpeth without
35 something of the nature of / phlegm having come out of his body.

With this we mean to say that, with respect to remedies, it is not necessary to write as many treatises as we must do with nutritive substances and foods in relation to the conservation of health. It could almost be said that every man needs a text written for himself; for this reason kings and magnates are accustomed to ask their doctors to write a text for them with the diet they should follow. So I
40 myself, the translator of this book, / ask forgiveness for having meddled in an art which is not mine, translating this book, which is the first that has been written to date—although I know well that Rabbi Moises [Maimonides], of blessed memory, wrote a treatise on the theme.

In reality I have two reasons for doing so: the first is that this text is written taking into account the nature of [this] country, for its author is Master Arnau de Vilanova who was in Barcelona, in the service of the King of Aragon; the second is
45 that this book was written according to the customs and habits of the Christians, / in whose lands we live. The great personages always asked him to compose diets, and in order to lighten the task, he composed this work.

I have deemed it fit to translate it, thus adding it to the similar work by the sage I have mentioned [Maimonides], because the author was a great scholar; his wisdom was so great and his science of medicine so high, that he surpassed all his predecessors, to such a point that the common people considered him a prophet. The truth must be accepted for itself; and the truth must be received from him who speaks it.

APPENDIX C

Prologue to the translation by Joseph bar Judah ha-Sefardi of the Catalan version (by Berenguer Ça Riera) of Arnau de Vilanova's *Regimen sanitatis ad regem Aragonum*[c]

I saw that this guide [to good health] was excellent and beneficial for two reasons: on the one hand for the merit of the man who wrote it, and on the other for the excellence of the man for whom it was written.

5 With / respect to the man who wrote it, you must know that he was a Christian writer called Master Arnau de Vilanova, a great scholar and brilliant physician, who lived among us in this our land. He knew the natural things of these lands and their temperament, and most of the guides that the ancients had written in Arabia, the temperament of their land, their nature, their food, their drink, and their customs, which are not the same as the temperament of our land, our nature, our

[c] New York, Jewish Theological Seminary, Elkan Nathan Adler (ENA) Collection Bill. The manuscript is not described in Steinschneider, *Die hebraeischen Übersetzungen*. We have not been able to

ואמנם השנות הדבור בתרופות הוא לנסות הדברים תמיד, וכל אחד
ואחד מחכמי הרפואה ינסה ויתאמת לו הדבר, לא יתאמת לזולתו מהדברים
הנמשכים אחר פעולת הסמים, מצד צורתם הנקרא טבע נעלם, ויחבר כל
אחד ספר ממה שנסה, כמו בחכמת הכוכבים מהמבטים אשר נכתבו על זה
25 הצד כמו שמבאר באלמגסטי.

והנה אומר כי יספיקו לנו המאמרים המחוברים בעניני התרופות
והסמים מאשר המחוברים בעניני המזונות השומרים הבריאות, והם מקבלים
התחלפות על הרוב ואחרים על המעט, וזה כי רוב הדברים השומרים
הבריאות, יקיפו באישים המתחלפים התחלפויות רבים, אין תכלית להם
30 והתרופות [ו]הסמים הפועלים על דרך שבארנו, כדמות כוללים, לא יתחלף
ענינם כי אם על צד הפחות; ויתר דמיון זה כי כבר נוסה שהאשקמוניאה
תמשוך המרה הכרכומית והטורביט ימשוך הבלגם; וזה הניחו הראשונים ולא
נראה לעולם מי שלקח אשקמוניאה שלא יצא מגופו תחלה מה שידמה
לכרכומות וטבעה, ולא מי שלקח הטורביט שלא יצא מגופו מה שהיה מטבע
35 הבלגם.

כונתנו בזה שלא נצטרך לחבר בעניני התרופות מאמרים שונים
כהצטרכנו בעניני המזונות ומסעדים בהנהגת הבריאות כמעט שאומר יצטרך
בזה לכל איש ואיש מאמר מחובר לעצמו, ולכן מדרך המלכים ואנשי המעלה
לצוות רופאיהם לחבר אליהם מאמר מסדר הנהגתם. ע"כ מעתיק זה הספר
40 אומר מתנצל בהיותי אוחז אומנות שאינה שלי להעתיק זה המאמר הראשון
המחובר עד היום, גם כן ידעתי חבר בזה הרס" ז"ל.

אמנם לי בזה שני טעמים, האחד כי זה המאמר נעשה לפי טבע
הארץ כי מחברו היה משטרא ארנבט דוילה נובא ועמד עם המלך דארגון
בברצלונא, השני כי המאמר הזה חבר לפי מנהג הנצרי וחוקותם אשר אנחנו
45 שרויים על ארמתם, ונבקש להם תמיד מגדוליהם לסדר להם הנהגתם ולמען
הקל הטורח חדש מאמר, בזה ראיתי להעתיקו מצורף אל זה החכם הנז[כר]
אשר חברו, חכם מאד וגדלה חכמתו, ונתעלה בחכמת הרפואה מכל אשר היו
לפניו עד שההמון חשבוהו לנביא והנה יקובל האמת מצד עצמו.
קבל האמת ממי שאמרו.

APPENDIX C

הקדמה ליוסף בר יהודה הספרדי

הנהגת הבריאות לחכם הנצרי משטרו ארנב דוילא נובא

בראותי הנהגה הזאת רבת המעלה והתועלת לשני פנים, האחד לפי
ערך האיש המחבר אותה והשני לפי גודל מעלת האיש אשר חוברה אליו. אם
5 לפי ערך האיש המחבר אותה, דע כי המחבר הזה היה נצרי, שמו מש" ארנב
דוילה נובא, חכם גדול ורופא מובהק היה בינינו ובארצנו זאת. ידע טבעי
הארצות האלה ומזגם ורוב הנהגות הבריאות אשר לראשונים חוברו בארץ
הערב, ומזג ארצם וטבעם ומאכלם ומשתם ומנהגם בלתי שוה למזג ארצנו

identify this translator. He may have belonged to the Jewish community of Castelló d'Empúries in Catalonia.

10 food, our drink, and our customs. On the great dignity of the man / for whom it was written, you must know that he was the powerful lord King of Aragon, and may his kingdom be exalted and extolled!

It is well known that this author [Arnau] applied all his effort, all his capacity of understanding, and all his science in order to be able to compose this guide with honesty and great discernment, with fine and subtle reflection, and with the aim of making it useful, as is fitting to the great dignity of the king who asked him for it. Thus when I perceived its excellence and utility I felt stimulated, I, Joseph bar 15 Judah ha-Sefardi (may I be remembered in the next world), / and I translated it from the Christian language to the Hebrew language—in spite of being conscious of my slight intelligence and my defective knowledge of both languages—so that it should be beneficial to us, unfortunate Jews, who live in exile in the land of the Christians.

APPENDIX D

Prologue to the translation by Leon Joseph of Carcassonne of Gérard de Solo's *Practica super nono Almansoris*[d1]

Thus says Leon Joseph, the translator, who lives in Carcassonne. Many years ago I directed my attention toward the study of and research into the profane sciences, which are several in number and nature; in number they are as many as 5 the days of the week, and each one has its own subject matter. / I had placed my confidence in them in order to undertake various pieces of research. In my eyes, the merits of these sciences were above all praise. This explains my zeal and wish to know them and to master them. I therefore followed in the footsteps of the learned men of our own times, as well as those of the recent and distant past who have had such concerns, so that they should illuminate my way with the light of their intelligence and understanding; I said to myself that they must have achieved 10 the level that had been acquired by those perfect men / long since disappeared.

But I realized that the lack of knowledge that they, and some of my people at this time, found themselves submerged in was great and immense, and that their words concerning the profane sciences were like "the words of a sealed book."[d2] Thus I said unto myself: perhaps my wish is greater than my intelligence, and the defect is mine for not being able to fathom the concepts, and it is my own lack of intelligence, the weakness of my spirit and my ignorance, that do not allow me to understand their words, and not their ignorance and their lack of knowledge, for 15 they are the most learned among all men / and I am nothing but the lowest of their class. I was perplexed until, after many attempts, I understood the truth of the situation and "the thing once again was full of sense, after having been without any meaning,"[d3] for I perceived that said lack [of knowledge] on the part of one sector of [our] nation was by no means strange. Its cause was not unknown and I was not unaware of the [Talmudic] law which referred to it,[d4] and I was surprised by what I saw, and felt considerable uneasiness because of this misfortune.

Then I heard a voice telling me that there was not one single cause, but many, 20 for the lack and absence of this knowledge / among some of our scholars. Sciences defeated them because their subject matter is more rational than in the bosom of our people, and they [the sciences] are as far from them as east is from west, and all the more so from the fundamentals of the Torah and of religious faith.

For this reason, some of our scholars refrained from studying them, from learning about them, and from investigating their aims; but others among them, who by

[d1] MS Paris, Bibliothèque Nationale, Héb. 1123. The Hebrew prologue was published in Hebrew by A. Neubauer and E. Renan, "Les écrivains juifs français," *Histoire littéraire de la France,* Vol. XXXI (Paris, 1893), pp. 770–778. The highly defective transcription was accompanied by a French

וטבענו ומאכלנו ומשתנו ומנהגנו. ואם לפי גודל מעלת האיש אשר חוברה
אליו, דע כי האיש הזה היה האדון רם ונשא התנשא מלכותו מעלה מעלה 10
מלך ארגון ירום הודו. ויודע כי המחבר הזה שם כל כחו והשגתו והחכמתו
לעשות ההנהגה הזאת ביושר ובתשיה גדולה ובעיון דק וטוב ובתכלית
התועלת, כפי מה שראוי לפי גודל מעלת המלך אשר בקש אותה ממנו. ולכן
בראותי מעלתה ותועלתה נתעוררתי, אני יוסף בר יהודה הספרדי, זלה"ה,
והעתקתי אותה מלשון נצרי אל לשון עברי עם היות כי ידעתי קוצר דעתי 15
וחוסר השגתי בשתי הלשונות למען תהיה לתועלת לנו היהודים האמללים
הגולים בארצותהם.

הנהגת הזאת נחלקת ליח׳ השערים.

APPENDIX D

הקדמה לליאון יוסף

אמר ליאון יוסף המעתיק, היושבי קארקאשונה: הנה מאז פקחתי
עינים לדרוש ולתור בחכמות החיצוניות אשר הם במין ובמספר רבות,
במספר ימי שבוע זאת, ולזה אין כולם כאחר בגדרות.
ונקשרה תוחלתי בהם לדעת לחקור מהם חקירות מתחלפות; גדלו 5
מעלות החכמות ההם בעיני בעל כל התהלות. וזה מרוב כספי ותאותי לעמוד
על דרישתם והתשקקי על חקירתם. והלכתי אחרי עקבות חכמינו הנמצאים
היום גם מאתמול גם משלשום לחשוב מחשבות ויאורו עיני בהם אור השכל
והלבבות באמרי: הגיעו הם במה שהגיעו [ה]שלמים שכבר חלפו והלכו להם
זה ימים. 10

ומצאתי היות החסרון הנמצא בהם ולקצת אומתי בדור הזה גדול
ועצום, ודבריהם בחכמות החצוניות כדברי הספר החתום ואמרתי: אולי כספי
גדול מהשגתי, והחסרון מגיע מצדי לעומק המושג וקצור שכלי ודלותי ודקותי
וסכלותי מהשיג ענינם לא מסכלותם וחסרון ידיעתם, ויחכמו מכל האדם,
והנני מכללם שריד. והיתי כמחריד עד שאחר האימוץ הרב עמדתי על אמתת 15
הענין הזה, ושב דבר שמן אחר שהיה רזה, כי ראיתי שאין על החסרון המגיע
לקצת האומה ראוי חהפ[לי]א. ואין הסבה בזה בלתי ידוע ולא נתעלמה ממני
הלכה, ונפלאתי על המראה ואשתומם רשעה חרא. ואמצא את קול מרבר
סבת חסרון הידיעה וההעדר מקצת חכמינו, איננה אחת אבל רבות. ומפני זה
אליהם החכמות רמות כי בהם דברים שכליים, מחוק המוננו כרחק מזרח 20
ממערב רחוקים ומשרשי התורה והאמונה על אחת וכמה.
ומפני זה מנעו קצת חכמינו מלעיין בהם ומלרעתם ומעמוד על ענינם;

---

translation of certain fragments. We have considered it opportune to produce a new edition from the original Hebrew, together with an English translation of the entire text.

[d2] Isa. 29:11.

[d3] See Judah Halevi, *The Kuzari (Kitab al-Khazari): An Argument for the Faith of Israel,* trans. from the Arabic by Hartwig Hirschfeld (New York: Schocken Books, 1964), Pt. 1, para. 28.

[d4] See Babylonian Talmud (henceforth **BT**), *Baba Kamma* 82b.

the mercy of God studied them with the intention of selecting what was fit to eat from what was only fit for throwing away, applied their intelligence to this research—"he found a pomegranate; he ate the fruit within, and the peel he threw
25 away"[d5]—and decided to do so in secret / and unknown to others, "in the clefts of the rocks and in the secret places."[d6] They had no right to propound this science in the squares and streets, or to discuss it, to show themselves to be favorable toward it, nor to conduct public debates with the aim of reading the complete truth, for knowledge of the truth can only be attained by means of the contrary. And all this [they did] for fear of the tongues of the foolish among the people, for they are but few and the others form the majority. Also for fear of those learned in the Torah, who proscribed the superfluous from our souls; whose power over wise men
30 comes not from strength nor from the breadth of their knowledge, / but rather from the strength of their hands and from their many ruses that make the mass of the common people believe whatever they think, that is to say, that these sciences and those who are concerned with them are sundered from the community of those who possess the Torah. Such are the virtues of those who insult and tell lies about learned men!

I saw what they wrote about *The Guide of the Perplexed* and what they did [to its author] in those times, even though he was an eminent man and most expert in the Torah, a thousand, nay even ten thousand, times more than they were.[d7] Rea-
35 son / and truth were on his side and his deeds support this; as a result, divisions increased and the lack [of learning] grew considerably among one part of our community [. . .],[d8] for the following reason: the majority of books concerning these sciences were written by men of other peoples, such as the Greeks, the Ismaelites, or the Christians, and other believers in religions that are different from the religion of the Jews, and those books reached us through translators. It is possible
40 that some of them did not have / a full and perfect command of the language or languages and translated one word by another because of the image and likeness; or they may have omitted one thing or added another because they lacked a deeper knowledge of the language in question; or the books from which they were translating may have contained errors and these then passed into the books that were being translated as a result of a lack of understanding or through exhaustion. And the first error was followed by another even more serious one, as those who studied with those books built "with the line of confusion and the plummet of chaos."[d9]
45 / When they realized their error and noticed their doubt, they redoubled their efforts to clarify the text or texts just as they were written, and on these weak and unstable foundations they built their castles. Anybody who endeavors to write a commentary and elucidate those texts moves further away from the truth and never finds it, and this is never suspected. I know that this is what happened to a
50 wise man of my nation who explained certain / expressions of the books of the learned Avicenna on the basis of an initial translation that was really distant from the truth at many points, with the result that he wrote a book that was acclaimed correct and held up as an example and model. Likewise, another scholar, after the former, explained the words of Avicenna according to his own opinion and the text that he had.[d10] And the people followed him, believing that he was right and that his words were meaningful. Until God sent us the learned rabbi Joseph Lorki, who
55 illuminated / our eyes with his latest translation, that which is closest to the truth. Then the commentators realized that the first translation was of no value, the matters were clear and simple, and they themselves were confused and ashamed.

Behold, then, that we saw with our own eyes that learned Jews had empty hands and lacked arguments. If those wise men were alive in our day, they would also be embarrassed today and ashamed of their opinions before God most awful and most mighty. We should surely not be surprised at this, for the cause is the corruption /
60 of the translations.

On seeing the obstacle that these causes represented and aware that the afore-

[d5] BT *Hagigah* 15b.
[d6] Song of Sol. 2:14.

גם קצת מהם אשר האל חננם לעיין בהם לברור האוכל מתוך הפסולת וכל
אחר לשכלו חקר רמון מצא תוכו אכל קליפתו זרק הוכיחו לעשות זה
25 ב[ה]סתר ובהצנע, בחוחים ובמחבואות לא היו רשאים להרביץ החכמה
בשוקים וברחובות ולהתוכח בהם להראות פנים להם ולהושיב ישיבה ברבים
להמציא האמת על בוריו כי לא יודע האמת כ"א בהפכו, וזה מפחדם מלשון
המון הסכלים, כי הם מתי מעט והנה הרבים, מצורף לזה מיראת התורנים
אשר מיתר החרימו מנפשנו אשר ידם תקפם על המעיינים לא בכחם ורב
30 ידיעתם רק לחוזק ידיהם, ורוב נכליהם להיות עדת ההמון ההוא נשמע אליהם
בחשבם היות החכמת ההם, והמעיינים בהם יוצאים מכלל בעלי התורה. אלה
מרות החסרים והמדברים על החכמים תועה.

והנה ראיתי הכתוב על ספר מורה הנבוכים ואת אשר עשו לו בימים
הראשונים עם היותו גאון ונסמך התורה יותר מהם לאלפים ורבבה, והצדק
35 והאמת אתו ומעשיו מוכיחין אותו ו[מפני] זה גדל החסרון והפליג, ההעדר
בקצת האומה אחר ההשתתפות [בהיותה] על זה סבה: כי רוב ספרי החכמות
ההם חברום חכמים מבני העמים כיונים ובני ישמעאל או הנוצרים ויתר
המאמינים אשר דתיהם [שונות] מדת היהודים.

והגיעו הספרים אלינו ע"י המעתיקים ואולי הם או קצתם לא היו
40 בקיאים בלשון ההוא או הלשונות על השלמות. והחליפו מלה אחת באחרת,
בצלם או בדמות, או חסרו מהם או הוסיפו לחסרון ידיעתם בלשון ההוא,
ידיעה בעצמות, או נפל בספרים אשר העתיקו ממנו טעות ומפני זה נפלה
השגיאה בספרים ההם וקיצור ההשגה והלאות. ונמשך מהטעות ההוא טעות
אחר גדול מזה כי יבנו עליו קו-תוהו ואבני-בוהו המעיינים בספרים ההם
45 בהרגישם בטעות ההוא ובספק ההוא בהעמיקם לבאר הלשון או הלשונות
כמו הם כתובים ועל יסוד רעוע וחלוש מגדליהם בונים וכל אשר יתקרבו
לפרש ולבאר הלשונות ההם יתרחקו ועל האמת לעולם לא יפגשו ובדבר הזה
אין בו חשש.

והנה ראיתי זה בחוש ובמוחש, שקרה לאחד מחכמי אומתי שביאר
50 לשונות מספרי החכם בן סינא בהעתקה הראשונה, האמתית הרחוקה מן
האמת במקצת המקומות עד שחבר מהם ספר, קראו ראוי להיות והעמידו
למופת ולאות גם חכם אחר קם אחריו וביאר לשונות בן סינא כפי דעתו
והלשון הנמצא אתו. וילכו העם אחריו עד שהיו לומדים מספרו בחשבם,
נמקו עמו וטעם לדבריו עד ששלחנו האל החכם ר" [יו]סף לורקי, אשר האיר
55 עיננו עם העתקתו האחרונה, האמיתית בקרוב ואז נמצאו המבארים ראשונה
ככלום, ויתבררו ויתלבנו הדברים ונשארו בושים ונכלמים.

והנה ראינו בענינינו כי אין כל מאומה בידיהם ונסתמו טענותיהם.
ואם היו החכמים ההם חיים כהיום, הנה כמונו היום, יבושו ויחפרו מדעותיהם
ודבריהם מאל נורא ואיום; אמנם אין זה מן הפליאה כי מאת שבוש
60 ההעתקה היתה להם נסבה.

ובראותי אלה הסבות המונעות, וידעתי לימוד החכמות הנזכרות אשר

---

d7 The allusion is to the so-called Maimonidean controversy. See note 12 above.

d8 The meaning of the two words in the original here is hopelessly obscure.

d9 Isa. 34:11.

d10 Leon's first reference is to the translation of Avicenna's *Canon* by Nathan ha-Me'ati, produced in Rome between approximately 1279 and 1283. His second reference is perhaps to Zerahiah Gratian, who carried out another (unfinished) translation of the *Canon* in Barcelona or Rome between 1270 and 1290.

mentioned sciences were known among the Christians—[God] is awful and marvellous; who could make him sustain us in our poverty!—I said to myself: I shall study their language [that of the Christians] a little, I shall attend their schools and houses of study, I shall follow their footsteps so that I am able to make use of whatever I might learn from their words and a little of the truth of their books,
65 both for myself and for any other person; / I shall study what the most perfect among them studied, if time allows me to and no accident or hindrance should prevent me from doing so, and if my intelligence is able to cope with this and is not bewildered. Consequently, I found great benefits in this, because in general their discussions on these sciences do not stray from the subject matter; they leave out
70 nothing when it is a question of / debating the truth or falsehood of a proposition [*lit.*, certain thing]; they are very rigorous concerning the questions and answers of a debate, which are linked together in such a way as eventually to bring out the truth by means of an analysis of opposing points of view, "like a lily among thorns."[d11] The way of studying that they practice today as regards these sciences is the same as that followed by our own scholars in earlier times in their study of the science of the Torah. I refer to that saying of the Talmud [on the scholars of Pumbedita who were able] to make an elephant pass through the eye of a needle,[d12] or as Rabbi Dosa said: "I have a younger brother who is a devil, who calls himself
75 [Jonathan]; with his talent he is able to make whatever / was prohibited permissible. And he has more than three hundred arguments to prove that the rival of a daughter may marry the brothers."[d13] And those conclusions of Rabbi Meir and his companion that the religious law that has to be followed was not that of the latter but the restrictive measures of the former. And the school of Shammai, as it appears in the appropriate place. If these and other sciences should remain beyond our reach, it is not because our intelligence is inferior to that of the Christians, for we possess the same capacity for understanding; indeed, it is the circumstances that have kept us apart, and the memory of them has been erased as a result of
80 exile and repression, and in the same way their pleasure, splendor, / and wealth [have been erased]. For this reason, the Christians have continued to advance in the profane sciences while we have continued to lose ground as a consequence of distress and oppression.

When I lived among the Christians, I was of an inferior condition in their eyes, for there is none of our nation who is honored in their eyes except him who is a physician and who cures them of their ills; in such a case, he sits at the table of
85 kings and remains standing before them, whether he be of humble birth or / of high rank, owing to his knowledge of medical science, which follows that of physics and law, for physical science has as its end man and his form, and it is here that medicine begins, the object of which is man.

I understood this and I took it as my model, and I said unto myself: I shall go in search of Jewish physicians and I shall beg them to have mercy on me and to teach me their science in exchange for a small payment or free of charge, for the money
90 of my purse has been exhausted. Thus I went forth to seek them and I found / that they were lacking nothing except that one science, for the two reasons that have already been explained and for another two, which make four. Not one less, nor one more. The majority of those who practice the art of medicine among my people do not intend to fathom the depths of their object; even if they were able to go into it in greater depth, they do not wish to tire their intelligence with this, and they mutter: this science is not a true science, but rather a skill; and with a "sound
95 of tumult"[d14] everybody wishes to turn it into a means of existence. / They make an ideal of it, and they say that a learned physician is he who asks large amounts of money and does not heal free of charge; this man, called an expert physician, with his deceits and frivolities makes it appear as if he knows or, sometimes, he introduces himself as if he were so-and-so, son of so-and-so. How many physicians do we see, lacking understanding, who have gathered treasures of gold and silver, without ever having passed a single examination; whereas others who have always been in pursuit of books have obtained knowledge, but have not earned a penny? For this and other reasons, they have not whiled away their time in the study of /

בין הנוצרים ידועות, נורא מאד ונפלא ומי יכילנו שעמד על ענינו ויקיפנו.
אמרתי בלבי: אלמד מעט מידיעת לשונם ואשב בישיבותיהם ובית עיונם
ואלך אחרי עקבותם למען אוכל להועיל לעצמי ולזולתי בהשיגי דבריהם
וקצת מאמתת ספריהם. 65

ואשקיף במה שהשקיפו השלמים מהם אם הזמן יספיק בזה [ולא]
ימנעני מונע ומקרה ואם השכל יכול להשקיף עליו ולא יטרידהו [הפ]חז ממנו
ואליו ומצאתי התועלת בזה גדול ורב, להיות רוב משאם ומתנם בחכמות
ההם, לא סרו במה שראוי לעיין בהם, דבר מהם לא יחסר בהיותם
מתפלו[פ]לים על אמתתם ואף בהעמדת שקריותיהם ומדקדקים בשאלותיהם 70
ותשובותיהם על צד הויכוח להוציא האמת ממרכז הפכו בבארם כל דבר
בשני הפכים כשושנה בין החוחים. והנה חק עיונם היום באלה החכמות כחק
חכמינו בדורות הקדומות בלמדם חכמת התורה, רוצה לומר תלמודה דמעילי
"קופא דמחטא", כמאמר דוסא, אח קטן יש לי בכור, שטן שמו, יודע להתיר
האיסור בחידודו. ויש לו שלש מאות וכמה תירוצים על צרת הבת שמותרת 75
לאחים. ובפלפול ר" מאיר וחבירו שאין הלכה כמותו רק בגזרותיו ושמאי
ובית דינו כמו שנרא זה במקומו והנה נמנעה ממנו ידיעת החכמות בכיוצא
בזה לא שיהיה שכלנו למטה משכליהם כי גם לנו לבב כמו הם, אמנם
מנעונו הסבות אשר קדר זכרונם על גלותינו ולחצנו ורוב תענוגם הודם
והמונם. ומפני זה המה עלו בחכמות החיצוניות מעלה מעלה. ואנחנו ירדנו 80
מטה לרוב הטרדות המעיקות.

וכאשר ארך זמן עמידתי עמהם והייתי שפ[ל] בעיניהם כי אין
באומתינו מי שיהיה מכובד בעיניהם אם לא יהיה הרופא אשר ירפאם
מחוליהם והוא יעלה על שולחן מלכים ולפניהם יתיצב יהיה שפל רוח או
בקומה נצב, יען ידע בחכמת הרפואה אשר היא אחרית הטבע ומשפטה, כי 85
סוף חכמת הטבע הוא האדם ותמונתו ובו תתחיל הרפואה מפני כי הוא
נושאה. והבנתי זה ולקחתיו למשל ואמרתי: אלך אל היהודים הרופאים
ואשאל מהם ואבקש מאת כבודם, יחמלו עלי וילמדוני מחכמתם בשכר מה
או בחנם כי אזל הכסף מכילי והנה תם. והלכתי אחרי עקבותם ולא מצאתי
בהם חסרון נרגש רק בחכמה ההיא לשתי הסבות הקדומות ושתים אחרות 90
שהם ארבע. ואל תוסיף בהם מאלה לא תגרע והוא שרוב כלל אומתי
המתעסק[ים] במלאכת הרפואה אין בדעתם להגיע עד תכליתה, גם כי יוכלו
לעיין בה לא ירצו להטריח בזה שכלם כי פוערים פיהם במאמרם: אין זאת
החכמה, חכמה כלל, אבל תחבולה וקול המולה להוציא כל איש ממנה המחיה
ויהי להם למשל וילחשו בלחשם לאיזהו רופא חכם השואל ממון רב ולא 95
בחנם ירפא והאיש ההוא, רופא אומן יקרא אשר בשקריו והבליו היותו יודע,
לפעמים יראה היותו מזה בן מזה. וכמה רופאים רואים אנחנו הנעדרי הבינה
אצרו אוצרות זהב וכסף מבלי בחינה ומהם עמדו תמיד אחרי הספרים
ואליהם ההשקפה ולא השיגו ממון אפ[ילו] פרוטה. זאת ועוד אחרת לא יעיינו

---

d11 Song of Sol. 2:2.
d12 BT *Baba Mezi*ᶜ*a* 38b.
d13 BT *Yebamoth* 16a.
d14 Ezek. 1:24.

100 medical books, but in their leisure time they drink and lie together in the flesh and
do other unworthy acts, such as reading books of fables, fantasies, and vanities.
Not one of them wishes to be learned in the field of medicine so as to avoid
entering Gehenna.

When I realized all this, I said to myself: there are no learned and expert men in
our region. And I went in search of wise Christians so that they should guide me in
such matters and so that they should show me the way, and so that they should
105 show me / old and new volumes. Among them I saw the works of the learned
physician Gordon, who was outstanding on the majority of subjects, both in theoretical study and in practical matters, and who achieved great fame. His works
proliferated and his prestige increased more than that of his brethren. Some of his
works have come down to us, such as the *Shoshan ha-refuʾah* [*Lilium medicine*]
and the *Haqdamat ha-yediʿah* [*Tractatus de prognosticis*]; but the translations of
these works are very defective, a shaky ladder. The Jews who translated [the
110 *Lilium*] in these lands / gained access to it by means of the vernacular; they expressed the Latin in Romance, adding that not every Latin word may be translated
in the vernacular tongue, and then they expressed it in the sacred language. In
spite of that, the greater part of this book is understandable to us, with the exception of some points which I myself have hastened to correct and improve as was
necessary so that the reader should be able to read it with ease; the same I have
done with the book *Haqdamat ha-yediʿah,* the translator of which paid hardly any
attention to the real meaning of the words. Seeking among their books, I found two
115 / new books, "two golden pipes,"[d15] exemplary, excellent and worthy, which the
people of our nation do not know. They are the books by the learned Gérard de
Solo and by the master Jean de Tournemire, of whose [existence] I had known for
ten years, but which I had been unable to acquire, neither in Montpellier, even
though it was their place of origin, nor in Avignon, nor in other important places,
since there were but very few [copies], and since the learned men of Montpellier /
120 anathematize and excommunicate anybody who should sell to those who are not
Christians. In order to acquire them, I invested all the funds that I was able [to
spare], without reflecting on the consequences. And the books reached my hands
this year, which is the year ninety-four according to the Christian calendar [1394].
The truth is that I paid double the price for them, owing to the yearning and desire
that I felt for them. I saw their excellence, and that they explained things and
answered matters in a suitable way. He who acts with them is as if he were grasp-
125 ing a column; / he will fear nothing, not even ten thousand physicians who might
come with proofs and disputes, for they are pure flour. No men of their equal can
be found in living memory. I believe that they did not fall behind and that they
chose the best [aspects] of medicine.

Behold, thus, that I begin the book of Gérard, based on Book 9 of the *Book of
Almansor* [by al-Rāzī, or Rhazes], short in length but great in quality, stronger
130 than a rock. It provides suitable and opportune answers. / Its art is a safe art. It
provides reasonable judgment on all things, which should be praised. Nothing is
left out and nothing is forgotten. Therefore, the decision to translate it was taken
calmly. When God wishes that I should finish it, I shall undertake Tournemire's
book; in his time he was at the head of all the scholars of Montpellier, who were
placed beneath him. I saw him with my own eyes and I spoke to him. He was a
man who was pleasant to talk to. His behavior was not like that of the other
scholars of his generation, who scorned those Jews who practiced the art of medi-
135 cine, / for he guided them as far as he was able. May his soul rest in the treasure
where the souls of the pious among gentile people live![d16] There is no book that is
comparable to it for the beauty of its science. It speaks of the limbs and of the
faculties, and of the illnesses that occur in them. It asks and answers, it explains
and clarifies. He teaches like a father to a son, true and proper words, which are
pleasing and agreeable to whoever listens to them.

I took the decision to translate these books, not for myself, but for the people /
140 of my nation, both those who live alongside me today and those who will come
after me, and who do not know the language of the Christians at all. When they

100 בספרי הרפואה, רק בעתות הפנאי בעברם עליהם בעתות השתיה והמשגל
והדברים אשר להם גנאי בקראם על ספר מספרי המשלים, וההזיות וההבלים.
אין איש מהם רוצה להיות ברפואה חכם למען לא יכנס בגיהנם.

ובהרגישי את כל זה אמרתי: אין נבון וחכם באיקלימנו זה. הלכתי
אחרי עקבות החכמים הנוצרים והיישירוני בדברים והראוני הדרכים וגלו אלי
105 הספרים חדשים גם ישנים. וראיתי ביניהם דברי החכם הרופא גורדו אשר
ברוב דבריו תשובתו בצד הפליג לעשות בעיון ובמלאכה ולכתר שם טוב
זכה רבו ספריו; וגדלו יקרותיו נתעלה על כל אחיו. והנה הגיעו אלינו קצת
ספריו והוא שושן הרפואה והקדמת הידיעה. אמנם הגיע לנו בהעתקה מאד
חלושה, סולם רעוע. הנה כי היהודים אשר העתיקוהו בארצות האלה לא
110 השיגוהו רק באמצעות הלעוזות לעזו הלטין והוסיפו בטענות כי אין כל מלה
אשר בלטין תוכל להיות לעוזה ובלשון הקדש להשיבה, אמנם רוב הספר
ההוא על כל זה מובן אצלנו, ואם לא במקומות מה אשר טרחתי אני להגיהו
ולתקנו כהלכתו למען ירוץ קורא בו וכן עשיתי בספר הקדמת הידיעה אשר
המעתיקו באמתות הלשונות לא שעה. ובחפשי אמתחות ספריהם מצאתי שני
115 ספרים חדשים לשני צנתרות הזהב ומשלים, מה מאוד הם טובים וראויים
אשר מבני אומתי נעדרים. המה ספר החכם גיראבט דישולא וספר החכם
מאישטרי יואן די טורנאמירא אשר זה עשר שנים שמעתי [ש]ומעם ולא
יכולתי להשיגם לא בהר ואם הוא המקום אשר ממנו חוצבו, ולא באויניון
ולא ביתר המקומות הטובים כי הנה במציאות מעטים, מצורף למה שחכמי
120 ההר היו מחרימים ומנדים כל מי שימכרם לזולת הנוצרים, והוצאתי בהשגתם
ממונות כפי כחי. ולא הרגשתי במי ומי והגיעו לידי בזאת השנה שהיא שנת
תשעים וארבעה למספר הנוצרים לחשבונם, וחי האמת קניתים בכפלים
משיווים לחשקי בהם ובתאותי ורצוני בם. וראיתי הפלגת טובם מראים פנים
לכל הדברים ומשיבים כהלכה לכל הענינים והפועל עמהם שיעמוד נשען
125 בעמוד ובצר לא יירא מדבר ומרבבות עדת הרופאים אם יבואו אליו
בנסיונות ובויכוחים להיות סולת נקייה לא קדם כמותם בזכירה. ואחשוב כי
לא יתאחרו והיותר טוב מהרפואה בחרו.

והנה מתחיל אני בספר גיראבט הבנוי על הספר התשיעי מהאלמנצור
אשר מעט הכמות ורב האיכות, חזק מצור, משיב כראוי ומגיד כהלכה
130 ומלאכתו מלאכה בטוחה, ונותן טעם לכל דבריו טעם לשבח, לא נעלם ממנו
דבר ולא נשכח ולכן שקטה ההסכמה להעתקתו. וכאשר יגזר השם בהשלמתו
אתחיל אחריו ספר טורנו מירה אשר היה בימיו על כל חכמי ההר לראש והם
עמדו תחתיו; ראיתיו בעניני ודברתי עמו, חסיד היה בדברו, לא היה חקו
כחוק חכמי דורו לבזות היהודים אשר במלאכת הרפואה מתעסקים יען היה
135 מישירם בכחו. בצרור חיי חסידי האומות תהיה צרורה נפשו, וספר אחד לא
דמה אליו ביופי חכמתו. דבר באיברים וכחותיהם והחליים המתהוים בהם,
שואל ומשיב, מברר ומלבן כאב המלמד את הבן חרש אמרים נכוחים
וישרים ימתקו ויערבו לשומעים.

והנה עלה על רוחי להעתיק הס[פרים] ההם לא לעצמי אבל לזולתי
140 מהעם, אשר עמי היום, ואשר יבואו אחרי הבלתי יודעים מאומה בלשונם כי

[d15] Zech. 4:12.
[d16] See 1 Sam. 25:29.

examine it and read it, they will see the perfection, the beauty, and the order of these books, they will bless me because of them and they will remember me for having been the reason why they should find them and the reason why they are able to compare themselves with the physicians in proofs and arguments, and this is a second life.

I believe that the fact that these books talk about things Christian is without significance unless [Jewish readers] take great pleasure in drinking wine, eating, drinking, and taking part in [Christian festivities], for, in that case, these translations will bring them no profit whatsoever; / it will be as if they had never been made and the yearning of their hearts will vanish. To tell the truth, if they read them many times, they will obtain different advantages, and they will feel calm and safe and will sleep peacefully in their beds. They will not fear him who attacks them in order to beat them with the stick of medicine, since they will be able to fight back without fear of the multitude, even if they are excommunicated. They will be able to put aside and put away all their other medical books, and should these two books be prohibited / they will be able to count on my explanations.[d17]

Thus, the aforementioned book of Gérard de Solo is Book 9 of the *Book of Almansor* [*Practica super nono Almansoris*] divided into the same parts and chapters as al-Rāzī divided his work. If this author [Gérard] does not point this out at the beginning of the book, as all authors of books are accustomed to do, it is because he trusted that everybody would understand that. I have prepared an index to these chapters, from the first to the last, so that everybody can find / what he wishes whenever he wants and so that he does not mix his thoughts or stray from the point. I have also thought to add in some chapters questions and answers following the disputes that I have found recorded, scattered here and there, some attributed to this same learned author, others to another eminent and authoritative scholar; others I drew up as well as I was able to when I could find nothing suitable that discussed the problem of that chapter in order to sharpen the intelligence and debate it and understand it in a suitable fashion. /

In the case of each question that is not from the book you will find at the beginning ALYH [*Amar Leon Yosef Ha-ma*^{c}*tiq* = Here speaks Leon Joseph, the translator], which are the initials of my name, so that you may know it is an addition. If you find any doubtful expression, do not reproach me or censure me in your hearts until you have checked it in another book, if God, in His mercy, has allowed you to learn the language of the Christians, or until you have consulted a learned physician. For the copy that I have translated contained errors, even though I have checked it twice / or thrice, [comparing it] with other books, and "all of them have been woven into the same cloth,"[d18] that is to say, it seems that they are all of the same origin, that all of them come from the same place. In those passages that turned out to be doubtful, there will be [the sign of] a small hand stretched out above them or some other sign at the side to indicate to whoever should be studying it that these passages have not passed unnoticed and that he can check them as well.

And to the community of physicians I say: Brothers, when you read the prologue of the author and see that he joins in his entreaty his king and his God and that / from one he obtains three and those different,[d19] I shall submit that I translated word for word, and do not get annoyed nor rail against me since you will have no other reasons for complaint after the prologue. My intention is that the book should reach you without suffering omissions or mistakes, so that you can understand it all and so that you can obtain benefits from it and sing its praises. If you do not wish to read the prologue, begin to study at the beginning of the book. I did not wish to remove any of its contents, rather / I wanted to add something to it.

I shall write [the name of] compound medicines in the Latin language, after having translated them into our language, with the help of God, so that those who are starting to practice the art of medicine can do so without effort or weariness. They will feel learned and intelligent, veterans in the art and perfect, even though they are but beginners. Without ever having seen the leading figure of the [medical] art, they will not be scorned by those crowned [with glory]. They will have no

בעברם עליהם וקראו ויראו שלמותם ויופי סידורם יברכוני בגללם ויזכרוני בלבו[ם] להיותי סבת חיותם ושיוכלו עמוד נגד הרופאים בנסיונות ובויכוחים ואלה חיי שניים ידעתי כי עם אלה הספרים דבריהם נוצרים לאין אם לא ימשכו את בשרם ביין לאכול ולשתות ולחוג חגות כי אז לא יועילו להם
145 ההעתקות, יען יהיו להם כמו שלא הועתקו וממורשי לבותם נתקו. אמנם אם יקראו בהם פעמים רבות יוציאו מהם תועלות מתחלפות ויעמדו בהשקט ובבטחה וישנו במטתם במנוחה. לא יפחדו מקול נוגש בשבט הרפואה יכם כי יוכלו ללחום מלחמותיהם לא ייראו מהמון אף כי יהיו מחורמים וכל שאר ספרי מלאכת הרפואה יוכלו להסתיר ולהעלים ועם שני אלה אם יחרימו
150 תחרים ומפני זה בירותי בספרם.

והנה הספר הזה הנקר[א] גיראבט דישולא הוא הספר התשיעי מספר האלמנצור נחלק לאותם החלקים והפרקים אשר חלק ראזי את ספרו, ואם לא זכרם החכם הזה בתחילת ספרו כחק כל מחבר ספר מהספרים הנה סמך על המבינים. וערכתי אני הפרקים ההם זה מראש ועד סופו למצוא כל איש
155 דבר חפצו כאשר ירצה ולא יבלבל מחשבתו ולא יטה. גם אמרתי: להוסיף בקצת פרקים שאלות ותשובות על צד הויכוחים מצאתים כתובים הנה והנה מפוזרים יוחסו קצתם אל החכם המחבר או לחכם אחר מטיב ומאשר וקצתם אחדש אני כפי יכלתי, כאשר לא אמצא איש עתי יתוכח בעיני הפרק ההוא לחדד השכל, ולפלפל ולדעת הפרק על אמתתו.

160 והנה תמצא בכל שאלה אשר לא תהיה בספר עצמו כתובה כתיבה בראשו אליה", והוא שמין שמי בראשי אותיות, למען תדע כי הן נוספות ואם תמצא לשונות מה מספקות אל תאשימני בלבך ואל תבזני בנפשך, עד שתדקדקו בספר אחר, אם האל חנגך לדעת לשונם או שאל לרופא חכם. כי הספר אשר ממנו העתקתי בו טעיות ואם דקדקתי עם ספרים אחרים שנים
165 או ג" פעמים וכלהו מחתינהו בהדא מחתא נראה שכולם ממחצב אחד חוצבו וממקום אחד יצאו. והנה במקומות אשר יפול בו הספק ידו נטויה עליו או סימן אחר שבה כמו אל צדו להעיר לב המעיין ולהודיעו כי לא נעלמו מעיני ותוכל לדקדקו אחרי. ואת עדת הרופאים, אחי, כאשר תקראו בפתיחת המחבר ותראו אותו קושר קשרים מתפלל במלכו ובאלוהיו ולמעלה יפנה
170 מאחר משלש ושונה ואני העתקתיו מלה במלה, אל תדברו עלי חרה ואל תלינו עלי תלונה, כי לא ישאר אליכם תרעומת אחר ההקדמה, כי רעתי רצויה יגיע לכם הספר בלי השמטה וחסרון יפול בו, למען תשיגוהו כולו ותוכלו לחתור מדבריו ולהשיב על מהלליו. ואם לא תרצו לקרות פתיחתות, תחילו לעין בתחלת ספרו ולא רציתי לחסר דבר ממה שדיו תחת רציתי
175 להוסיף בו והנה אכתוב כל הרפואות מורכבות בלשון לטין אחר העתיקי אותם בלשוננו, בגבורת השם למען אועיל למתחילים להתעסק במלאכה יוכלו לסדר בלי עמל ויגיעה. יראו עצמם חכמים ונבונים, זקני המלאכה, ושלמים גם כי יהיו מתחילים. לא ראו מאורות במלאכה כל הימים, לא יבזו אותם העטרים ולא יצטרכו לעיין בספר שמות הסמים אם ידע הרפואות על

---

d17 The meaning of this passage is unclear in Hebrew, and it is difficult to translate.
d18 BT *Berakoth* 24a.
d19 An allusion to Gérard de Solo's invocation of the Trinity in the proemium to his *Practica*.

180 need to consult any other book of medicine provided they know the remedies / by heart, just as is written: "[God] will keep your foot from being caught";[d20] they will be considered important physicians, which is what I hope to achieve.

God knows what my heart desires. I ask Him to help me and to guide me in my translation and to give me strength to bring my task to conclusion and to translate the other work as well. And that He should show me the wonders of His Torah, and that He should send us His Messiah. Amen.

APPENDIX E

Prologue to the translation by Abraham ben Meshullam ben Solomon Abigdor of Bernard Alberti's *Introductorium in practicam pro proiectis*[e1]

Abraham ben Meshullam ben Solomon Abigdor says: When I was young my desires were those of a young man, and equally my thoughts and my urges; not that I wished to study the art of medicine because of a yearning for knowledge of
5 the science, but rather just the name itself was enough for me, / so that I would be called Master and earn those undeserved profits as is customary nowadays among the majority of the men who are engaged in this art, especially those of our people.

When I grew up, and reached the grade of manhood, my desires were those of a man, and equally my thoughts and my urges. I decided to study and to go deeply into this science from every side; and I wished to do so in order to attain true
10 knowledge, not for anything / subject to destruction and corruption. Thus I went up to the mountain, to Montpellier, the renowned city, to hear the science of medicine from the mouths of Christian doctors and scholars, and there I found numerous books and useful commentaries on the fundamental principles of that science, and if the Lord should prolong my life, I shall translate some of them. I found useful books, which lead us straight to the art [of medicine], and from among them I specially chose this one called *Mabo ba-melakhah* [*Introduction to*
15 *the Art*], which is a very useful guide / for anyone who wants to devote himself to this art with confidence and rectitude. The questions of which this book treats are all explained very clearly, as are the remedies that it includes, and that according to the time in which we live, the climates and the temperaments. [With its help,] the physician will be able to apply them with surety. I have translated it from the Christian language to that of our people, at the request of some of my friends and colleagues.

May the Lord protect me from all errors and make me worthy of translating this
20 [book], as well as other books, / and may He prolong my days and the days of those who translate books of medicine, which are great things [?], and the days of all brothers of Israel [?].[e2] Amen.

The author says: This book, called *Introduction to the Art,* was written for the sake of those who have been instructed in the science of medicine but have not yet been trained in its use. It is founded on the first chapter of Book 4 of Avicenna['s *Canon*] and was composed by the distinguished scholar Bernard Alberti, a promi-
25 nent figure in the honorable academy that is in the city of Montpellier, / at the request of his friends and pupils. These are the contents of his words: At your request, my friends, who have toiled and labored with me in the study of this science, the science of medicine, I decided to write for you a useful though small book concerning the art of medicines. I shall follow the path of the first chapter of Book 4 of Avicenna. I shall arrange, in an orderly way, the remedies that in my
30 opinion are well known, efficient, and proven. I pray that this book / may not be trivial in your eyes, since I deem it to be of great utility. . . .

[d20] Prov. 3:26.

[e1] The Hebrew text (MS Halberstam 406, fol. 74) was edited by Moritz Steinschneider, "Medizinische Handschriften im Besitz des Herrn Halberstam," *Magazin für die Wissenschaft des Judentum,*

180 פה כבתוב וישמר רגל מלכד. ויחזיק[והו] לרופא חשוב וזאת היא כונתי. והאל
הוא היודע מה הנכסף בלבי וממנו אשאל יהיה כעזרי ויישירני בהעתקתי,
ויתן לי יכולת להשלים מלאכתי ולהעתיקו הספר האחר עמו ויראני נפלאות
תורתו וישלח לנו את משיחו אמן סלה.

APPENDIX E

הקדמה לאברהם בן משולם אביגדור

אמר אברהם בן משולם בן שלמה (בן) אביגדור בהיותי נער (היו)
תשוקותי תשוקות נער, ומחשבותי מחשבות נער, ותאותי תאות נער,
וחשקתי ללמוד מלאכת הרפואה לא מפני כוסף ידיעת החכמה, אך הספיק
5 לי בשם לבד כדי שאקרא רב, ארויח הרווחים ההם המגונים כמו שהוא היום
מנהג כולל ברוב האנשים המתעסקים במלאכה הזאת ובפרט (וכל שכן ב)
מבני אומתנו. וכאשר גדלתי והגעתי למדרגת האנשים היו תשקותי תשוקות
איש ומחשבותי מחשבות איש ותאותי תאות איש ונתתי אל (את) לבי
לדרוש ולתור ולחקור בזאת החכמה מכל צד לחשקתי בה מצד ידיעת האמת
10 לא לתכלית כלה ונפסד, ואעלה ההר"ה הוא העיר ההוללה מונפשליר לשמוע
חכמת הרפואה מפי נבוני הנוצרים וחכמיהם (מפי חכמי הנוצרים ונבוניהם)
ומצאתי שם ספרים רבים וביאורים מועילים (בשרשי החכמה הזאת ואם
יאריך השם ימי אעתיק מהם קצת. וספרים מועילים) מיישירים במלאכה
ומכללם בחרתי בייחוד (ביארתי) זה הספר הנקרא מבוא במלאכה מועיל
15 ומיישיר מאוד למי שירצה להתנהג במלאכה הזאת באמונה וביושר, והענינים
אשר יכללם הספר הזה כולם מבוארים וגלוים והרפואות הנכללות בו כפי
העת אשר אנחנו בו (והענינים) והאיקלימים והמזגים, יוכל הרופא להרגילם
בבטחון. ולבקשת קצת אוהבי וחברי העתקתים מלשון אדומי אל לשון בני
עמי. האל ישמרני מכל שגיאה ויזכני להעתיקו ולהעתיק ספרים אחרים
20 ויאריך ימי וימי (מעתיקי רפואות דברים כבירים עם) כל ישראל חברים (אמן).
(אמר המחבר) זה הספר נקרא מבוא במלאכה חובר בעבור
המשכילים בחכמת הרפואה ועדין לא הורגלו בשמושה ונבנה יסודו על האופן
הראשון מהספר הרביעי (מאבן סיני) וחברו החכם המעולה ברנרט (ברנט)
אלביריט האחר המיוחד בישיבה הנכבדת אשר בעיר מונפשל"ייר לבקשת
25 אוהביו ותלמידיו וכה (וחביריו וזה) היו (sic) תורף דבריו, לבקשתכם ידידי
אשר עמלתם ויגעתם עמי בלמוד החכמה הזאת (חכמת הרפואות) ראיתי
לחבר לכם ספר קטן ומועיל במלאכת הרפואות אדרוך בו דרך האופן
הראשון מהספר הרביעי מא"ס אתקן בו הרפואות הידועות המועילות (אשר
ראויות) המורגלות לפי דעתי ואני משתחוה (!) אליכם שלא יהיה זה הספר
30 נקל בעיניכם, כי הוא רב התועלת בעיני. ואם ימצא איש (על צד הקנאה)
יורה חיצי סכלותו נגדי ראוי לכם לעמוד נגדו בכל מאמצי כחכם אחר
שרציתי לעמוס עלי זה המשא הכבד בעבורכם ולתועלתכם.

1883, *10*:157–168, on p. 165. We have left the text as Steinschneider had it, including the use of parentheses for brackets.

[e2] The passage is difficult to decipher and translate.

History, Novelty, and Progress in Scholastic Medicine

*By Chiara Crisciani**

RECENT TRENDS in the historiography of scientific thought have stressed how the images of science held in different historical periods have affected the specific researches of both individuals and groups of scientists.[1] For modern historians the expression "image of science" encompasses many components, including notions of the appropriate ethical commitments, social expectations, and mental attitudes of the scientist as well as such concepts as the possibility of progress in science, its proper transmission, and its place among other human activities. These elements bear no direct relation to either the solution of specific scientific problems or the construction of definite theories; rather, they serve to define the kind of problems, conceptual tools, and goals a scientist can choose from, and the permissible scope and structure of the research undertaken.

Analyses of images of science have proved useful for understanding the specific features of some forms of scientific thought that are very different from those of modern science. In the case of Western medieval "science"—that is, *scientia,* rationally ordered, supposedly certain knowledge about the natural world—such features include the closed, albeit dynamic, structure of scholastic learning; the tendency of this system of knowledge to differentiate sharply between speculative truth and operative utility; and its problematic attitude toward the "new" and toward the progressive growth of knowledge.[2] On this last issue —ideas about scientific progress in late medieval and early modern culture— A. C. Crombie, Paolo Rossi, and A. G. Molland have written valuable studies.[3]

* Dipartimento di Filosofia, Università degli Studia di Pavia, 27100 Pavia, Italy.
I should like to thank J. Agrimi and F. Gallone for the help they gave me during this work.

[1] See Paolo Rossi, "Problemi e prospettive nella storiografia della scienza," *Rivista di Filosofia,* 1972, *63*:123–125; Rossi, *Immagini della scienza* (Rome: Editori Riuniti, 1977), esp. pp. 7–10; Thomas S. Kuhn, *The Structure of Scientific Revolutions* (Chicago: Univ. Chicago Press, 1962); and Kuhn, *The Essential Tension: Selected Studies in Scientific Tradition and Change* (Chicago/London: Univ. Chicago Press, 1977).

[2] On this huge subject I will confine myself to citing Alexandre Koyré, "Du monde de l' 'à peu près' à l'univers de la précision," in *Etudes d'histoire de la pensée philosophique* (Paris: Gallimard, 1961); Eugenio Garin, *Medioevo e Rinascimento* (Bari: Laterza, 1966), esp. pp. 29–34, 203–206; John E. Murdoch and Edith D. Sylla, eds., *The Cultural Context of Medieval Learning* (Dordrecht: D. Reidel, 1975), esp. the editors' introduction, pp. 1–30, Murdoch's "From Social to Intellectual Factors: An Aspect of the Unitary Character of Late Medieval Learning," pp. 271–339, and Guy Beaujouan's "Réflexions sur les rapports entre théorie et pratique au Moyen Age," pp. 437–477; Edward Grant, "Aristotelianism and the Longevity of the Medieval World View," *History of Science,* 1978, *16*:93–106; L. M. De Rijk, *La philosophie au Moyen Age* (Leiden: E. J. Brill, 1985), esp. pp. 95–103; Rossi, *Immagini della scienza;* and Rossi, "Aristotelici e moderni: Le ipotesi e la natura," in *Aristotelismo veneto e scienza moderna,* ed. Luigi Olivieri, Vol. I (Padua: Antenore, 1983), pp. 125–154.

[3] Alistair C. Crombie, "Some Attitudes to Scientific Progress: Ancient, Medieval, and Early Modern," *Hist. Sci.,* 1975, *13*:213–230, rev. in *Il concetto di progresso nella scienza,* ed. E. Agazzi (Milan: Feltrinelli, 1976); Paolo Rossi, "Sulle origini dell'idea di progresso," *ibid.,* pp. 37–87; Rossi, *I filosofi e le macchine (1400–1700)* (Milan: Feltrinelli, 1962), Ch. 2, "L'idea di progresso scientifico";

The present essay seeks to apply this type of analysis to the learned medicine of the thirteenth to fifteenth centuries. My aim is to examine whether and how scholastic physicians dealt with the problem of the growth of medical knowledge. How did they construe the changes that had taken place or were taking place in their discipline? Did they interpret those changes as the accretion of new things or as deeper understanding? Did the changes appear to them simply as an unbroken line continuing into the future, or also as a recovery of knowledge once possessed by mankind and then lost? What kind of history had medicine undergone, in their eyes? And what progress could take place? Starting from these questions, I want to show that Latin physicians in the late Middle Ages adhered to an image of science that encompassed ideas of the "new" and of increase by accretion, even though such ideas raise difficulties and require special care. In this image of science, changes were perceived and even advocated, but they were acknowledged as progress in only a limited and particular way.

Consequently, my argument concerns representations (of progress) and images (of science), rather than the analysis of real advances and effective novelties. Unquestionably, such advances and novelties are to be found in the medicine of the period. Physicians gave new interpretations to authoritative texts, dealt with old diseases in new ways, and tried to identify new pathologies. To choose only one example from the present volume, Danielle Jacquart's analysis of the different therapeutic options of three contemporary physicians bears witness to lively movement. I will endeavor to show how such novelties appear in physicians' and surgeons' representations of their discipline and its development, and to point out the criteria used to incorporate the "new" into the framework of already existing notions. As will be seen, the image of medicine offered by these physicians reveals how difficult it was for them either to appreciate advances as such or to insert them into the body of medical knowledge.

The texts on which my discussion is based range from the end of the thirteenth century to the middle of the fifteenth, come from different regions of Europe, and fall into several genres of medical writing. Obviously, over such a long period and such a wide scope learned medicine was neither unchanging nor monolithic. There are, for example, significant differences between the commentaries on works of ancient medical authorities written by Taddeo Alderotti at Bologna in the late thirteenth century, the treatise on pharmaceutical theory produced by the Montpellier master Arnau de Vilanova around 1300, the manual of surgery written by Henri de Mondeville at Paris in the early fourteenth century, and the *practica* (a general survey of practical medicine) written by Michele Savonarola at Padua in the third or fourth decade of the fifteenth century. However, these texts are, in one important respect, alike: all are connected in a more or less direct way with the institutional organization of medical knowledge that took place in the universities and that often goes under the name of medical scholasticism. Commentaries were produced directly for, or in close connection with, university teaching. The compendia known as *practicae* are texts whose purposes were halfway between teaching and professional practice; the examples I

---

and A. G. Molland, "Medieval Ideas of Scientific Progress," *Journal of the History of Ideas,* 1978, *39*:561–577. The broader topic of the idea of progress in medieval culture seems still worth examining. See also Guy Beaujouan's report on research to be undertaken on this theme, in *Bulletin de Philosophie Médiévale,* 1988, *30*:20–36.

shall examine were all written in Latin by physicians who were university trained and university teachers. And although surgery was not taught in all universities and certainly not taught exclusively in universities, the Latin surgical texts considered here were intended precisely to promote, for surgery too, a style of didactic transmission shaped explicitly along the lines established in universities for medicine.[4]

Hence, notwithstanding the differences between universities, the regional or national trends that can be perceived in late medieval medicine, and the length of time spanned, these texts are markedly homogeneous, at least in their more general features, owing to the influence of the institutional university setting.[5] Indeed, all these authors agreed precisely on the image of medical science, which was for them a *scientia* fully mastered only by the learned, divided into two parts (theory and practice), and structured in epistemological levels ranging from the most general explanations down to the most particular interventions. All of them referred to many of the same canonical authoritative texts and used the same techniques and rules in expounding their knowledge. Most important, all of them shared the epistemological criteria and naturalistic world view offered by scholastic Aristotelianism. And finally, all of them maintained, if in a somewhat stereotyped way, the image (deeply rooted in Christian anthropology) of medicine as a providential gift bestowed by God on mankind.

Yet within this generally stable cultural context, there were varied views about the conditions and ways of growth of medical knowledge. For instance (to look at two writers active in the first two decades of the fourteenth century), Pietro Torrigiano's opinion on the increase of medical knowledge differed from Henri de Mondeville's ideas about the improvement to be pursued in surgery, as will become apparent. In my view, such differences do not depend mainly on the specific formation, place, or time of each author; rather, attitudes to the "new" and to the possibility of progress vary according to the epistemological levels and literary genres of scholastic medical knowledge. In other words, these attitudes appear different depending on whether we examine commentaries, compendia of practical medicine, surgical writings, or texts on pharmacological therapy.

I. THE LATE MEDIEVAL VIEW OF THE HISTORY OF MEDICINE

Medieval Christian anthropology regarded human nature, including corporeal nature, as determined by sin. Because of the Fall mankind lost immortality, among other gifts, and therefore entered time, knew death, and lost physical perfection. Thus weakness, frailty, progressive decay of the body, and diseases are "natural" (that is, normal) features of fallen man. In medieval culture, this picture

[4] See Jole Agrimi and Chiara Crisciani, *Edocere medicos: Medicina scolastica nei secoli XIII–XV* (Milan/Naples: Guerini, 1988), esp. pp. 5–20; also Nancy G. Siraisi, "Some Recent Work on Western European Medical Learning, ca. 1200–ca. 1500," *History of Universities,* 1982, *2*:225–238; and Siraisi, "Reflections on Italian Medical Writings of the Fourteenth and Fifteenth Centuries," in *History and Philosophy of Science: Selected Papers,* ed. Joseph W. Dauben and Virginia Staudt Sexton, *Annals of the New York Academy of Sciences,* 1983, *412*:155–168.

[5] On the durability of the university setting, the interlacement between continuity and change in early Renaissance medicine, and the relation between notions of progress and real change in the medicine of that period see Andrew Wear, Roger K. French, and Iain M. Lonie, eds., *The Medical Renaissance of the Sixteenth Century* (Cambridge: Cambridge Univ. Press, 1985), esp. the editors' introduction.

affected the ways in which medicine was considered. Indeed, medicine, along with other techniques, was viewed (at least from the twelfth century) as a gift of God—that is, a tool that enables human beings somehow to remedy the state of bodily necessity by which they are burdened in their earthly life.[6] This view was also a commonplace with physicians themselves, as the following remark by Arnau de Vilanova (d. 1311) shows: "I hold that medicine is a science [*scientia*] that was established by Divine Providence to be of help to men against the hardship and necessity that mankind met because of the transgression of the first father."[7]

Medicine thus entered the world and time along with disease and thereafter could have a history. For medicine was not conceived as an instant and total gift of a miraculous power, but rather as the donation of a capacity to develop a form of wisdom—as a possibility that entailed and required coming into being. The conditions of medicine's rise in time (sin, gift) are outside of time, but it unfolds as a discipline in the course of time, undergoing periods of latency or scientific flowering. However, in the frequent summaries provided by scholastic physicians of the vicissitudes of their discipline, the times of God, of myth, and of the medical sects of antiquity were not always arranged in linear succession.

Let us consider, for example, Nicolò Bertucci (who worked in Bologna and died in 1347) and Nicolò Falcucci (who worked in Florence and died in 1412). Each wrote a sort of encyclopedia of practical medicine at a time when scholastic medical knowledge was well consolidated in terms of its organization in branches and the acquisition of authoritative texts, but before the historical and critical requirements of Renaissance medical humanism had been established.[8] Consequently, these two authors could and did assume a broad thesis about medicine and its development. Both of them opened their works with a kind of general evaluation of medicine (under the heading *De commendatione medicinae*) that included discussions of its history.[9]

Falcucci, starting from the divine gift ("for God revealed many things about this science to his prophets and his men"), thought that medical knowledge had originated simultaneously among the wise men of all peoples: "Indians and Latins, Persians and Greeks . . . and Arabs and Jews." From them, however, derived only the principles of the art, which were subsequently developed and articulated in the texts of prestigious authors and in the chains and layers of

[6] On this theme see L. M. De Rijk, "Some Notes on the Twelfth-Century Topic of the Three (Four) Human Evils and of Science,Virtue, and Techniques as Their Remedies," *Vivarium,* 1967, *5*:8–15.

[7] Arnau de Vilanova, *Contra calculum,* in Arnau, *Opera omnia* (Basel, 1585), col. 565FG; see also Pietro d'Abano, *Conciliator* (Venice, 1565), *differentia* 9, fols. 14Db–15Gb.

[8] Nicolò Bertucci, *Collectorium medicine utilissimum . . . Bertucii bononiensis . . .* (Lyons, 1518) (hereafter **Bertucci, *Collectorium***); and Nicolò Falcucci, *Falcutii Nicolai Sermones medicinales* (Pavia, 1481–1484) (hereafter **Falcucci, *Sermo***). References to Falcucci's *Sermones* are to the copy owned by the Pavia University Library, in which the type of numeration varies from section to section and is absent in some sections; in the following references, I either use the numeration that is available or refer to the text by *Sermo* and chapter.

[9] On historical accounts of the development of medicine in antiquity see Wesley D. Smith, "Notes on Ancient Medical Historiography," *Bulletin of the History of Medicine,* 1989, *63*:73–109. See also Brian P. Copenhaver, "The Historiography of Discovery in the Renaissance," *Journal of the Warburg and Courtauld Institutes,* 1978, *41*:192–214. For useful hints about the ideas of cultural development held during the Middle Ages see Edouard Jeauneau, "Nani gigantium humeris insidentes," *Vivarium,* 1967, *5*:79–99; and Gregorio Piaia, *Vestigia philosophorum: Il medioevo e la storiografia filosofica* (Rimini: Maggioli, 1983).

commentaries that constituted the very full—perhaps too rich—medical disciplinary tradition.[10] This tradition was therefore the fruit of the initial gift from God, but also of reason (*ratio*) and personal experience (*experimentum*). These last were the means by which "medicinal science was discovered and subsequently carried forward." The first discoverers, however, did not leave written texts and seem to have limited themselves to *experimenta*.[11]

Bertucci, too, was sure of the very ancient origin of medicine. According to his reconstruction, "the authors of medicinal science were the first of all," and medical research was introduced well before the "principles of other sciences were to some extent found."[12] He also believed that these first researchers were empirics rather than reasoners; in any case, as he and others pointed out, for the most part all that remains of them are luminous but uncertain traces—specifically, their fame, which evaporates in the distance of myth.[13] "Their writings have not reached us," Bertucci hypothesized, "on account of the interruption of study as a result of the Flood."[14]

But the origin of medicine was also considered from another standpoint, one both historical and epistemological. Besides an origin in divine gift, medicine was generated and developed from—or, better, owed its discovery (*inventio*) to—humbler and more prosaic beginnings.[15] As in other arts that arose under the impetus of necessity (both biblical condemnation and daily goad), the discoverers were those who realized, not only through revelations and dreams but also through luck or "example or imitation," the efficacy of certain remedies and procedures.

Iacopo da Forlì (d. 1414), Matteolo da Perugia (d. before 1473), and other fourteenth- and fifteenth-century commentators on the first Hippocratic aphorism insisted on these themes.[16] That is, they presented the discoverers of medicine as having found out remedies by chance or as having observed and imitated nature, adapting to human needs some instinctive forms of self-healing practiced by animals. Such discoveries became customary practice: repeated because of their efficacy, they were passed on from generation to generation through usage and became immutable. The remark of Hugh of St. Victor (d. 1141) that "all the sciences were first in usage before they were in art" is valid for medicine as well.[17] From this viewpoint, then, as Pietro Torrigiano remarked, adducing Aristotle in his support, the origins of any art, including medicine, are "rough and

[10] Falcucci, *Sermo* 1, fols. 2ra, 2rb; and see also Giovanni Matteo Ferrari da Grado, *Practica* (Pavia, 1497), fol. 2rb.

[11] Falcucci, *Sermo* 1, Ch. 8, "De inventoribus seu auctoribus medicinae," fols. 6vb–7vb.

[12] Bertucci, *Collectorium*, fol. 2ra; see also Pietro d'Abano, *Conciliator* (cit. n. 7), *differentia* 1, fols. 3Aa–4Cb, and *differentia* 5, fols. 9HFa–bEF.

[13] See, e.g., Gasparino Barzizza, *Oratio habita in funere Jacobi de Turre Foroliviensis* . . . , in Barzizza, *Opera*, Vol. I (Rome, 1723), p. 24; and Iacopo da Forlì, *Medicina scientia preclarissima*, as ed. in Agrimi and Crisciani, *Edocere medicos* (cit. n. 4), pp. 263–273.

[14] Bertucci, *Collectorium*, fol. 2rab. There is also a specific reference to the Flood in *Practica Petrocelli Salernitani*, in *Collectio Salernitana*, ed. S. De Renzi (Naples, 1852–1859), Vol. IV, p. 188.

[15] The theme of *inventio* is dealt with particularly in the *De sectis* of Galen and, above all, in the commentaries on the Hippocratic *Aphorisms*, beginning with Galen's own. See also *La préface du De medicina de Celse*, ed. and trans. P. Mudry (Lausanne: Arts et Metiers, 1982), pp. 53–54.

[16] See Iacopo da Forlì, *In Afforismos Ypocratis expositiones* . . . (Padua, 1477), *quaestio* 1, fol. 2rab; and "Matheolus of Perugia's Commentary on the Preface to the Aphorisms of Hippocrates," ed. and trans. Pearl Kibre and Nancy G. Siraisi, *Bull. Hist. Med.*, 1975, *49*:426–427.

[17] Hugh of St. Victor, *Didascalicon* 1.12, ed. C. H. Buttimer (Washington, D.C.: Catholic Univ. Press, 1939).

boorish," and every science "in its beginning is crude and immature." Thus Torrigiano noted that the ancient authors of medicine "ordered into books . . . only the accidents that they sensibly grasped in the patient, only . . . numbering illnesses."[18]

In the eyes of medieval medical writers, the fruits of divine bounty, the fabulous wanderings of myth, the repetitive fixity of usage, and the rough improvements—concrete but only accidental—due to sense perception constituted a sort of prehistory of medicine; they were perceived as the preconditions of its origin as *scientia*. Whether medicine was considered to arise from below through empiricism and usage, or to descend from on high through revelations, all accounts of its history—presented as an itinerary at once temporal and epistemological—hailed an author in whom the two directions flowed together and in whom a decisive epistemological shift took place at a precise moment. This was Hippocrates, "first finder of medicinal science," as Bertucci called him,[19] discoverer par excellence.

Hippocrates incarnated, as it were, the divine generosity; he is a personified gift. Pietro d'Abano (d. 1316) stated that "God, mercifully caring for mankind, may have created Hippocrates as someone without flaws to perfect medical knowledge." Pietro was here recalling an ancient belief about the special role of Hippocrates, and many others would repeat this praise. Giovanni Matteo Ferrari da Grado (d. 1472) expressed it thus: "Since human infirmities had become so many, God Almighty, driven by mercy to mankind wasted by such diverse diseases, brought Hippocrates to light."[20] Hippocrates was a man but a special one, for he concluded the mythical genealogy of the discoverers of medicine, who were held to be, and venerated as, gods or demigods. It was he who, after the Flood—or at least after that period of interruption of about five hundred years in which the art lay dormant[21]—called medicine back to light. Above all, he was the author par excellence—that is, "the most ancient writer of this art," the first whose works have reached us. Pietro d'Abano, Bertucci, and Falcucci all stressed the special place and role of Hippocrates.[22] Thus the proper origin of medicine as *scientia* was represented as linked from the beginning to the presence of texts, and *scientia* was seen as strongly related to the activity of writing down what had been discovered.

The historical and epistemological break that took place with Hippocrates was identified as the transformation of medicine from discovery, usage, and art as yet without rules to *scientia,* doctrinal discipline, scientific art, and tradition. This was the opinion held by Taddeo Alderotti (d. 1295) and Pietro d'Abano, among

[18] Pietro Torrigiano, *Plusquam commentum in Parvam Galeni artem* . . . (Venice, 1557), fol. 5H; his reference is to Aristotle, *Metaphysics* 1.10, 993a15. See also Matteolo da Perugia, *De laudibus medicinae,* in *Tre orazioni nuziali* . . . , ed. A. Messini (Rome: Istituto Grafico Tiberino, 1939), p. 40; and Pietro d'Abano, *Conciliator* (cit. n. 7), *differentia* 1, fol. 3Cb.

[19] Bertucci, *Collectorium,* fol. 1rb.

[20] Pietro d'Abano, *Conciliator, differentia* 1, fol. 3ABa; and Ferrari da Grado, *Practica* (cit. n. 10), fol. 2ra.

[21] On the latency of medicine see also Pliny, *Naturalis historia* 29.2; Celsus, preface to *De medicina* (cit. n. 15), pp. 17, 58; and Isidore of Seville, *Etymologiae* 4.3.

[22] Pietro d'Abano, *Conciliator, differentia* 1, fols. 3Ba, 3Db, *differentia* 3, fol. 6Bb (here he refers to Galen, *De colera nigra*); Bertucci, *Collectorium,* fol. 2rb; and Falcucci, *Sermo* 1, fol. 7vb. Among others in antiquity, Celsus had already stressed the role of Hippocrates with particular emphasis; see his preface to *De medicina* (cit. n. 15), pp. 17, 63–67.

others.[23] Although they referred to Aristotle, who described the difference between art and *scientia* as an epistemological distinction bearing no necessary relation to any historical event, they thought that Hippocrates' work (and thus a particular moment of history) implemented the shift in medicine from art to *scientia*. In their view, those who came before Hippocrates had made use, as it chanced, sometimes of *experimentum*, sometimes of reason, and had trusted fortuitous examples; investigation was disorganized and unsystematic; the art appeared *in fieri* and "in motion, since then the habit [of scientific knowledge] was not perfected and lay quiet in the soul."[24] Some of these characteristics were still present in the work and procedures of Hippocrates, in that he was described as using logic by instinct and without having learned it, perhaps through divine will and inspiration. Nevertheless, with him there supposedly took place a shift from random procedures to the cogency of logic, from the fluidity of oral communication to the stability of written transmission, which thus established a disciplinary tradition.

From Hippocrates then—and without any more interruptions—medicine as "an almost perfect science" was introduced and the continuity of an equally "perfect tradition" began, as Pietro d'Abano and Iacopo da Forlì attested. All other physicians, Hippocrates' successors, received this legacy, which, in the view of late medieval Latin medical writers, could truly be called *scientia*: an ordered stillness, an articulated but structural whole of levels of knowledge, methods, doctrines, and techniques of transmission. According to Nicoletto Vernia (d. 1499), in the place of investigation that "at one time before it was reduced to science was extremely inconstant," an utterly constant discipline was imposed; according to Bertucci, this discipline took the lead among the others by "its certitude of knowledge or efficacy of demonstrating," both of which made it stable ("fixed and immobile").[25] In a sense, at the very moment of the appearance of medical *scientia* in history, with Hippocrates, its history seems to end. A unique temporal break separates a period of nonhistory of the science (when in fact the latter is not yet fully established) from another period in which it is perhaps equally impossible to talk of history of the science, given that what now begins is the history of its transmission. The task of expounding and transmitting *scientia* predominates over discovery; *scientia* comes to coincide with both doctrine and tradition. In the "historical" descriptions mentioned above and in many other passages, it is no accident that the accounts provided contrasting pictures of Hippocrates and his successor Galen: the former was termed discoverer and finder, the latter defined as teacher and considered the model of the faithful commentator.[26] Together, the two are obviously *principes medicine* and the very

[23] "Medicina dicitur ars dum est in fieri, et scientia dicitur dum est in facto esse": Taddeo Alderotti, *In C. Gal. Micratechnen commentarii* . . . (Naples, 1522), fol. 6va. "Alii vero iam fere ab eo perfectam accipientes, ut sic plus eam habent scientiam appellare": Pietro d'Abano, *Conciliator, differentia* 3, fol. 6BCb. See also Ferrari da Grado, *Practica*, fol. 2va.

[24] Aristotle, *Posterior Analytics* 2.19, 100a. The use made by scholastic physicians of this passage deserves further investigation; see Agrimi and Crisciani, *Edocere medicos* (cit. n. 4), pp. 40, 151–153. The same passage of Aristotle is cited by Ferrari da Grado, *Practica*, fols. 2rb–2va; and by Nicoletto Vernia, "Quaestio an medicina . . . ," in *La disputa delle arti nel Quattrocento*, ed. Eugenio Garin (Florence: Vallecchi, 1947), p. 116.

[25] Pietro d'Abano, *Conciliator, differentia* 3, fol. 6Bb; Iacopo da Forlì, *Medicina scientia* (cit. n. 13), p. 269; Vernia, "Quaestio," p. 116; and Bertucci, *Collectorium*, fols. 1rb–1va.

[26] See, e.g., Torrigiano, *Plusquam commentum* (cit. n. 18), fol. 5E; "Matheolus of Perugia's Commentary" (cit. n. 16), p. 426; and Iacopo da Forlì, *Medicina scientia* (cit. n. 13), p. 269.

greatest authors, since, as Alderotti pointed out in praise of his own role as a commentator, "science is upheld in two ways: in one way from full discovery, in another way from its explanation."[27] Equal in dignity and authority, they embrace two diverse roles in the "historic" rhythms of medical science: Hippocrates closes the series of author-discoverers; Galen begins that of the author-commentators.

II. CONCEPTS OF TRADITION AND PROGRESS IN MEDICAL COMMENTARIES

A great many learned medical texts are commentaries—that is, expositions of tradition. Did the commentators believe that their exegeses of the texts of the authors entailed changes, accretions, and progress? They were certainly aware that the medical textual tradition had undergone changes and that in this sense it had a history. First of all, the medical authors—those who had consigned wisdom and truth to written texts—had succeeded one another in time; therefore they were often presented in a chronological series that alluded also to doctrinal connections between them or to regional groupings.[28] Moreover, it was recognized that such texts had been transformed over time, owing to successive geographic and, consequently, linguistic "translations." Indeed, the commentators were well aware of the many facets and problems of the medical *translatio,* a word that means moving both from one nation to another and from one language to another. A case in point is Taddeo Alderotti's vigorous criticism of Constantinus Africanus as a translator.[29] In any case, medical authors and texts, coming from different epochs, countries, and languages, were innumerable. As Bertucci put it, "For no other science is found written with the pen of so many writers."[30]

The antiquity of the early medical writings together with their great number made tradition appear a secure guarantee of the dignity, stability, and scientific continuity of medicine. But the very multiplicity of the texts could make the tradition itself appear, as in Falcucci's words, a *congestio*[31]—that is, a mass of material that is, at first sight, rich but confused, disordered, and not directly usable. To late medieval Latin medical writers, it seemed as if the truth, which is in principle transparent and luminous and certainly contained in these authoritative texts, was rendered implicit and opaque precisely because of the means by which it had been transmitted through history. Therefore truth had to be recovered using the pure light of exposition. Alderotti stressed the indispensable role of the commentator: "Written *scientia* that is not irradiated with explanation does not perfect the understanding."[32] The tradition thus involves two textual levels: that of the works of the authors, and that, even richer, of the many expositions by the commentators, composed to give a clear and current interpretation

[27] Taddeo Alderotti, *Expositiones in arduum Aphorismorum Ipocratis . . .* (Venice, 1527), fol. 1ra (referred to hereafter as "commentary on the *Aphorisms*").

[28] Such is the presentation in Bertucci, *Collectorium,* fols. 1ra–2va. See also Ferrari da Grado, *Practica* (cit. n. 10), fol. 2rb; and Falcucci, *Sermo* 1, fol. 2va.

[29] Alderotti, commentary on the *Aphorisms,* fols. 1ra, 40va. For a similar criticism of incorrect translations see Arnau de Vilanova, *Explicatio super canonem vita brevis,* in *Opera omnia* (cit. n. 7), col. 1681E. (This is the same work cited in the article by Michael McVaugh under the title *Repetitio super canonem vita brevis;* the variation in title is found in different early editions of Arnau's works.)

[30] Bertucci, *Collectorium,* fol. 2rb.

[31] Falcucci, *Sermo* 1, fol. 2rb.

[32] Alderotti, commentary on the *Aphorisms* (cit. n. 27), fol. 1ra.

of the authors' words. Commentaries too, in their turn, are disposed in genealogical strata, always recalled by each new commentator as essential instruments for his own work of analysis of the text of the author.

It is well known that scholastic exposition, while always defined as "faithful," can in reality be aggressively interpretative when confronting authority, in medicine as in other disciplines.[33] Medical commentators' interventions consist not only of corrections of earlier interpretations (such corrections, in turn, being then expected and invited for the commentator's own work of exegesis), but also of *addenda, additamenta, ampliationes,* or *complementa*—active additions to the text of the commented author. These terms occur, for instance, in the commentaries by Ugo Benzi (d. 1439) on the *Aphorisms,* and those by Ferrari da Grado and Giovanni Arcolano (d. 1458) on the *Nonus Almansoris.*

As Torrigiano noted in his commentary on Galen's *Techne,* the work of the commentator really involves "not only explication but application . . . as the need occurs." Torrigiano did not intend to confine himself to repeating the content of the Galenic text; rather, "often digressing, we have interposed some material . . . sometimes indeed revealing [Galen's meaning], but sometimes providing confirmation with a more detailed account of the causes, and sometimes adding extrinsic material."[34] As a result, his commentary would earn the title *Plusquam commentum,* and Torrigiano would earn fame for the "*new* and *unheard-of* opinions" therein expressed.[35]

Innovations, additions, complements: do these entail a concept of cumulative progress? If they do, it is a special concept of progress that must be interpreted in the framework of the rather complex interaction among authority (*auctoritas*), truth (*veritas*), and explication (*expositio*) that defines scholastic exegesis.[36] Certainly the relation between author and commentator is not one of simple repetition; it involves conscious ambivalence in which respect is intertwined with interpretative force. The commentator is well aware of his necessary function (without which the author would be dumb) and often describes his work in terms of an accretion ("I add," "I shall abbreviate," "I complete"). He knows he is introducing innovations. But he does not intend to represent his contribution as a cumulative outgrowth of new knowledge or as a change from the inner truth of theory. He and the author are in fact linked by the same atemporal, immutable value of truth—truth that is certainly embedded in the author's text but that the vicissitudes of history may have obscured. The commentator finds the truth again and brings it back to light.

Once again the dominant idea—at least at the level of representation, which is our present concern—is that of the actualization of an already defined potential, the recovery of a truth that has been given but that, with time, has become hidden. This truth will be explained, clarified, steadily deepened by generations

[33] See Agrimi and Crisciani, *Edocere medicos* (cit. n. 4), Chs. 3, 4, and bibliography cited there.

[34] Torrigiano, *Plusquam commentum* (cit. n. 18), fols. 10D, 1BC.

[35] Filippo Villani, *Vite di illustri fiorentini,* in *Croniche di Giovanni, Matteo e Filippo Villani* (Trieste, n.d.), p. 438.

[36] On this topic I will confine myself to citing A. J. Minnis, *Medieval Theory of Authorship* (London: Scholar Press, 1984); Garin, *Medioevo e Rinascimento,* esp. pp. 203–206; De Rijk, *La philosophie au Moyen Age,* esp. pp. 95–103 (both cit. n. 2); and Franco Alessio, "Conservazione e modelli di sapere nel Medioevo," in *La memoria del sapere: Forme di conservazione e strutture organizzative dall' antichità a oggi,* ed. Paolo Rossi (Bari: Laterza, 1988), pp. 118–120.

of commentators succeeding one another in history, reaching with continual effort toward an objective that appears predetermined and definite and that neither changes nor grows. Changes, deviations, and gaps that time or the limits of the author himself may have introduced do not affect the truth itself but only the extent to which it is either hidden or explicit. In such a representation the innovations introduced by the commentator appear as deepenings of interpretation of a given truth. In sum, it is a matter of *promotio veritatis,* which in this context, it seems to me, should be read as unveiling, explanation, and above all diffusion of truth.

Thus, in the medical textual tradition, the image of *scientia* formed by the process that began with Hippocrates and continues through the work of the "proven authors" and "approved expositors" seems to be that of a well-ordered system, definite in its extent and methods, its texts, and its fundamental theories. But the knowledge in question is not immobile, sterile, or purely repetitious: it shows variations, internal dynamics, and possible decadence and deviation, and it must favor development. Within it there are different epistemological levels and strong crosscurrents. However, its dynamism occurs within a system that is presented as already substantially laid down, defined once and for all, closed. Medical knowledge is mobile, certainly, but in essentials not modifiable.[37]

III. PROGRESS IN COMPENDIA OF PRACTICAL MEDICINE

I mentioned earlier an epistemological as well as a historical development of medicine. In medieval medical writings one finds an epistemological arrangement in which the "historical" phases correspond approximately to the various hierarchical levels of scholastic medical knowledge: theoretical *scientia,* practical *scientia,* art, *experimentum,* and intervention. Many scholastic medical authors outline this scheme, expressed very clearly by Falcucci, when dealing with the problem of the relationship between the theoretical and the practical parts of medicine.[38] The line along which these levels are organized appears for the most part hierarchical and unidirectional from high to low (that is, from theoretical reasoning to operative interventions), in contrast to the historical line, which—as we have seen—was described as a movement from usage and empiricism toward *scientia.* Nonetheless, however these authors describe medicine's past, the "historical" process appears duplicated and systematized within an epistemological dynamic that is rich in transitions and relations, but relatively atemporal, that is, independent of time.

A closer consideration of the role of discovery (*inventio*), to which medieval commentators on the first Hippocratic aphorism often give special attention,[39] enables one to understand the effect of this duplication better. For these

[37] On these themes see Chiara Crisciani, "Exemplum Christi e sapere: Sull'epistemologia di Arnaldo da Villanova," *Archives Internationales d'Histoire des Sciences,* 1978, *28*:245–292, esp. pp. 276–279. Also see Andrew Wear, "Galen in the Renaissance," in *Galen: Problems and Prospects,* ed. Vivian Nutton (London: Wellcome Institute for the History of Medicine, 1981), esp. pp. 241–244; Rossi, "Aristotelici e moderni" (cit. n. 2); and Molland, "Medieval Ideas of Scientific Progress" (cit. n. 3).

[38] Falcucci, *Sermo* 1, fols. 5rb–5va.

[39] See, e.g., "Matheolus of Perugia's Commentary" (cit. n. 16), pp. 426–428; see also Torrigiano, *Plusquam commentum* (cit. n. 18), fol. 32rv; Petrus Hispanus, *Commentarium singulare . . . super librum dietarum universalium Isaac* (Lyons, 1515), fol. 20r–v; and Falcucci, *Sermo* 1, fol. 7va.

commentators "discovery" came to assume at least two meanings. In the first, it referred to the transition of medicine toward *scientia*—an episode that took place once and for all when the rich and varied incoherence of discovery was incorporated into established doctrine and discipline. The other meaning referred to a lively dynamism, to the multiplicity and diversity of forms of acquisition of data and knowledge that—within this framework of established and unmodifiable doctrine—occurred continuously in medicine at the level of operative art and intervention, on the model of that unique historical event. Thus, what is recorded by "history" is always repeated in epistemology. This can also mean that when systematizing the epistemological relation between reasoning and *experimentum*, or between *scientia* and art (*ars*), care is taken to root this systematization in a particular moment in the discipline's past; the correctness of this system is, as it were, guaranteed by explaining its "historical" genesis.

As I noted above, this epistemological systematization entailed setting epistemological levels within medical knowledge. In fact, from the end of the thirteenth century authors often described medical knowledge as a diversified system ranging from the most general theoretical explanations to the most particular forms of operational intervention.[40] The part of medical knowledge that dealt most closely with practical intervention was defined as "practical part," "operational *scientia*," "scientific art." Indeed, it partook of features of *scientia*, being rational knowledge. But its rationality was not destined to remain abstract. On the contrary, it was shaped as a set of rules, directions, and regular recurrences that were intended to lead to practical intervention and useful operations. Discussions about the relationship between these two parts of medicine often stressed that medicine as art depended on medicine as *scientia*. Physicians believed, however, that their knowledge was bound to be simultaneously rational and useful, and they acknowledged that this level of medicine as operational art was, owing to its proximity to experience, necessarily more open to the lively but disordered richness of discovery (*inventio*). It seems as though this epistemological level had preserved those features of flexibility and mobility typical of the "prehistory" of medicine as pictured in the "historical" accounts of its development.

Latin works on practical medicine of the thirteenth to fifteenth centuries constitute a very wide and diverse genre:[41] they include treatises on diagnosis and therapy, collections of clinical cases and recipes, large treatises (*summae*, or practical compendia) about diseases and therapies organized from head to foot, and texts halfway between commentary and compendium. In the last class we find *practicae*, commentaries (produced in university teaching) on authoritative canonical texts on practical aspects of medicine—for instance, the *Nonus Almansoris* by Rhazes; in these works the commented text is likely to be clarified by "modern" additions.

[40] See Nancy G. Siraisi, *Taddeo Alderotti and His Pupils: Two Generations of Italian Medical Learning* (Princeton, N.J.: Princeton Univ. Press, 1981), esp. Ch. 5; Per-Gunnar Ottosson, *Scholastic Medicine and Philosophy* (Naples: Bibliopolis, 1984), esp. Chs. 1, 2; and Agrimi and Crisciani, *Edocere medicos* (cit. n. 4), Chs. 1, 2, 5.

[41] On this genre see Luke Demaitre, "Scholasticism in Compendia of Practical Medicine," *Manuscripta*, 1976, *20*:81–95; Michael R. McVaugh, "Two Montpellier Recipe Collections," *ibid.*, pp. 175–181; Andrew Wear, "Explorations in Renaissance Writings on the Practice of Medicine," in *Medical Renaissance*, ed. Wear, French, and Lonie (cit. n. 5), pp. 118–145; and Agrimi and Crisciani, *Edocere medicos* (cit. n. 4), Chs. 5–8.

The ways in which these texts on practical medicine are built seem explicitly cumulative. Authors of *practicae* declare by the very titles of their works their intention to collect and aggregate: the practical works of Guglielmo da Brescia (d. 1326) and Iacopo Dondi (d. 1359) are both called *Aggregator,* and Bertucci entitled his work *Collectorium.* The writers also announce their intention to add to information collected from texts whatever insights and information the experience of many people and the expertise of professional practitioners may offer. For instance, in the *Thesaurus pauperum,* Petrus Hispanus (d. 1277) claims that he wants to collect faithfully what he has been "able to find in books of ancient physicians and masters and of moderns who make use of experience [*experimentatores*]." In the *Breviarium* attributed to Arnau de Vilanova, the author claims he wants to collect "all that he has experienced and whatever in his age he has seen experienced by any master, or even simple men and women, and by empirics."[42] Such writers propose to collect this information according to an order that makes it rationally explicable and easily memorizable—that is, functional in professional practice.

It should be emphasized that "collecting" as used by the authors of the *practicae* means selecting and bringing together not only hitherto-unknown information but also information that is hard to find, either because it is scattered through too many lengthy and disorganized texts or because it has not yet been organized in written form.[43] Here there does seem to be acceptance of a kind of "progress," in the sense of integration and even accumulation of innovations, data, cases, and recipes that are expected continually to enrich the doctrinal patrimony of medicine. But since the authors of the *practicae* believed that the transition from a situation that was plural and *in fieri* to the unitary order of science and doctrine *in esse* had already taken place in medicine, they perceived anything brought in from below as a sprouting of new discoveries and data that could not modify the scope of the discipline or the definite state of theory. The "new" could confirm and enrich theory, but to succeed it needed to be controlled, transferred from the realm of causal to that of doctrinal information through specific criteria.

The criteria that serve to select among what has been experienced or reported are reason and authority.[44] In this way a compatibility between the "new" and what has already been acquired is established, and often enough the new is neutralized, in the sense that it is made homogeneous with, and hence assimilable to, what is already attested. Thus recipes or accounts of clinical cases are more likely to receive endorsement if, in addition to the witness of personal experience, similar instances can also be found in an authoritative text, or if they appear explicable by valid arguments, or if they are accredited by the consensus of a community of experts. The reciprocal and circular reinforcement of these criteria of selection reduces the progressive or cumulative significance of collecting. In fact, what is accepted is that which is already known by other means, and what is accumulated is that which already has a predetermined epistemological space to which to return.

[42] Petrus Hispanus, *Thesaurus pauperum,* in Petrus Hispanus, *Obras medicas,* ed. Maria Helena Da Rocha Pereira (Coimbra: Univ. Coimbra, 1973), p. 79; and Arnau de Vilanova, *Opera omnia* (cit. n. 7), col. 1055C.

[43] E.g., Bertucci, *Collectorium,* fol. 1ra; and Ferrari da Grado, *Practica* (cit. n. 10), fol. 2rb.

[44] As one example see Falcucci, *Sermo* 1, fols. 6va–7ra, on the relation between these criteria.

So that novelties may be more easily accepted, an institutional point of view is often put forward in compendia of practical medicine (here I am using "institutional" in a very broad sense). This form of integration and control of the "new" takes place through the organization of the allegedly new data and discoveries into written and sanctioned form. For instance, Arnau de Vilanova claims that "it is necessary that every researcher write down what has been discovered. . . . Indeed, in this way adding something to what has been already found is easier than always finding again all that needs to be known."[45] Furthermore, this process makes the data homogeneous, publicly transmissible and teachable like the patrimony of knowledge that medicine already possesses. In other words, if data, discoveries, and recipes keep their absolute and unrelated singularity, they remain empirical and, as such, irrational, useless, and constantly in need of rediscovery. By contrast, *experimentum*—the valid experience of an individual practitioner—is defined as such to the extent that it can enter into a network of relations with reason and authority, into a series with other *experimenta,* and into texts elaborated according to the established criteria of collecting.[46] Things that are not susceptible of being treated in this way may well be "new" (or, better, "wonderful"—*mirabile*), but they remain outside *scientia*. The therapies and the casual and ritual formulas of old wives and rustics that are not translated into written *experimenta,* the procedures to which no rational cause is assigned, the revelations that cannot be subsumed under reason, and the successes obtained solely by chance and luck remain on the plane of empiricism and resemble archaisms.[47] They are amorphous; that is, they are not part of that historical and epistemological dynamic that would render them significant in terms of *scientia* and thus capable of being integrated and used.

The texts of practical medicine are undeniably open to the acquisition of new findings, even if they also pay the usual attention to the ancient authors. The importance ascribed in these texts to collaboration among modern practitioner-authors and to the role of the community of experts in legitimating the findings of its members, generation after generation, bears witness to these writers' sense that medical knowledge can grow, that progress can occur. However, the representation of medical knowledge as a whole and the criteria according to which this integration of the "new" takes place suggest that acquisition of new findings is considered more an epistemological transformation than an accretion of data. Moreover, although the greater closeness to the level of experience and intervention that is typical of medicine as art entails an acknowledgment of lively change, the change is perceived as occurring within a closed system—the system that is defined once and for all by the theories, the methods, and the texts of medicine as *scientia*.[48]

[45] Arnau de Vilanova, *Explicatio* (cit. n. 29), col. 1679 (but see also cols. 1679–1703). The necessity of writing down what has been experienced is also stressed in the *practicae* and especially in the commentaries on the first Hippocratic aphorism.

[46] See McVaugh,"Two Montpellier Recipe Collections" (cit. n. 41); and Agrimi and Crisciani, *Edocere medicos* (cit. n. 4), Ch. 8, section 3.

[47] See Pearl Kibre, "The Faculty of Medicine at Paris: Charlatanism and Unlicensed Medical Practice in the Later Middle Ages," in Kibre, *Studies in Medieval Science* (London: Hambledon Press, 1984), no. 13, pp. 1–20; and Jole Agrimi and Chiara Crisciani, "Medici e *vetulae* dal Duecento al Quattrocento: Problemi di una ricerca," in *Cultura popolare e cultura dotta nel Seicento* (Milan: Franco Angeli, 1983), pp. 144–159, esp. pp. 149–150.

[48] For this peculiar concept of dynamism in *scientia* in the Aristotelian tradition see, among others,

IV. CHANGE AND PROGRESS IN SURGICAL TEXTS

According to some of the authors under discussion, all branches of medicine do not follow the same temporal rhythms in their emergence, even if all follow, it would appear, the same historical and epistemological scheme. Thus, Falcucci likened medicine in its entirety to a tree "in which there are at once flowers and fruits, some green and some ripe," a strange tree in which different time patterns coexist, grafted together into the formal synchrony of a single model.[49]

The picture of the development of surgery presented by learned Italian and French surgeons writing in Latin between the mid-thirteenth and mid-fourteenth centuries certainly fits Falcucci's description. I will consider here principally Bruno Longoburgo or Longobucco (d. ca. 1286), Teodorico Borgognoni (d. 1298), Lanfranco (fl. 1290–1296), Henri de Mondeville, and Guy de Chauliac (d. 1368), and the surgical portions of some of the *practicae*.[50] Almost all these authors describe a sort of "history" of surgery, trying to reconstruct its genealogy—its origins, vicissitudes, and divisions of opinion.[51] The care taken by these authors in reconstructing the past of their discipline was a first step toward raising its cultural dignity, although they knew much more effort would be required.

In fact, these writers on surgery recognized with devastating clarity the immaturity of surgery as a discipline. They suggested that perhaps it had only recently fallen into this miserable situation. According to Guy de Chauliac, its decline had taken place, not because of a divinely caused natural disaster or a mythical centuries-long latency, but because it had become separated from the vigorous and central trunk of *physica;* in itself, this separation was necessary, but it had been brought about in imprudent and unproductive ways.[52] From Hippocrates to Avicenna surgery and medicine (*physica*) had been united, but later, "on account of either carelessness or excessive engagement with curing, surgery was separated [from medicine] and left to the hands of mechanics"; as a result, many forms of intervention "which belong to surgery" are now "on account of our pride and disdain left to barbers and women."[53]

At the same time, and associated with this separation, which meant surgery's degradation toward empiricism, was the loss of the essential textual basis of the

---

Francesca Calabi, "La crescita del sapere scientifico secondo Aristotele," *Atti della Accademia delle Scienze di Torino,* 1976–1977, *111*:169–199; Mario Vegetti, *Il coltello e lo stilo* (Milan: Il Saggiatore, 1979), esp. pp. 88–89; Patricia Reif, "The Textbook Tradition in Natural Philosophy, 1600–1650," *J. Hist. Ideas,* 1969, *30*:17–32; Charles B. Schmitt, "Experience and Experiment: A Comparison of Zabarella's View with Galileo's in *De motu,*" in Schmitt, *Studies in Renaissance Philosophy and Science* (London: Variorum, 1981), no. 8, pp. 80–138; and Rossi, "Aristotelici e moderni" (cit. n. 2), esp. pp. 141–149.

[49] Falcucci, *Sermo* 1, fol. 5rb.

[50] The texts used are *Cyrurgia Guidonis de Cauliaco et Cyrurgia Bruni; Teodorici; Rolandi; Lanfranci; Rogerii; Bertipaliae* (Venice, 1498) (hereafter cited by the individual author's name and a short title); and Henri de Mondeville, *Die Chirurgie des Heinrich von Mondeville . . . ,* ed. Julius Leopold Pagel (Berlin, 1892).

[51] See, e.g., Bruno da Longoburgo, *Cyrurgia magna,* fol. 83ab; Mondeville, *Chirurgie,* pp. 135, 137–139; and Guy de Chauliac, *Cyrurgia,* fols. 2vb–3rab.

[52] Guy de Chauliac, *Cyrurgia,* fol. 2vb.

[53] Lanfranco, *Cyrurgia,* fol. 168va; Bruno, *Cyrurgia magna,* fol. 83ab; Guy de Chauliac, *Cyrurgia,* fols. 2vb–3rab. See also M. C. Pouchelle, *Corps et chirurgie à l'apogée du Moyen Age* (Paris: Flammarion, 1983), esp. pp. 27–28; and Tiziana Pesenti, " 'Professores chirurgie,' 'medici ciroici,' e 'barbitonsores' a Padova nell'età di Leonardo Buffi da Bertipaglia (m. dopo il 1448)," *Quaderni per la Storia dell'Università di Padova,* 1978, *11*:1–38.

discipline. According to Falcucci, "Its science perished . . . and all that remained of it were brief descriptions," for the most part badly written and incomprehensible.[54] This was not surprising, since practice had passed into the hands of illiterates, idiots, and rustics. Nor was it surprising, therefore, that the transmission of such notions (scarcely describable as knowledge) had become a kind of teaching carried on purely by rote, in closed family circles, often with jealous precautions of secrecy. This was the dreary picture drawn by late medieval surgeons of surgery's immediate past.[55]

Two characteristics emerge from these historical accounts. First, the events in surgery's past were set entirely in recent times—in historic, almost contemporary times—with no mythical or divine background. Surgery was treated as a specialization that was at once newer and more primitive than other branches of medicine, at least in its present and ascertainable configuration. And second, surgery's course was portrayed as downward, its history one of successive interrelated losses that had brought it back to that prescientific phase in which medicine as a whole had been before Hippocrates. Usage, static and repetitive, dominated; authoritative texts were absent or filled with lacunae; the discoveries of empirics remained as such, casual and not reduced to writing. It was true that recent surgeons of importance and their disciples had progressively corrected the most serious faults (this was, for instance, Henri de Mondeville's opinion),[56] but much remained to be done. It was now necessary to bring surgery (or bring it back) to a form of organization proper to *scientia*.

This is the thrust of the insistent appeals in some fourteenth-century Latin surgical works, notably those of Henri de Mondeville and Guy de Chauliac, for an instruction in the art that, without abandoning the hands-on training more indispensable in this than in other branches of medicine, would be based on sanctioned texts and on a preliminary knowledge of the general principles of medicine. This is the meaning of the urgent requests that the practice of surgery, precisely because it was tied to manual activity, should be directed by a body of rules and canons, certainly not as stringent as those of theoretical medicine, but stable because rational and written—and thus, it was claimed, able to direct a truly efficacious practice. This is what lies behind the criticism and refutations that Guy de Chauliac, for example, directed at the many empiric remedies that abounded in earlier surgical treatises, at the many blunders that made certain works "brutish," and at the "fables" of his thirteenth-century predecessors Ugo da Lucca and Teodorico Borgognoni.[57] Such crude archaisms would find no space in Guy's text; instead, he would make use of access to new translations of works of Galen, which had enriched surgical doctrine. And finally, the same desire for the organization of surgery as *scientia* lies behind the anxiety—particularly evident in Henri de Mondeville[58]—to make sure that whatever was acquired by direct experience, by assiduous reading of texts and tireless study, and

[54] Falcucci, *Sermo* 7.1; he refers here frequently to Albucasis.

[55] Henri de Mondeville strongly criticized hermetic forms of training, and so, by implication, did Teodorico Borgognoni. See Mondeville, *Chirurgie* (cit. n. 50), pp. 12, 39; and Teodorico, *Cyrurgia*, fol. 166ra.

[56] Mondeville, *Chirurgie*, p. 138.

[57] Guy de Chauliac, *Cyrurgia*, fols. 2vb, 3ra, 13rb, 15vb, 69rb.

[58] See esp. Mondeville, *Chirurgie*, pp. 124–125, 332–333.

through reasoning should be systematized according to a definite order and committed to writing.

The transition of surgery from prehistory to history, along with its corresponding accession to the dignity of *scientia*, is indeed taking place in these texts; the transformation is driven by a vigorous impulse that appears—here too—directed toward a predetermined goal, a model to be striven toward and arrived at but not improved upon. The personal formation of Teodorico Borgognoni, as he himself describes it, is symbolic of this attitude. For a certain time he saw and followed the really very expert cures of his father and master, Ugo, but he could not "comprehend or learn [them] to the full"; only when he turned to the texts of Galen could he give a rational and doctrinal sense to the things that he had learned through usage. Thus, Teodorico perceived—in a significantly closed circle of usage and tradition—that his father's particular interventions were to be found in Galen, but there joined to explanatory reasons and in a scientific systematization provided by the text of the ancient author.[59]

These descriptions of the "history" of surgery also serve to confirm several themes I have insisted upon above. The development of medical knowledge is represented as filled with movement and life, but it is also, as it were, a blocked development in the sense that it is directed toward predetermined goals following a stable scheme in which the dynamics of epistemological relations dominate and direct supposedly historic events. Only one event is really allowed: the transition from nonscience to science—and vice versa, as in the case of surgery, which was once a science, was subsequently degraded into usage and empiricism, and is "now" acquiring (which simultaneously means recovering) disciplinary dignity. One may note that the more surgery emerges as a new and relatively autonomous discipline, the more it imitates, at any rate in formal writings on the subject, the disciplinary structure of medicine.[60] It seems, indeed, as if the development of potential cannot take place except according to an already realized model or in imitation of a preexisting form.

Once a definitive systematization has been established, the lively integration of new material (which does not, however, alter the configuration of the discipline) appears in surgery too. Indeed, the surgical authors under discussion claimed that now, after having taken steps to restore the dignity of the discipline as *scientia*—for instance, through writing texts and standardizing the training of future surgeons—it was necessary to deal with problems and cases of which the ancients were not aware and which the empirics could not even bring into focus. Thus, it was the "new" that must be dealt with, as we can perceive from the following words of Henri de Mondeville: "It seems absurd and almost heretical to believe that the glorious and sublime God gave Galen such a sublime ingenuity with the pact that nobody after him would find anything new. . . . Has God not given each of us, like Galen, his own natural ingenuity? In fact, our ingenuity would be really poor had we always to resort to already found things."[61] The

[59] Teodorico, *Cyrurgia*, fol. 106ra.

[60] See, e.g., Mondeville, *Chirurgie*, pp. 70–71. In this passage Mondeville defines the epistemological status of surgery, taking up the same questions traditionally discussed in defining medicine as a discipline: whether it is *scientia* or art, whether it is theoretical or practical, and so on.

[61] Mondeville, *Antidotarium*, a treatise appended to his *Chirurgie*, on p. 508. The treatise was appended to supply surgeons in particular with information on drugs.

statement is typical of Mondeville's strong emphasis on the improvements that the "moderns" were invited to pursue in medicine.

However, although the usual criteria—authority, reason, experience—for the integration of new material are certainly valid for surgery, the relation between them appears more uncertain and flexible in surgical texts than in the compendia of practical medicine: in the surgical treatises, they sometimes reinforce one another and sometimes stand in contrast. This state of affairs occurs precisely because surgery as a subject of medical knowledge is less systematized, very close to intervention (*opus*), and endowed with a somewhat exiguous tradition. The choice in integrating new material becomes easier when, for instance, an *experimentum* or an instrument is upheld by reasoning or when texts and personal experience agree, but this happens only rarely.

Thus it is not unusual to find very diverse opinions: alongside someone who repudiates a procedure because, he says, he has proved its inefficacy by experience, one finds others who recommend it because a prestigious author (Albucasis, for instance) records it with approval. The process of choosing what can be embodied in surgery as a discipline is therefore uncertain and oscillating, but also dynamic and open. By the same token, the dialectic between old and new, between tradition and progress, is extremely lively in surgical writing, precisely because of the involvement of surgeon-authors in the work of raising the dignity of their discipline.

Many pages of Henri de Mondeville are representative of this dialectic.[62] His work is rich in expressions of admiration for the progress the moderns are making in surgery by continually adding cases and forms of intervention unmentioned by the ancients. There is also bitter and disgusted criticism of the excessive confidence in the ancients and the attachment to custom that all patients and a good many of his colleagues display.[63] But however suggestive these remarks may be, and certainly they permit us to recognize here a specifically cumulative meaning of progress, one ought not to forget the context in which Mondeville's really vehement innovative thrust was placed. In fact, alongside his enthusiasm for the "new," we find an equally obsessive demand for order, for the enclosure of the otherwise uncontrollable riches of innovation within the compactness of a rational and written system (which he himself intended to provide with his own text).[64]

V. ANCIENTS AND MODERNS IN PHARMACOLOGICAL LITERATURE

Although compilations of medical recipes constitute a separate genre of medieval medical literature, knowledge about drugs was considered part of medicine as an operative art; therefore treatises on practical medicine contain pharmacological sections. In what follows, I shall consider works devoted exclusively to pharmacology as well as the pharmacological portions of such general works as the

[62] This attitude of Mondeville's has many facets and deserves fuller analysis than space permits here. His most interesting pages on tradition and progress are to be found in *Chirurgie*, pp. 11–12, 61, 69, 138–143, and esp. 144; and in *Antidotarium*, pp. 507–508 (where he recalls the metaphor, subsequently also used by Guy, of the moderns as dwarfs on the shoulders of giants), and p. 511. See also Pouchelle, *Corps et chirurgie* (cit. n. 53), esp. pp. 16–17, 52.

[63] Mondeville, *Antidotarium*, p. 507.

[64] Mondeville, *Chirurgie*, pp. 332–333.

Conciliator of Pietro d'Abano, the *practicae,* and books on surgery. "Pharmacology" is here treated as a partially separated subject in order to underline some aspects of the field (for instance, the relation between the idea of decadence of the human organism and the consequent unreliability of ancient prescriptions) that seem particularly relevant to our central topic.

The uncertainty, the flexibility, and a readiness to accept coupled with a sense of alarm in the face of the "new" that we noted above are even more evident in the case of pharmacology, which seems one of the most anomalous and troubled branches of medicine. In the first place, and in contrast with surgery, it is doubtful whether we can even say that the discipline was institutionalized or in the process of becoming so, whether in terms of teaching or practice. Certainly medicine was traditionally divided into *diaeta, potio,* and *chirurgia* (that is, cures by diets, drugs, and instruments), and in this sense knowledge about drugs was always considered a consistent disciplinary field. But in the context of university teaching and the corporative organization of the medical profession in the late Middle Ages, the status of pharmacological knowledge and the recognition of professional competence in this area seem fluid and variable.

Pharmacology nevertheless relates, in both its merits and its limitations, to some well-established aspects of the disciplinary image of medicine already referred to. Above all, knowledge about drugs was grounded in a specific divine gift: God had infused herbs with therapeutic virtues for the benefit of fallen mankind. Moreover, the remedies supposedly found in the "prehistory" of medicine by discoverers through chance (*inventio*) and by animals through instinct were for the most part pharmacological. Also, the contemporary customary practices that learned physicians denounced as deplorable and attributed to old wives, rustics, and empirics mostly concerned pharmacological remedies. In this instance it seemed more urgent than ever to develop these gifts, to pass from usage to doctrine, and to channel casual empiricism and rote performance of practices that were sometimes even superstitious into rational rules.

And this, in fact, the ancient authors were considered to have done: authoritative texts on this subject were not lacking. But thirteenth- to fifteenth-century learned physicians found these texts scarcely usable for a variety of reasons, as attested in their own writings on pharmacological subjects. Besides terminological difficulties—more troubling and dangerous here than in other subject areas—a major problem was that so many authors had written on this subject and it was very difficult to reconcile one with another. This was especially true with respect to ingredients and doses, given that each author referred primarily to the herbs, drugs, and measures in use in his own time and region. The difficulties of translation that resulted were real, not limited to terminology, and were virtually insurmountable by late medieval physicians. A further obstacle was that some pharmacological texts were excessively schematic and compressed, either because they followed a pattern of hermetic precepts or, more prosaically, because they had been produced in the context of daily contact with pharmacists, disciples, and colleagues and were thus merely abbreviated summaries to facilitate memorization.

Late medieval interpreters appear to have been deeply sensitive to another basic fault. They noted that when ancient authors made an attempt to impose a more complex and scientific pharmacological theory, for example, the theory on the procedure of assigning "degree" to medications, the result was delusory: the promises of Galen and Dioscorides, the founders of this theory, remained only promises; and now scholars wandered uncertainly, unable, in the words of Pietro d'Abano, to "find an authentic author with whose position they agree."[65] To these scholastic and theoretical difficulties, which diminished the force of authority, were added considerable practical difficulties as well. Medieval physicians and pharmacists were aware that the different geographical origins of the texts meant that they often described unrecognizable herbs and irreproducible recipes, and that the ancients' prescriptions were of uncertain applicability to modern illnesses. Guglielmo da Saliceto (d. ca. 1280) compiled a list of prescriptions found in Avicenna that, although useful in Avicenna's time and place, were in Guglielmo's own time and place of little efficacy or even harmful.[66] Furthermore, and more important, in the course of history infirmities had multiplied, diversified, and become complicated to an infinite extent (as, e.g., both Ferrari da Grado and Michele Savonarola asserted)[67] through the progressive physiological decadence of mankind caused by the Fall, so that it was now impossible to use for current ills remedies designed for individuals who had been much stronger and more robust, and perhaps also simpler. Some fourteenth-century physicians therefore held that the only safe course was to deal very carefully with the "dogmas of the ancients" or, even better, to avoid them altogether. According to Arnau de Vilanova, "Many remedies of the ancient physicians were healthful, which are now deadly, just because of the deficiency of human nature. Thus let the modern physicians be careful, because the revolutions of the world, separating the ages, distort the statements of the ancients." And according to Falcucci, "The physicians must distinguish the statements of the ancients according to the flow of the ages. In fact, nature does not remain unchanged, but lapses into deficiency from time to time." But, as he and others warn, care is needed with modern remedies too: "Do not, O physician, be fond of modern remedies."[68]

In truth, in this field it was necessary to distrust everything and everyone: ancients and moderns, the vulgar and the learned. In this situation, with no rules accepted by the majority[69] and with no authors on whom it was possible to rely completely or whose instructions were directly usable, the ancients suggested remedies that seem occult because they had become incomprehensible, the vulgar proposed irrational empiric remedies, and the moderns prescribed fanciful and inharmonious preparations of which they boasted complacently.[70] And yet to

[65] Pietro d'Abano, *Conciliator* (cit. n. 7), *differentia* 138, fol. 195Ca; see also the anonymous text edited in Michael R. McVaugh, " 'Apud Antiquos' and Medieval Pharmacology," *Medizinhistorische Journal,* 1966, *1*:18. On the pharmacology of the late Middle Ages see the exhaustive introduction by Michael R. McVaugh to Arnau de Vilanova, *Aphorismi de gradibus,* in Vol. II of *Arnaldi de Villanova Opera medica omnia,* ed. L. García-Ballester, J. A. Paniagua, and M. R. McVaugh (Granada/Barcelona: Univ. Barcelona, 1975).

[66] Guglielmo da Saliceto, *Summa conservationis* (Venice, 1490), fols. a5va, b4va.

[67] Ferrari da Grado, *Practica* (cit. n. 10), fol. 2rb; and Michele Savonarola, *Practica maior* (Venice, 1559), dedication, unnumbered folio.

[68] Arnau de Vilanova, *Contra calculum* (cit. n. 7), col. 1568EF; and Falcucci, *Sermo* 1, fol. 10rab, quotation in *Sermo* 2, fol. a6va.

[69] See, among others, Savonarola, *Practica* (cit. n. 67), fols. 11vb, 14ra.

[70] There is a great deal of testimony that could be quoted here. See, e.g., Lanfranco, *Antidotarius* (the fifth *tractatus* of his *Cyrurgia magna* [cit. n. 50]), fol. 206va; Arnau de Vilanova, *Aphorismi de gradibus* (cit. n. 65), p. 196; Savonarola, *Practica,* fols. 9ra, 13ra; Guy de Chauliac, *Cyrurgia* (cit. n. 50), fols. 51va, 56ra; and Antonio Guainieri, *Opus preclarum ad praxim . . .* (Pavia, 1518), fol. 243ra.

have faith in oneself, that is, in only one person, was considered virtually impossible. In fact, claimed Bruno Longoburgo, if what someone, a single operator, did were correct, "the ancients . . . would have already said [it] in their books." If someone disagreed completely with the most famous ancients, his ideas would be rejected, "even though," in Arnau de Vilanova's words, "the reason for the error of his position is not in any way apparent; in fact, it is more likely that this one person, given that he was . . . equal to them in skill and experience, is deceived by false imaginings than that the ancients erred."[71]

Well-balanced texts such as Arnau de Vilanova's *Explicatio* and the first *Sermo* of Falcucci provide at least some suggestions, and very strong cautions, on how to orient oneself in this unreliable and dangerous terrain so as not to fall into "error and labyrinth," as Teodorico Borgognoni put it. The recipes of the vulgar may be collected, according to Arnau, provided "they are not manifestly repugnant to reason"; the practitioner who adopts them must remember, however, that they have a regional validity and origin, so that they can be exported and extrapolated only with difficulty.[72] But in this subject not everything found in books, even those written by ancient authors, will be able to satisfy the physician immediately. Therefore, in Falcucci's view, the modern physician "ought for a long time to think over the *experimenta* of the ancients"; he must not insist only on written texts, but "with skillful ingenuity he must satisfy the propensity toward the new . . . for he is of wretched ingenuity who uses only things that have already been discovered."[73]

And yet just at this point, where it appears indispensable to "relinquish custom,"[74] where the push toward innovation is a recognized necessity and is, moreover, founded directly on medical-anthropological theorizing about the decadence of the organism, their very manner of writing betrays the scholars' uncertainty, their sense of insecurity and alarm: lists of opinions ("some . . . others . . . I") proliferate, or series of contrasting views about a drug or a cure are simply reported and not resolved. Insecurity is also manifest in the way the criteria of reason, authority, and experience seem no longer able to sustain one another. It is expressed in continual oscillation, often by the same author, between mistrust of and lingering faith in the recommendations of the ancients; and between irony and caution, not without expectancy, toward the suggestions of the moderns.[75]

All this seems to confirm that, even when it is necessary to accumulate innovations, it is difficult to undertake the task, given the absence of just that closed and guaranteed disciplinary framework within which alone integration appears

[71] Bruno, *Cyrurgia magna* (cit. n. 50), fol. 91ra (the statement concerns both surgical and pharmacological treatment); and Arnau de Vilanova, *Explicatio* (cit. n. 29), col. 1682FG.

[72] Teodorico, *Cyrurgia* (cit. n. 50), fol. 106ra; and Arnau, *Explicatio,* cols. 1680–1681AD.

[73] Falcucci, *Sermo* 1, fols. 10ra–10va. For wavering estimations of the value of the pharmacological prescriptions of the ancients see the following works by Arnau de Vilanova: *Antidotarium,* cols. 385E–386EFD; *De vinis,* col. 588C; *Speculum,* col. 62C; *De simplicibus,* col. 379D; *Contra calculum,* col. 1568A; and *Breviarium,* col. 1291E; all in *Opera omnia* (cit. n. 7). Also, Falcucci, *Sermo* 2, fol. a6va; and Savonarola, *Practica* (cit. n. 67), passim.

[74] For shifting attitudes toward what is customary see, e.g., Bernard de Gordon, *Lilium medicinae* . . . (Venice, 1498), 5.8; Teodorico, *Cyrurgia,* fol. 117ra; Mondeville, *Chirurgie,* p. 144 (both cit. n. 50); and Savonarola, *Practica,* fol. 9vab.

[75] Many pages of the *practicae* of Savonarola and Guainieri show this attitude. See also Falcucci, *Sermo* 1, fols. 10ra–10vb.

possible: there appears to be no accepted norm or form according to which innovation can be processed. However, norms and forms of a different type emerge. These are not merely disciplinary but also social, having to do with the scientific-professional community and the competence of the experts who are its members. In the case of pharmacology, perhaps even more clearly than in that of operative medicine, the community of experts and rational practitioners plays an important role. In fact, the community's approval, secured by the circulation within it of suggestions for remedies and recipes, appears in many cases the most secure criterion, simultaneously flexible and relatively authoritative, on which an individual can rely. Such a stress on the role of the community implies a further insistence on the necessity of the written word—since only if the new discoveries are written down can they be evaluated, tried out, and corrected by colleagues—as well as a decisive emphasis on the tasks of collaboration and exchange among the members of the group.[76]

Of course, the values of collaboration, openness, control of innovation, readiness for correction, publicity, and circulation of results are often associated with the concept of progress in a branch of knowledge.[77] But despite the clear presence of these values in late medieval descriptions of pharmacology and surgery and of practical medicine more generally, there is a very special idea of progress—shown, for example, in the writings of Henri de Mondeville. Mondeville stressed the idea of progress tied to exchanges between generations and within communities of experts.[78] But, we must note once more, this progress is closed. In fact, Mondeville compared the community of experts to a flock of sheep in a sheepfold; the sheepfold certainly has a door, but it is also strongly encircled by fences and defenses.[79]

VI. CONCLUSION

I have sought to show that scholastic medicine was characterized by a concept of increment, but one that was not progressive in the sense of continual accumulation; by the indefinite integration of new material, but within a closed doctrinal and disciplinary framework; and by a lively dynamism that paradoxically did not involve radical changes. As a final clarification and example of these ideas, it may be useful to recount two metaphors with which some scholastic medical authors described the way their knowledge developed.

One of these metaphors is organic and agricultural. Hippocrates, as repeated by many authors following Galen, "was a good farmer who first sowed this

[76] On these themes, more developed analyses can be found in Luke Demaitre, "Theory and Practice in Medical Education at the University of Montpellier," *Journal of the History of Medicine and Allied Sciences,* 1975, *30,* 103–123; Demaitre, "Scholasticism in Compendia" (cit. n. 41); McVaugh, "Two Montpellier Recipe Collections" (cit. n. 41); and Agrimi and Crisciani, *Edocere medicos* (cit. n. 4), esp. pp. 185–188, and Ch. 8, section 3.

[77] See Rossi, "Sulle origine"; and Molland, "Medieval Ideas of Scientific Progress" (both cit. n. 3).

[78] Mondeville, *Chirurgie,* pp. 12, 61, 64, 67, 69–70, 124, 138–139, 332–334. Similar orientations can also be found in Arnau de Vilanova, *Explicatio* (cit. n. 29); and Savonarola, *Practica* (cit. n. 67). "The sense of intellectual community and intellectual exchange" and "the sense of progress and discovery" in the development of medical teaching and research can be perceived also in the work of Tommaso del Garbo, as Katharine Park points out in *Doctors and Medicine in Early Renaissance Florence* (Princeton, N.J.: Princeton Univ. Press, 1985), esp. pp. 206–209.

[79] Mondeville, *Chirurgie,* pp. 64–65.

art";[80] his seeds produced rich harvests of knowledge because someone later irrigated them and made them grow. Like every organic metaphor, this one carried the idea of movement and growth, of development in time. Usually, in such metaphors, old age or decadence is said to follow; it is no accident that the authors of the texts that have been discussed do not dwell on this, being content, it seems to me, to point out the necessity of growth (which had, however, already taken place). But this growth cannot go beyond the organic form that was predetermined and potential in the seed. Otherwise, the organism that is thus well developed would lose its physiognomy and identity.

The other metaphor is architectonic and artisanal. To indicate the necessary accretions and improvements called for in science, Henri de Mondeville and Guy de Chauliac have recourse to the image of a house, a fine building: workers continually labor to repair things that are broken, to replace parts, and to straighten the walls, but the house is "already begun and completed."[81] Its structure is definite, and the activities of the workmen modify neither its form and dimensions nor its substantial solidity. Between them, these metaphors epitomize the peculiarities ultimately inherent in any conception of medical progress held by scholastic authors.

[80] Bertucci, *Collectorium,* fol. 2rb; and Matteolo da Perugia, *De laudibus* (cit. n. 18), pp. 39–40. See also Owsei Temkin, "Alexandrian Commentaries on Galen's *De sectis ad introducendos,*" in Temkin, *The Double Face of Janus and Other Essays in the History of Medicine* (Baltimore/London: Johns Hopkins Univ. Press, 1977), p. 192.

[81] Mondeville, *Chirurgie,* pp. 11, 70; and Guy de Chauliac, *Chirurgia* (cit. n. 50), proem, fol. 2ra.

Theory, Everyday Practice, and Three Fifteenth-Century Physicians

*By Danielle Jacquart**

MORE THAN AT ANY PREVIOUS period, medical authors during the first half of the fifteenth century alluded in their general works to the conditions of their actual practice by describing—besides pathological cases—remarkable events or characters, local habits, or the prevalent mood of their contemporaries. Some of them seemed almost to turn into "ethnologists," observing with curiosity and attention the surrounding world.[1] Their medieval predecessors had occasionally incorporated anecdotes or remarks on the customs of the day, but to a lesser extent. This larger opening onto contemporary life no doubt suggests an evolution in the way physicians viewed their function. The renewal of the genre of the *practica* in the fourteenth century with Bernard de Gordon's *Lilium medicine* or John of Gaddesden's *Rosa anglica* had already shown an intention to connect the theoretical basis of medicine with the particular conditions of everyday practice. At the end of the thirteenth and the beginning of the fourteenth century, scholastic debates had revealed a new approach to the relation between theory and practice which involved an emphasis on the *particularia,* the experience of which is necessary to the part of medicine that cannot be taught but must be acquired through habit (*ars assuescibilis*).[2] The emphasis on the *particularia* was best illustrated by the collections of *consilia,* that is, descriptions of cases written for didactic purpose. In the first half of the fifteenth century this emphasis became stronger: at the same time that Michele Savonarola was composing his *practica,* one of his Paduan colleagues, Antonio Cermisone, wrote hundreds of *consilia* intended to illustrate different parts of Avicenna's *Canon.*[3] The increasing number of such accounts, which were previously used to illustrate theoretical issues, suggests a more acute awareness of the difficulty of applying authoritative general rules to individual cases, and also a closer attention to the human and social context, an awareness and attention that could lead to adaptations in the traditional means of therapy. It can then be asked how physicians, aware of these difficulties, made their choices. My purpose in the present study

* Ecole Pratique des Hautes Etudes, IVe Section, 45, rue des Ecoles, 75005 Paris, France.

[1] See Danielle Jacquart, "Le regard d'un médecin sur son temps: Jacques Despars (1380–1458)," *Bibliothèque de l'Ecole des Chartes,* 1980, *188*:35–86.

[2] On the renewal of the genre of the *practica* see Luke Demaitre, "Theory and Practice in Medical Education at the University of Montpellier in the Thirteenth and Fourteenth Centuries," *Journal of the History of Medicine and Allied Sciences,* 1975, *30*:103–123; and Demaitre, *Doctor Bernard de Gordon: Professor and Practitioner* (Toronto: Pontifical Institute of Medieval Studies, 1980). Among the many studies of scholastic debates on the relation between theory and practice see the recent account of Jole Agrimi and Chiara Crisciani, *Edocere medicos; Medicina scolastica nei secoli XIII–XV* (Naples: Guerini e Associati, 1988), pp. 21–47.

[3] For a list of Antonio Cermisone's *consilia* see Tiziana Pesenti, *Professori e promotori di medicina nello Studio di Padova dal 1405 al 1509* (Padova: Lint, 1984), pp. 72–91.

is to point out the link between theory and practice in three fifteenth-century authors who were also famous practitioners by examining several themes: their ideas about plague, a disease that every physician at this time had to deal with in his practice; their resort to other sciences or techniques, such as astrology, magic, and alchemy; and their account of a common illness with a complex conceptual background—pleurisy.

Although all three were in fact contemporaries, Antonio Guaineri and, even more, Jacques Despars are usually studied by historians in the context of the Middle Ages, while Michele Savonarola is considered a figure of the Renaissance. Savonarola's attachment to the Ferrara court, the relations he established with artists and humanists, and Theodore Gaza's translation of his works into Greek all contribute to confer on him the stature of a Renaissance scholar; the kind of medicine that he professed was nevertheless "medieval." The three doctors had similar careers: all three were university professors and appointed physicians at a court. Antonio Guaineri studied at the University of Padua under Biagio Pelacani da Parma and Iacopo da Forlì. He taught at Pavia from 1412, then at Chieri. He was made physician of the duke of Savoy in 1427, and he died, probably at Pavia, after 1448.[4] Born at Padua in 1385, Michele Savonarola was made master in medicine in 1413 and taught at the university. He became physician at the court of Este in Ferrara, where he died after 1466.[5] Jacques Despars, born at Tournai around 1380, was *magister regens* of the faculty of medicine at Paris from 1411 to 1419. He visited Italy, probably in 1418. Between 1420 and 1450 he lived alternately in Cambrai and Tournai. He was called several times to the court of Burgundy. He died in Paris in 1458.[6] The three physicians' courtly responsibilities led them not only to treat the mighty but also to observe ordinary people, about whom they had to advise the princes. The two Italian authors chose the genre of the *practica,* and it was *in practica* that Savonarola taught at the University of Padua. They both showed some aversion to scholastic debates. Antonio Guaineri apparently did not excel in them and avoided them systematically, although he put some *dubia* (or debate topics) in his medical works. While Michele Savonarola placed *dubia* at the end of each chapter of the *Practica maior,* in the prologue he expressed reservations about logical argumentation:

[4] On Antonio Guaineri's life and works see Ernest Wickersheimer, *Dictionnaire biographique des médecins en France au Moyen Age,* 3 vols. (Geneva: Droz, 1979), Vol. I, pp. 34–35; and *ibid.,* Vol. III, *Supplément,* ed. Danielle Jacquart, p. 23. Guaineri's different treatises on specific topics were often published in complete collections during the sixteenth century; all references here to **Guaineri** are to the edition of **Lyons, 1525.** See also H. R. Lemay, "Anthonius Guainerius and Medieval Gynecology," in *Women of the Medieval World: Essays in Honor of John H. Mundy,* ed. Jules Kirshner and S. F. Wemple (Oxford/New York: Basil Blackwell, 1985), pp. 317–336.

[5] On Michele's life and works see Tiziana Pesenti, "Michele Savonarola e Padova: L'ambiente, le opere, la cultura medica," *Quaderni per la Storia dell'Università di Padova,* 1977, *9/10*:45–101; and Pesenti, *Professori di medicina* (cit. n. 3), pp. 187–196. This study is based mainly on Michele Savonarola, *Practica canonica de febribus* (Lyons, 1560), written before 1439 (cited hereafter as **Savonarola, *Practica canonica***), and on Savonarola, *Practica maior* (Venice, 1547), written between 1440 and 1446 (cited hereafter as **Savonarola, *Practica maior***). A recent edition is J. Nystedt, *Libreto de tute le cosse che se manzano: Un libro di dietetica di Michele Savonarola, medico padovano del secolo XV: Edizione critica basata sul Codice Casanatense 406* (Stockholm: GOTAB, 1982), reviewed by Jole Agrimi in *Aevum,* 1984, *58*:358–365.

[6] Despars's main work, a commentary on Avicenna's *Canon* (Books 1 and 3 and Book 4, Fen 1), was published at Lyons in 1498 (cited throughout as **Avicenna, *Canon,*** and **Despars, *Commentarium;*** the numbers of the two works correspond exactly). On his life and work see Wickersheimer, *Dictionnaire biographique,* pp. 326–327; and Jacquart, *Supplément,* pp. 134–135 (both cit. n. 4).

"I do not hate dialectical arguments, but I consider deserving a heavy punishment those who wish to get the title of physician while burning to show off their long beard in debates of this kind." Although this statement is something of a topos, since Galen had already mocked the sophists' long beard, it nevertheless attests a firm stand on this field.[7] Jacques Despars's main work does not belong to the same genre. A commentary upon Avicenna's *Canon*—which deals with the whole range of theoretical and practical issues—traditionally contains some scholastic questions. Despars, however, restricted them to the exposition of the theoretical outlines of Book 1 and to the general definition of fever in Book 4. The portions devoted to the description of diseases and to treatment have no scholastic questions and are very close to the genre of the *practica* as exemplified by the work of Guaineri and Savonarola.

I. WHAT IS PLAGUE?

It will perhaps appear superfluous to return to the subject of plague: everything seems to have been said about the attitude held by men of learning after the Black Death, about their confusion, their occasional lack of courage, and their hesitation to admit the notion of contagion, even though it was acknowledged by civil authorities and common opinion. Fourteenth- and fifteenth-century physicians wrote much on this subject; twentieth-century scholars have written even more.[8] However, it would be hard to characterize a fifteenth-century practitioner without attempting to discern his attitude toward one of the major problems he had to face.

Plague was a constant presence during the first half of the fifteenth century, in Italy as well as in France. Antonio Guaineri devoted a treatise to the subject; in it he refers explicitly to the epidemic that raged in 1402 (at Pavia?), and to another in 1416 at Chieri. Before writing the treatise, that is, before 1440, he could have seen the disease at Chieri in 1420, at Pavia in 1431, and in Savoy in 1435.[9] Michele Savonarola had even more experience with the plague: it raged at Padua in 1428, and in 1432 and throughout the period it scarcely left the nearby city of Venice. He himself mentions the disease brought by Greeks en route for the Council of Ferrara.[10] Patients died within three days, spitting blood. The author imputed what we can today identify as pulmonary plague to the malignancy of the water that the travelers had drunk at "Trebizunda" (Trebizond). Jacques

[7] Galen, *De differentiis pulsuum* 2.3, in *Claudii Galeni Opera omnia,* ed. K. G. Kühn (Leipzig, 1821–1833), Vol. VIII, pp. 571–576. For a translation of this passage in French see Paul Moraux, *Galien de Pergame: Souvenirs d'un médicin* (Paris: Les Belles Lettres, 1985), pp. 84–88.

[8] Among the fundamental studies, I note here Karl Sudhoff, "Pestschriften aus den ersten 150 Jahren nach der Epidemie des Schwarzen Todes (1348)," *Archiv für Geschichte der Medizin,* 1911, *4*:191–222, 389–424; 1912, *5*:36–87, 332–396; 1913, *6*:313–379; 1914, *7*:57–114; 1915, *8*:175–215, 236–286; 1916, *9*:53–78, 117–167; 1917, *11*:44–92, 121–176; 1925, *17*:35–39, 256–257. See also A. Campbell, *The Black Death and Men of Learning* (New York: Columbia Univ. Press, 1931); A. Coville, "Ecrits contemporains sur la peste de 1348 à 1350," in *Histoire littéraire de la France,* Vol. XXXVII (Paris: Imprimerie Nationale, 1937), pp. 325–390; and J. N. Biraben, *Les hommes et la peste en France et dans les pays européens et méditerranéens,* Vol. I: *La peste dans l'histoire* (Paris/The Hague: Mouton, 1975), Vol. II: *Les hommes face à la peste* (1976). I will refer to other studies on specific topics.

[9] Antonio Guaineri, *De peste* 2.2. My locating of epidemics is based on Biraben, *Les hommes et la peste,* Vol. I, pp. 378–379, 395–396.

[10] Savonarola, *Practica canonica* 9.4.

Despars, who wrote the part of his commentary devoted to fevers between 1441 and 1446, refers to the Parisian epidemic of 1400. He was not yet a physician in 1400, but other allusions show that he practiced in time of plague, undeterred by the risks. He reports how some patients suddenly died after having consulted him; among his habits during epidemics was to have his mouth bathed with vinegar and absinthe "before coming down from my room to examine the infected persons' urines and advise visitors."[11] Plague was present at Paris in 1418–1419, and in Flanders and Artois in 1438–1439—that is, when Despars was living in these regions.

The two Italian authors begin their explanation of plague with a survey of the vocabulary. Several words were used: *pestis, pestilentia, febris pestilentialis, lues, contagium, epidemia, undimia, mortalitas*. The proliferation of synonyms was a major difficulty in fifteenth-century medicine, since the authors based their knowledge on translations made, at different times and in different regions, from Greek, Arabic, or Hebrew. In the case of plague, texts from Latin antiquity were also important. Did all these words express the same meaning? Savonarola distinguished *pestis* from *pestilentia:* "*Pestis* is a venomous disposition, occurring in the air, which causes the disease called *pestilentia.*" Although he enumerated and defined the different words, he minimized the significance of this activity by quoting Avicenna: "We [the physicians] do not decide from words, but from intentions and differences." In the *Canon* this sentence does not concern pestilential fevers, but the inward senses or cerebral faculties. In his commentary upon this passage, Despars claims that in his day terminological confusion was greater than in Avicenna's: for example, according to some authors the *imaginativa* and the *fantastica* constitute a single sense, according to others they have to be distinguished; sometimes the *imaginativa* is described as the faculty that keeps the sensible forms, whereas the *fantastica* composes and divides them, sometimes the contrary is said. He concludes that "there are such discrepancies in the use of words that today it is impossible to speak properly about those faculties." Different words rarely coincide with the conceptual nuances.[12]

Despite this mistrust of words, all three authors paradoxically continued to follow Isidore of Seville's example, that is, to try to discover through etymology the true meaning or power of the word (*vis vocabuli*). Never extinguished during the Middle Ages, the Isidorean tradition seems to have been reinvigorated in the fifteenth century. All three authors' assertions about terminology were partly derivative of earlier writers, but they also reflect a degree of personal opinion. In comparing the etymologies they proposed, one can find both similarities and divergences.[13]

[11] On the Parisian epidemic: Despars, *Commentarium* 1.4.29; cf. Jacquart, "Le regard d'un médecin" (cit. n. 1), p. 65. For description of sudden deaths: *Commentarium* 4.1.4.1; cf. Jacquart, "Le regard," p. 66. For precautions taken by Despars: *Commentarium* 4.1.4.4. The words "infected" and "infection" are used to translate *infectus* and *infectio* with their medieval meaning, that is, "contaminated," without reference to the modern notion of infection. On the ancient meaning see M. D. Grmek, "Le concept d'infection dans l'Antiquité et au Moyen Age: Les anciennes mesures sociales contre les maladies contagieuses et la fondation de la première quarantaine à Dubrovnik (1377)," *Rad Jugoslavenske Akademije*, 1980, *384*:9–55.

[12] Guaineri, *De peste* 1.1; Savonarola, *Practica canonica* 4.1: and Avicenna, *Canon* (and Despars, *Commentarium*) 1.1.6.5.

[13] The following quotations are from Guaineri, *De peste* 1.1, and Savonarola, *Practica canonica* 4.1. The etymologies put forward are based on Simon of Genoa, *Synonyma,* and Isidore of Seville, *Etymologiae* 4.6, respectively.

Contagium comes from *contangendo,* either because it pollutes what it touches, or because from the contact of one, it reaches the others very easily. . . . *Epidemia,* according to Simon of Genoa, is like *superadveniens,* for it comes from above. It is an acute disease affecting a great number of people in a brief *contagio,* or coming from above, i.e., that it seems to have a divine origin . . . . An epidemic is a common disease, mortal or not, touching numerous people in a definite space of time. It is also called *pandimon.*—ANTONIO GUAINERI

Isidore says that plague is a *contagium,* because, after having caught one, it goes quickly to others. . . . I explain *epidemia* in the following way in the hope of agreeing with the authorities and the *vis vocabuli:* it is a poisonous and contagious mutation, having its root above the sky and occurring in the air; it is able to cause in human bodies diverse diseases that are usually fatal. Therefore any poisonous disease that is usually fatal and that comes from the superior world is called *epidemia;* perhaps the ancients understood this, as they frequently asserted that plague is sent by God's wrath. Indeed, *epidemia* comes from *epi,* which means "above," and *dimos,* "people," for coming from above it infects people. . . . *Undimia* differs from *epidemia,* because one comes from an inferior root, the other from a superior root—MICHELE SAVONAROLA

Savonarola emphasizes the heavenly origin of epidemic to a larger extent than does Guaineri. According to both, plague is called contagious because it touches a great number of people in the same period of time.

Despars also quotes the usual etymology for *epidemia* ("coming from above"), but without commenting upon it.[14] He prefers to stress the word *pandimon,* which, according to Oribasius and Simon of Genoa, was used to describe the spread of the same disease among a large number of people. Unlike Savonarola, Despars did not consider an epidemic disease usually mortal: he quotes as an example the *reumata,* which raged twice in his time and which killed only the weakest among the elderly. Although he does not provide any specific explanation for the word *contagium,* Despars, when commenting upon the chapter Avicenna devoted to the nine diseases that could be passed from one man to another, shows clearly what he meant by *morbus contagiosus.*

Avicenna enumerates nine contagious diseases [*morbi contagiosi*]. They are called contagious because they infect the patient's surroundings and because they are said to go from one man to another. It is not the same disease that goes from one to another, for it is impossible for an accident to do this, but it generates a similar disease among those who surround the patient for a long time and are similarly predisposed.[15]

Despars judiciously observes that it is not the disease itself that is transmitted, but its ability to generate a similar one. Like all supporters of contagion in the Middle Ages—including the fourteenth-century Andalusian physicians—he attributed this phenomenon, in the case of plague, to the transmission of putrid vapors that spread out of the patient's body through both breath and pores, infecting the house, the neighbors, the village, and the town, as well as clothes and other objects.[16] Even if they arrived at different conclusions, Savonarola and

[14] Despars, *Commentarium* 4.1.4.

[15] *Ibid.,* 1.2.1.8.

[16] On the Andalusians Ibn al-Khatīb and Ibn Khātima see Manfred Ullmann, *Islamic Medicine* (Edinburgh: Edinburgh Univ. Press, 1978), pp. 86–96. Since pulmonary plague alone is contagious from person to person, while bubonic plague is transmitted to human beings by fleas, it was judicious to mention, as did Despars and several other authors, the possibility of contamination not only by breath but by the contact of things and clothes.

Despars, like almost all their contemporary colleagues, agreed to attribute the start of plague to the corruption of the air. But according to Guaineri, this was not the only possible origin. He saw plague as above all a poisoning: the complexion is altered by "something" that acts by virtue of a quality, a substance, a specific form, or an occult property. Food can also bring the disease: "The immediate cause of plague is a poisonous infection of food or air."[17]

On the subject of the factors that predispose a person to plague Savonarola and Despars go in quite different directions from the same starting point. Following a Galenic line, Avicenna had said that clean bodies are less exposed, whereas bodies that have been weakened, by sexual intercourse for instance, or whose pores have been dilated by too many baths are more vulnerable. This assertion gave Savonarola the occasion to introduce some advice of a moral and religious nature. Cleanness of the body follows purity of the soul: "Since clean bodies are the most protected in time of plague, endeavor at first to purify your soul."[18] Commenting upon the same sentence of the *Canon,* Despars begins by mentioning the noxiousness of public baths. He then quotes from *De differentiis febrium,* in which Galen put forward some of the ideas taken up by Avicenna. Despars summarizes the Galenic passage, dividing it into three propositions: clean bodies do not suffer from pestilential air, one can suppose that there exist in the air some seeds of plague, and bodies filled with superfluities are more exposed. Although he comments in a detailed way upon the first and the third propositions, he contents himself with paraphrasing and quoting the second: "Secundo supponit quod in aere nos circumdante sint aliqua semina pestilentie dicens: *Supponatur autem in exemplo secundum circumdantem nos aerem inferri quedam pestilentie semina.*"[19] This Galenic assumption did not suggest any comment to him, but he did not think it right to omit it, even if it seemed irrelevant to his subject.

The same belief in transmission by air led Savonarola and Despars to different stands on the notion of contagion, mainly because they did not attribute the same role to causes "coming from above." Savonarola asks a question often posed in writings on plague: "Why does it commonly happen that in a town in which

[17] Guaineri, *De peste* 1.1: "Et ideo his visis dico quod pestis est morbus contagiosus immediate a nutrimentorum vel aeris venenosa infectione proveniens. Potest enim humana complexio duobus modis infici venenose. Primo mixti alicuius alteratione a qualitate seu tota substantia vel forma specifica hunc inducendi effectum virtutem habentis. Et talium quedam ab extra, quedam ab intra possunt approximari, de quibus infra in secunda huius principali parte Deo auxiliante faciam mentionem. Secundo ab aeris seu nutrimentorum venenorum infectione. Et hoc potest dupliciter evenire. Uno modo a constellatione seu radiorum stellarum commixtione vel aerem vel nutrimenta naturaliter ad venenosam infectionem disponente. . . . Alio modo ex venenoso eidem permixto vapore. . . . Ex quibus patet immediata causa pestis que est nutrimentorum vel aeris infectio venenosa."

[18] Savonarola, *Practica canonica* 4.3; see Avicenna, *Canon* 4.1.1: "Et preparatio corporum ad illud in quo sumus de passione est ut sint plena humoribus malis, nam munda forsitan non patiuntur ex illo. Et corpora debilitata iterum sunt patientia ex eo sicut illa que multiplicant coitum et corpora dilatatorum pororum humida multe balneationis."

[19] Despars, *Commentarium* 4.1.1; the paraphrase is in roman, the quotation from Avicenna in italic. Avicenna probably relied on this passage of *De differentiis febrium* 1.6, but he omitted, no doubt intentionally, the allusion to the "seeds of plague"; thus he did not mention, as Galen did, the closing of pores, but instead their dilatation, more suited to explain the penetration of corrupted air. Despars recognized Avicenna's source and quotes the Galenic passage in full. On the problem of "the seeds of plague" in Galen see Vivian Nutton, "The Seeds of Disease: An Explanation of Contagion and Infection from the Greeks to the Renaissance," *Medical History,* 1983, *27*:1–34. Nutton shows that this idea was rarely taken up during the Middle Ages. We can note that, despite his wide erudition, Despars never refers to Niccolò da Reggio's translation of *De causis procatarcticis,* another treatise in which Galen mentions these "seeds."

inhabitants are dying of plague, some of the people who are locked up, like monks or prisoners, remain untouched?" He suggests several answers: since the air is thicker in a closed-in space, it is less receptive to infection; the monks' diet is usually good; prisoners eat much garlic ("the theriac of the poor"), which helps to correct the air, and so on. The opposite observation might also bring into question the influence of the superior world: it occasionally happens that cloistered monks or prisoners die, while the rest of the town is safe. Savonarola's explanation is astrological: just as a favorable ascendant is sought to build an edifice or to choose a piece of land, it can be admitted that this cloister or that prison was built under a bad ascendant, which predisposed it to receive the impregnation of noxious air more easily. He concludes that physicians have to try, by their own study or with the help of astrologer colleagues (*socii*), to know the superior cause of the disease.[20] Astrology was thus used not only to explain the starting of epidemics but also to justify the selectivity of infection. Guaineri, too, adhered to this kind of explanation, supplying more information on the precautions to be taken: whereas some recommend waiting forty days before reoccupying a previously infected house, and others (like Savonarola) a period of six months, he himself advised that it be left empty for at least three months, the necessary time for a *triplicitas* of the zodiac.[21]

Despars could not avoid mentioning the astrological cause, since it appeared in every tract on plague. However, he treats it only as a historical phenomenon, described by astrologers. He evokes the primary causes of plague in connection with the Black Death, alluding to the preaching of some theologians who asserted that great epidemics are sent by God, as in the age of the Pharaoh, to punish mankind.[22] As for the astrologers, they maintained that it was a fatal conjunction between Saturn, Jupiter, and Mars that had caused the 1348 epidemics. Despars's opinion was that physicians need not concern themselves with primary causes but only with the immediate ones, by means of which the superior powers act on human bodies. This clear stand fixed impassable borders between the roles and the respective competences of the priest, the astrologer, and the physician.

Their views on causality largely determined the kind of preventive advice our three doctors offered. In Guaineri's view, since plague was mainly a venom, acting through its specific or occult form, it had to be avoided with antidotes, with the whole range of *bezoardica* that he listed. He did not forget the primary causes and recommended prayer to God and the healer saints, such as Saint Anthony, Saint Sebastian, and Saint Christopher. He acknowledged as a proof of the divine origin of plagues the fact that they are sometimes foretold by holy apparitions from the Virgin or the saints, and he remembered that he had heard plague prophesied several months before the first deaths occurred in 1402 and 1415.[23]

Savonarola too recommended fasts and processions and even advised physicians to pray. By this kind of recommendation Girolamo Savonarola's grandfather was not merely expressing religious devotion or resignation; he was keeping in mind the general principles of a scientific medicine that acts on causes. When

[20] Savonarola, *Practica canonica* 9.1–6, quoting from 9.1.
[21] Guaineri, *De peste* 3.3.
[22] Despars, *Commentarium* 1.3.5, 4.1.4.1.
[23] Prescription of prayers to the healer saints: Guaineri, *De peste* 2.4; prophecy: *ibid.*, 2.2.

the causes have a divine origin, the only technique to be used is that of prayer.[24] He nevertheless considered flight the best way to escape the disease, especially when a member of one's family had been touched. In addition to astrological conditions, which affect the whole house, hereditary predispositions have to be taken into account. Savonarola noticed the injustice that the prescribing of flight implies: "Medicine is not a just invention, since it gives more protection to the mighty." Following his predecessors, he took care to deliver some preservative rules addressed to the poor, who were unable to flee.[25]

It was as much through his refusal to integrate celestial causes into medical science as through observation that Despars was led to accept contagious transmission. Flight was not the main advice he gave, nor did he prescribe either prayer or astrological measures. Like all his colleagues, he found himself deploring the physicians' incapacity: whatever they did, they were reduced to shame and slander. Whether they decided to treat with bleeding or with drugs, they attended more often to their patients' death than to their recovery, and "some prattlers would report that one or another has been saved without the physician's help, whereas others rapidly perished because of his advice." Aware that the fight against this disease was beyond the reach of medical art, Despars, instead of turning to celestial powers for help, trusted in *respublica,* in civil authorities. According to him, plague victims should be driven away, in the same way as lepers were, kept out of towns and isolated from healthy people. Such measures should be taken at the beginning of epidemics, before the vast number of victims makes this impossible. The entry of infected people or things into towns should be forbidden, and infected houses or rooms closed, the inhabitants evacuated, and all the implements left behind. Despars quotes as an example the duke of Milan, who had posted guards at the gates of towns in order to prevent the entry of travelers from infected places. It is in fact known that in 1399, Gian Galeazzo Visconti had instituted in Milan a kind of quarantine line with armed guards.[26] Surprisingly enough, it was Despars, a man from northern France, who praised the duke of Milan for his measures, whereas neither of our Italian authors referred to him. Even Pietro da Tussignano, one of the duke's advisers, showed in his *Consilium pro peste evitanda,* written around 1398, less firmness and conviction than Despars toward contagion. Tussignano's treatise permitted astrology to play its part and, much like Savonarola's, noted the predispositions of places according to their astral ascendants.[27] Guaineri, who dedicated his *De peste* to another duke of Milan, Filippo Maria, advised that conversation with people

[24] "Tu itaque ora et sanos et patientes orare instruas": Savonarola, *Practica canonica* 9.3. His treatise on plague in the vernacular also describes plague as a divine punishment: "La peste mandate foe nei populi per la ira de Dio a quelli"; see *I trattati in volgare della peste e dell'acqua ardente di Michele Savonarola,* ed. Luigi Belloni (Milan: Stucchi, 1953), p. 5.

[25] Savonarola, *Practica canonica* 9.3. The treatise in the vernacular also takes into account the social background of patients, mentioning "siropo per i zentilhomini, bevanda per li poveri": *I trattati,* ed. Belloni, p. 16.

[26] The *respublica:* Despars, *Commentarium* 1.2.1.8, cf. Jacquart, "Le regard d'un médecin" (cit. n. 1), p. 67. Physicians' incapacity and civil measures: Despars, *Commentarium* 4.1.4.3; see the passage quoted in Jacquart, "Le regard d'un médecin," p. 82. On the edict of Bernabò Visconti (1374) and the measures taken by Gian Galeazzo see Grmek, "Le concept d'infection" (cit. n. 11), pp. 40–47.

[27] "Quare moniales et incarcerati non moriuntur? Et per rationem participationum aspectus unius loci quia monasterium potest esse edificatum sub tali ascendente quod in uno tempore aer illius loci est magis corruptus quam aer alterius loci. Ideo quando unus moritur in una habitatione, alii debent recedere tanto citius quanto sunt generati ex uno patre." Pietro da Tussignano, *Consilium pro peste evitanda,* in Iohann von Ketham, *Fasciculus medicine* (Venice, 1500).

coming from infected places be avoided and thought that the *gubernatores reipublice* should forbid their subjects to talk to such people at close range. Shortly after this remark, lost among many others, he showed that he was very far from the notion of contagion when he recommended that plague victims be transported to healthy places, if possible to a town that had not been touched by epidemics, so that they could breathe an air that had not been poisoned.[28]

II. ASTROLOGY, MAGIC, ALCHEMY

The painful experience of the plague epidemics, which revealed the shortcomings of medicine with striking clarity, no doubt played a decisive role in the adoption by some practitioners of techniques borrowed from other arts. The resort to astrology, magic, or alchemy—always a temptation for doctors—seemed more necessary than ever, and led at times to the abandonment of rationality. Guaineri, Savonarola, and Despars represent, in the different positions they took on these three outside resources, the uncertainties and difficult choices that their profession faced.

The part the three physicians assigned to astral causes in the outbreak of plague reflects to some extent their differing attitudes toward astrology. The *impressiones celi* could intervene in their explanations as primary causes, that is, causes by definition exterior to the human body. Jacques Despars repeatedly stressed that physicians had to be concerned only with bodily causes, through which primary causes act. Similarly, attention should be paid only to the nearest causes, and not to remote ones.[29] His rejection of astrology did not ensue solely from this hierarchy within causality. The applications of astrological data to medicine did not seem to him very convincing. The determining of good times for bleeding gave him the opportunity to describe the different levels of involvement in medical practice that in his time the use of astrology could attain.[30]

According to Despars, among those who observed the celestial dispositions in order to determine the time for bleeding, many paid special attention to the phases of the moon. Several views were put forward: according to some, the first quarter was the most favorable time because of the predominance of hot and wet, both qualities of blood; according to others, it was the third quarter, during which the humors were supposed to be neither constrained nor boiling; still others followed the adage "The old moon seeks for the old, the new one for the

[28] Guaineri, *De peste* 2.1. Guaineri points out that even conversation with healthy persons who come from infected places can pass on the disease, as the closeness of sulfur can cause fire. Differentia 3.3: "Et si in aliqua propinqua villa in qua sanus aer existit infirmus statim accederet, vel faceret se portari, maxime si in illa domo nullus alius ex peste fuerit defunctus, securius adhuc esset."

[29] Despars, *Commentarium* 1.4.20: "Celum remotissima causa est dispositionum humanarum et medico sufficiat propinquiores causas cognoscere earum." *Ibid.* 4.1.4.1: "Ad curam pestilentialium egritudinum sufficit medicos scire causas proximas inferiores a quibus procedunt vel a quibus pendent nec eis curandum est de causis primis et remotissimis que nullatenus subiacent suis operibus nec aliquid imprimunt in corpora nostra sine medio causarum inferiorum." The starting point of this sentence is one of Avicenna's statements on the causes of the corruption of air: "Prima causa longinqua ad illud sunt figure celestes et propinque dispositiones terrestres." Despars nevertheless considers it necessary to make an exception among primary causes for God: "Deo solo excepto qui omnia quecumque voluit fecit in celo et in terra, in mari et in omnibus abyssis et nihil est quod resistere possit sue voluntati, quapropter iugiter exorandus est in pestilentiis." This statement is of a general nature; prayers are in no way included in his medical advice. Furthermore, he is against the worship of healer saints. See Jacquart, "Le regard d'un médecin" (cit. n. 1), pp. 70–71, 84.

[30] Despars, *Commentarium* 1.4.20.

young"; and there were some who relied on "Egyptiac days." The most astrologically minded physicians did not merely observe the moon but also took into account the course of the planets. A gradation could also be found among such followers of astrology, according to whether they were using the great almanac, supplying each year for every day the signs and aspects in which each planet was to be found; the middle almanac, showing the entry of the moon into the signs of the zodiac, the oppositions of the sun and moon, the favorable days for bleeding, and the favorable days and nights for prescribing purgatives; or the small almanac, which showed only the entry of the moon into the signs of the zodiac. The use of the "great" and the "small" almanacs is attested at the Parisian faculty of medicine; its register reports that in 1452 a candidate to the baccalaureate was graduated, despite the insufficient length of his studies, because he promised to offer every year to the faculty these two kinds of almanacs—"unum almanach magnum et unum parvum."[31]

Despars did not place himself in any of the numerous categories of medical astrologers he had enumerated. Astrological judgments appeared to him uncertain, variable and ambiguous; their authors were deceived by the diversity of the intermediary causes, by the innumerable dispositions able to impede the action of the celestial impressions. Despars was thus listing the arguments usually put forward by opponents of astrology such as Nicole Oresme, who in the fourteenth century had stressed the uncertainty of the calculations and the relativity of astral determinism, which can be shifted by fortuitous circumstances.

In the very years in which Despars was writing, the question of calculation of favorable days for bleeding and purgation was particularly topical. In 1437 an astonishing controversy on this subject took place at Paris between the master of the medical faculty, Roland l'Escripvain, and a theological student, Laurent Muste.[32] Despars wrote the relevant part of his commentary between 1436 and 1438, while he was living at Tournai; and he had probably heard about this controversy, especially since in 1437 Roland l'Escripvain entered the duke of Burgundy's service. It was perhaps the appearance of this new rival that led Despars to a slight concession to astrology, although he had previously questioned its usefulness. Considering the widespread craze for this art, among princes as much as among ordinary people, he conceded that practitioners might pretend to be astrologers in order to keep up their reputation. He advised young physicians to take into account the astral dispositions when determining the time for bleeding, if there was no risk in doing so. On the other hand, whenever the patient's state required it, they had to prescribe bleeding even if astrological conditions were unfavorable.[33]

In spite of this concession to fashion—intended to keep the patients' confidence—Despars remained firm as far as knowledge was concerned. While commenting a few years later on a passage in Avicenna devoted to putrid fevers, he examined the possibility that "the periodic motions of the humors which cause fevers might proceed from the sky, that is, from the dominant planets, the moon

[31] See Ernest Wickersheimer, *Commentaires de la Faculté de médecine de l'Université de Paris (1395–1516)* (Paris: Imprimerie Nationale, 1915), p. 193.

[32] See Thérèse Charmasson, "L'établissement d'un almanach médical pour l'année 1437," *Comptes Rendus du 99e Congrès National des Sociétés Savantes, Section des Sciences,* 1974 (pub. 1976), *5*:217–234.

[33] Despars, *Commentarium* 1.4.20.

governing phlegm, Mars bile, Saturn melancholy."[34] He offered three arguments against this view. First, even if one conceded that the motions of the humors would follow the slow or fast rhythm of the planets governing them, it would not explain why melancholic fevers recur every four days, choleric fevers every three days, and phlegmatic fevers daily. Second, the same periodicity should be noticed in a healthy state, which is not the case. Third, the sanguine fevers, too, should be subjected to paroxysms, following the course of Jupiter or the sun, whereas they are continuous. Having set forth these arguments—answered to some extent by remarks of Savonarola's that I shall mention below—Despars presented the influence of the planets upon the humors in a way that gave more emphasis to the analogy between their mutual qualities, and consequently their rhythms, than to any relation of cause and effect: "Melancholy, which is cold and dry, has the slowest motion among the humors, as Saturn, which is cold and dry, has the slowest motion among the planets, covering its circle and completing its revolution within thirty years." In theory, Despars was not much disposed to believe in the influence of astral dispositions upon physiological processes. In practice, although denying any validity to the resort to astrology, he allowed physicians to feign observing the course of the planets in order to accommodate their patients.

Michele Savonarola presented at length the influence of the stars on the motions of the humors in his *De balneis,* postulating several times that "everything here below is subjected, in its actions and its properties, to the influx and government of stars, and above all of planets, which are the noblest among the celestial bodies."[35] The daily motion of the humors follows the four stages of the course of the sun: blood is moved first, then bile, next phlegm, and finally melancholy. Since the motion of the planets is not measured in days but in years, Mars exerts its influence the most strongly every three years, Saturn every four years. The frequency of choleric and melancholic fevers follows these rhythms. Astrology also played an important part in his *Physionomia.*[36] It is nevertheless noteworthy that in the writings that Savonarola addressed more specifically to physicians, astrology, although mentioned, played only a minor part. In the *Practica maior,* dedicated to Sigismondo Polcastro, we find, besides unambiguous statements concerning plague, an allusion to the role of the planets in connection with the "latitude of complexions": in summer the dominance of one or another planet joins together with the influence of the sun to make the heat more or less intense. Another allusion concerns epilepsy, also called "lunatic illness"; it is supposed to follow the stages of the moon and the qualities inherent to them: for example, epileptics whose illness is caused by the wet quality have their crisis at the first or the last quarter.[37] Introduced with some discretion into the works he reserved for practitioners, or more widely commented on in the writings meant for other

[34] *Ibid.,* 4.1.2.1.

[35] Michele Savonarola, *De balneis* (Ferrara, 1485), fol. 102r. This passage was analyzed by Lynn Thorndike, *History of Magic and Experimental Science,* Vol. IV (New York/London: Columbia Univ. Press, 1934), pp. 209–211.

[36] The second part of the *Physionomia* treats the astrological basis of this art; see Thorndike, *History of Magic,* Vol. IV, p. 193; and A. Denieul-Cormier, "La très ancienne *Physiognomie* de Michel Savonarole," *Biologie Médicale,* 1956, *45*:i–cvii.

[37] Savonarola, *Practica maior* 2.1, 6.1.20. On Sigismondo Polcastro, who studied medicine and taught from 1426 at Padua, see Pesenti, *Professori di medicina* (cit. n. 3), pp. 167–170.

readers, astrology seemed to Savonarola capable of helping physicians understand health and disease in general, as well as in the case of a particular patient. It was one of the ways to investigate primary causes.

Antonio Guaineri's position on astrology was more complex. He regarded it as an imperfect way to investigate causes. Confident in astrological data as far as plague was concerned, he challenged their usefulness with respect to the eight-month fetus. The diversity of particular temperaments could falsify the general rule according to which the celestial dispositions would be unfavorable to the birth of an eight-month fetus: a woman with a sanguine temperament could be delivered more quickly and her baby could survive, even if the celestial influx was contrary. In order to know "the perfect cause" of the phenomenon, one had to consider both the astrological explanation and the reason put forward by medical science.[38]

Guaineri was more interested in astrology as an operating technique linked to magic. Numerous prescriptions in his works combine the observance of the stages of the moon or of the patient's nativity with the wearing of precious stones, the engraving of images, and the absorption of animal substances.[39] This kind of prescription was not given on just any occasion, but only when medical science proved to be powerless, as in cases of plague, poisonings, epilepsy, furious mania—that is, all the afflictions that Guaineri considered to be caused by a venom, by a specific and occult form, and not by an alteration of the bodily qualities. Whenever a disease could be compared to poisoning, it had to be treated "with a few ceremonies." To neutralize the possible indignation of the reader who might think that such advice looked like a witch's prescription, he justified his standpoint by referring to certain natural principles, which, as they were a matter for natural philosophy, did not have to be expounded within a *practica*.[40] Some allusions to the action of stellar rays allow us to suppose that Guaineri was inspired directly or indirectly by al-Kindī.[41] Astrology, especially

[38] Antonio Guaineri, *De egritudinibus matricis* 30: "Patet ergo secundum medicos quare in octavo natus non vivat, in septimo vero sic, quod si hanc cum astrologorum ratione coniunxeris quesiti perfectam causam habebis." Savonarola does not question the astrological explanation; following the *Problemata* attributed to Alexander of Aphrodisis, he puts forward the unfavorable influence of Saturn: Savonarola, *De balneis* (cit. n. 35), fols. 102r, 103r. On the eight months' fetus see Ann Ellis Hanson, "The Eight Months' Child and the Etiquette of Birth," *Bulletin of the History of Medicine,* 1987, *61*:589–602.

[39] On Guaineri's account of these procedures see Thorndike, *History of Magic,* Vol. IV (cit. n. 35), pp. 215–231; M. Préaud, *Les astrologues à la fin du Moyen Age* (Paris: Jean-Claude Lattès, 1984), p. 202; and Danielle Jacquart, "De la science à la magie: Le cas d'Antonio Guaineri, médecin italien du XVe siècle," *Médecines, Littératures, Sociétés,* 1988, *9*:137–156. On the general problem of the passage from astrology to magic and necromancy in the Middle Ages see Graziella Federici-Vescovini, "L'astrologia tra magia, religione e scienza," in Federici-Vescovini, *"Arti" e filosofia nel secolo XIV* (Florence: Enrico Vallecchi, 1983), pp. 171–193.

[40] See Thorndike, *History of Magic,* Vol. IV, pp. 229; and W. G. Lennox, "Antonius Guainerius on Epilepsy," *Annals of Medical History,* 3rd Ser., 1940, *2*:482–499. This passage in *De egritudinibus capitis* 1.7 is very ambiguous: "Omnes ergo egritudines venenositate participantes melius per talia sic communiter a nobis ceremoniis quibusdam adiunctis curantur quam aliter. Si que ergo in hoc meo opusculo deinceps ad modum vetularum descripsero, pensa non ab re me hic illa descripsisse, nam etsi tibi forsan precantationes appareant, nihil tamen est sine ratione positum quam tibi ubi opus esset ex principiis naturalibus assignarem. Sed cum hoc sit philosophi declarare huic meo opusculo esset impertinens."

[41] Al-Kindī, *De radiis,* ed. M. T. d'Alverny and F. Hudry, *Archives d'Histoire Doctrinale et Littéraire du Moyen Age,* 1974, *41*:139–260. Guaineri mentions stellar rays in connection with plague (*De peste* 1.1) and, more explicitly, puts forward the action of influxes in discussing the question, Why do some melancholic illiterates become learned? (*De egritudinibus capitis* 15.4).

Ptolemy's *Quadripartitum,* led him to consider the possibility of diabolical interventions: some persons, because of a particular state of the sky at the time of their birth, are in the devil's grip.[42] The importance he attached—while deploring it—to the competition to physicians from magicians or sorcerers suggests another reason for his attitude: the resort to procedures that gained the confidence of most people helped practitioners to act upon the patients' imagination, upon their *fantasia,* thus increasing the efficacy of the strictly medical treatment. For this reason, learned physicians needed to know what the old women knew. Apart from this realistic argument, it is obvious that, without explicitly admitting it, Guaineri was following Pietro d'Abano.[43]

The attraction that magic exerted over Guaineri went beyond questioning the therapeutic efficacy of this or that procedure. Any topic could be a pretext for evoking the subject of magic with obvious pleasure. In his account of hectic fever in *De febribus* he is prompted to ask whether it is possible to come back from senility to youth. He mentions the episode from Ovid's *Metamorphoses* in which Medea extracted Aeson's blood and put in its place some saps and herbal decoctions that could restore his youth with only this comment, "God knows whether it is true."[44] Searching for similar phenomena, Guaineri refers to the power attributed to Albertus Magnus, "who had made, with herbal saps during a particular disposition of the stars, a head that was able to speak." The origin of this legend—attributed also, with some variations, to Gerbert, Roger Bacon, and Robert Grosseteste—is to be found, as far as Albertus Magnus is concerned, in a passage of the *Meteora* describing blowers, a kind of vessel used as a bellows and usually made in the form of a human head. Unaware of its origin, Guaineri saw in this legend an opportunity to repeat the condemnation, often uttered in his works, of the mendicant friars, who in his view were responsible for spreading such beliefs.[45] The reader feels somewhat confused by Guaineri's attitude toward magic, sometimes indulgent, sometimes skeptical or violently unfavorable. While holding up to ridicule all that derived from the most obvious credulity, he showed himself greatly interested in the practices of astrological magic.

Despars agreed with Guaineri in condemning the mendicant friars' harmful role and their diffusion of belief in supernatural interventions, but he radically disagreed in his intransigence toward every irrational practice. He violently con-

[42] Guaineri, *De egritudinibus matricis* 1, quoted in Lemay, "Anthonius Guainerius" (cit. n. 4), pp. 328–329; cf. Ptolemy, *Quadripartitum* 4.3.

[43] See Pietro d'Abano's famous question on incantations in *Conciliator* 156 (Venice, 1476). Cf. Jole Agrimi and Chiara Crisciani, "Medici e *vetulae* dal Duecento al Quattrocento: Problemi di una ricerca," in *Cultura popolare e cultura dotta nel Seicento* (Milan: Franco Angeli, 1983), p. 155. Even Despars, otherwise skeptical about magical powers, acknowledges this action on the imagination of those who believe in these kinds of procedures: "Carmina non sanant sua virtute, sed spe, confidentia et imaginatione erroneis . . . bona imaginatio non sanat nisi spe, confidentia et consolatione consequentibus" (*Commentarium* 1.4.1). Nevertheless, he refrains from describing these procedures and advises learned physicians against using them.

[44] Guaineri, *De febribus* 3.3, discussing Ovid, *Metamorphoses* 7, lines 285–293 (Guaineri names Jason instead of Aeson, Jason's father).

[45] See the passage of the *Meteora* in which Albert mentions the implement called a *sufflator*: 3.2.17, ed. A. Borgnet (Paris: Vives, 1890), Vol. IV, p. 634. On the description of these "machines" and the origin of the legend see Lynn White, *Medieval Technology and Social Change* (Oxford: Clarendon Press, 1962), pp. 90–92; and White, *Technologie médiévale et transformations sociales* (Paris/The Hague: Mouton, 1969), pp. 110–112. In the *Image du monde,* the making of a speaking head able to pass on secrets is ascribed to Virgil; see J. W. Spargo, *Virgil the Necromancer* (Cambridge, Mass.: Harvard Univ. Press, 1934), pp. 132–133. On Guaineri's attitude toward the mendicant friars see Jacquart, "De la science à la magie" (cit. n. 39).

demned the competition exerted by magicians and sorcerers and refused to adopt any of their practices.[46] If he allowed physicians, in certain cases, to pass themselves off as astrologers, he accepted no compromise over magic. Savonarola, however, gave no clear statement on magic. His opinion on Pietro d'Abano's magical views, as well as his attitude to incantations, pseudoreligious formulas, or exorcising processes, was ambiguous. Concerning epilepsy, he noted the wearing of pious amulets as one of the remedies that had been passed on by "our Christian predecessors"; nor did he omit the traditional invocation to the Magi.[47] In the chapter on impotence, a favorite place for discussions of evil spells, he did not hesitate to list several magical remedies (for example, to behead a hoopoe at new moon and keep its heart), even if he confessed that he did not much believe in magic spells ("I have heard a lot, I have seen little").[48] *De vermibus* is even stranger, since it lists six spells intended for expelling intestinal worms.[49] Yet Savonarola expressed his skepticism, even his aversion to practices that went against natural principles and religious faith. He offered two reasons for enumerating the spells: popular opinion confirmed the efficacy of these remedies, which were used by a lot of people; and his reader would be amused by them.

Alchemy was another possible resource for physicians. In his search for remedies to palliate the deficiencies of the medical art, Guaineri could not avoid the subject. Several times he recommends remedies invented by a hermit who was led by his successes to abandon alchemy for medicine, but, as usual, claims that he lists them because of their success among common people. It is likely that he was in touch with alchemists.[50] It is well known that Savonarola also took some interest in this art, although he did not consider the transmutation of metals possible. His treatise *De aqua ardente* followed the tradition set by Taddeo Alderotti's *consilia* on alcohol, but it also gave Raymond Llull as an authority in the opening pages.[51] Savonarola described aqua vitae as the solar quintessence, having a marvelous power, especially to act on the heart, an organ placed under the influence of the sun.

Is it possible to detect in Jacques Despars a similar interest in alchemy? A sixteenth-century manuscript amazingly attributes to a certain "Jacques des Pars

[46] See Jacquart, "Le regard d'un médecin" (cit. n. 1), pp. 70–73, 84–85.

[47] Savonarola, *Practica maior* 6.1.20. Savonarola hesitated to believe that Pietro d'Abano was a necromancer. See *Physionomia* as quoted in Thorndike, *History of Magic,* Vol. IV (cit. n. 35), p. 188.

[48] Savonarola, *Practica maior* 6.20.32: "Et hanc curam plurimi vulgarium incantationibus et fascinationibus, ita meo tempore tractaverunt ut magis perfecti sint quam medici naturales et ego de eorum effectibus miratus rei diabolice quam divine ascripsi, quorum sectam in presentiarum non sequar naturalibus omnino accomodatus. . . . Accedo ad ea quibus obviatur maleficiis malorum virorum et malarum mulierum."

[49] Michele Savonarola, *De vermibus* (Lyons, 1560), pp. 801–802.

[50] Guaineri, *De egritudinibus capitis* 9.8: "Quidam heremita alkimista magnus plurima paralitica membra cum infrascriptis unctionibus curavit. . . . Unde alkimiam dimittendo medicus se fecit a quo experta plura habui. . . . Ab hiis unguentis cum vulgares in hiis magis quam in aliis in re fiduciam maiorem adhibeant discedere nescio." For other examples showing Guaineri's link with alchemy see Thorndike, *History of Magic,* Vol IV (cit. n. 35), pp. 230–231.

[51] For Savonarola and alchemy see Thorndike, *History of Magic,* Vol. IV, p. 213. In the last decades of the thirteenth century Taddeo Alderotti wrote seven *consilia* on the distillation of alcohol and on its use in medicine, describing it as *omnium medicinarum mater et domina;* see Nancy G. Siraisi, *Taddeo Alderotti and His Pupils: Two Generations of Italian Medical Learning* (Princeton, N.J.: Princeton Univ. Press, 1981), p. 301. Among the treatises falsely attributed to Raymond Llull, Savonarola probably refers in his *De aqua ardente* to *De secretis nature seu de quinta essentia,* which is, in fact, a revision of John of Rupescissa's *De consideratione quinte essentie.* See Robert Halleux, *Les textes alchimiques* (Turnhout: Brepols, 1979), p. 108; and Michela Pereira, "Filosofia naturale lulliana e alchimia," *Rivista di Storia della Filosofia,* 1986, *47*:747–780.

Lombart" one recipe "to congeal quicksilver with herbs," but this cannot serve as proof.[52] However, the few allusions that are scattered through his commentary allow one to suppose that he considered alchemy a reliable art. Commenting on an experiment of Avicenna's that consisted in determining, with the help of an alembic, whether bones were made of a substance moister than air, he not only mentioned the method for distilling medicinal waters but described at length the equipment necessary to alchemy. He compared the making and subsequent refining of melancholy to the cupellation performed by goldsmiths. In a sentence listing the different kinds of metals, he used the standard alchemical metaphors. The point at issue was the noxiousness of the waters with which metallic substances were sometimes mixed: "The kinds of metals because of which waters happen to contract their nature are quicksilver, lead, tin, silver, bronze, brass, copper, and gold, which is the noblest among all, as the king of metals, sulfur being their father and quicksilver their mother."[53] This quotation alone is not sufficient to show that Despars was devoting himself to alchemy; it may merely come from general knowledge. Nevertheless, he showed the utmost restraint with regard to alchemy, in contrast to his frequent condemnations of astrology and, of course, magic. We are allowed to suppose that, if he was interested in alchemy, it was not, like Guaineri, in order to integrate certain mineral remedies into medicine, nor, like Savonarola, to search for the help of celestial powers, but to understand the natural rules that govern matter. From this perspective, he might have been attracted to the art of transmutation. As far as medical practice was concerned, he did not intend to apply rules other than those derived from strictly medical theory, the only system of knowledge that he considered relevant to finding an efficacious therapeutics.

III. AN EXAMPLE OF ORDINARY DISEASE: PLEURISY

For a medieval practitioner, to apply medical theory meant, in most cases, to follow the rules given in the Greek and Arabic sources. Within the limits of a common system, that is, Galenism, the broad range of authorities available by the end of the Middle Ages provided physicians with a variety of explanations on specific topics, and thus with a variety of recommendations for prescription. From antiquity to the Renaissance, the etiology and treatment of pleurisy were among the most controversial issues in medicine.

Antonio Guaineri devoted a special treatise to this nosological entity, which had already been discussed in the *Corpus hippocraticum* and whose name seems to go back even earlier. According to the *Regimen in Acute Diseases,* "Now the acute disease are those to which the ancients have given the names of *pleuritis, peripneumoniē, phrenitis, lēthargos, kausos,* and such as are akin to these, the fever of which is on the whole continuous. For whenever there is no general type of pestilence prevalent, but diseases are sporadic, acute diseases cause many times as many as or more deaths than all others put together."[54] Although he did not actually quote this text, Guaineri probably remembered it as he noted the

[52] MS Orléans, Bibl. mun. 291, fol. 138.

[53] Despars, *Commentarium* 1.1.3.2; 1.1.4.1; and 1.2.2.1.16 (quotation).

[54] *Regimen in Acute Diseases* 5. I quote W. H. S. Jones's translation of Hippocrates, 4 vols. (Loeb Classical Library) (Cambridge, Mass.: Harvard Univ. Press, 1967), Vol. II, p. 67, but I keep the Greek terms of the diseases.

Patient suffering from pleurisy: an illustrated initial from a passage on pleurisy (note the words* pleuresis, pectore, costis, *etc.) in Avicenna's* Canon, *Vatican, MS Urb. Lat. 241, folio 142r. Courtesy of the Loren C. MacKinney Collection, University of North Carolina.

ravages caused by pleurisy. Instead of citing Hippocrates' authority, he referred to actual events. It seemed to him that pleurisy was more widely diffused and killed more people than in previous times, because of bad diets, or astral dispositions, or climatic variations. During the very year in which he was writing his treatise, which he did at the request of some colleagues, the disease had raged like plague in numerous places. This observation is surely related to Michele Savonarola's allusion to a particular pleurisy: "You must know that in our city of Padua, at Trevisa and Venice, in 1440, around March, some malignant pleurisy broke out, which was contagious and which caused numerous deaths."[55] According to him, eight out of ten patients died. Savonarola's perplexity arose mainly from the contagiousness of the disease, which had not been noted by the ancients. Among the different diseases that Avicenna had listed as passing from one man to another, Savonarola picked "putrid abscesses" as a category to which pleurisy could belong. Neither Guaineri nor Savonarola described this "pleurisy" precisely enough to allow us to determine the exact nature of the disease. Both observed its epidemic nature. Guaineri suggested the influence of an astral disposition. More strikingly, Savonarola mentioned the possibility of transmission by contagion, an explanation that he avoided in the case of plague.

The concept of *pleurisis*—which corresponds in modern medicine most often

[55] Savonarola, *Practica maior* 6.5.13; cf. Antonio Guaineri, *De pleuresi* 1 (Lyons, 1525) (hereafter **Guaineri, *De pleuresi***). This treatise is addressed to Antonio Magliani of Chieri, physician of the duke of Savoy.

to pneumonia, though sometimes to pleurisy[56]—had become more and more complex since antiquity. Medieval physicians referred principally to Avicenna and to Mesue, and to a lesser extent to Galen's *De interioribus* (*De locis affectis*). As he usually did, Guaineri tried to find the truth through the words, but they led him to an embarrassing ambiguity. After he had given the Arabic and Persian equivalents, *sosata* and *birsen,* which meant "abscess of the breast,"[57] he gave an anatomical analysis. Four parts are to be distinguished in the breast: membranes, muscles, the heart, and the lung. Muscles and membranes were the only parts to play a role in pleurisy. To the second category, membranes, belong the mediastinum, the diaphragm, and the *pleura,* "the membrane that covers the ribs inside," that is, the parietal part of the pleura. After this preamble, Guaineri set forth the usual distinction in medieval nosology between *pleurisis vera* and *pleurisis non vera,* which involved both the anatomic location of and the substance responsible for the inflammation. The distinction based on anatomy led him to note a difficulty in vocabulary: *vera pleurisis,* as its name says, occurs when the membrane called *pleura* is affected; the point was that *pleura* in Greek meant, according to numerous authorities, vaguely "the side." Keeping to this etymology, Guaineri proposed the following definition: "*Vera pleurisis* is an abscess of the side caused by bile or blood, provoking continuous fever, biting pain in the side, coughing, difficulty in breathing, and a 'serrated' pulse. *Non vera* is a *pleurisis* in which one of these conditions is missing, as I shall say at greater length below." Through the description of the signs of *non vera,* Antonio decided more firmly upon the meaning of *pleura.*[58] Under the heading of *non vera,* he placed the inflammations caused by a humor other than blood and bile or by flatulence, as well as the abscesses of the mediastinum, the diaphragm, and the intercostal muscles. The mention of these last muscles could have undermined the definition of *vera pleurisis* as "an abscess of the side": the problem was solved by admitting that the inflammation of the intercostal muscles provokes a less biting pain in the side. Between the Greek etymology and the precise meaning given by anatomists to *pleura,* Guaineri was unable to choose. However, his description shows clearly that he, like Mondino dei Liuzzi,[59] restricted *pleurisis vera* to the inflammation of "the membrane that covers the inside of the ribs."

More dependent on the authorities, Michele Savonarola and Jacques Despars did not adopt the same definition. Following Avicenna and Mesue, they thought that *pleurisis vera* affects not only "the membrane that covers the inside of the ribs" but also the mediastinum, the diaphragm, and the internal muscles. It is clear that the imprecision of the anatomical descriptions, which did not take into account the two membranes forming what we today call the pleura, contributed much to the confusion of the explanations.[60] It has to be added that the Latin term *pleura* did not appear, with reference to *pleurisis,* in the translations of the

[56] See M. D. Grmek, *Les maladies à l'aube de la civilisation occidentale* (Paris: Payot, 1983), p. 197.

[57] Guaineri, *De pleuresi* 1. The terms *sosata* and *birsen* are notably in Avicenna's *Canon* 3.10.4.1.

[58] Guaineri, *De pleuresi* 4.

[59] "Pleura autem est panniculus substantie durus et nervosus in quantitate magnus, cooperiens omnes costas internis et ideo colligantiam habet cum omnibus membris contentis in concavitate pectoris. . . . [Pleuresis] vera fit in hoc panniculo." Mondino dei Liuzzi, *Anatomia,* ed. E. Wickersheimer (Paris: Droz, 1926; rpt. Geneva: Slatkine, 1977), Ch. "De panniculis pectoris."

[60] See A. Baffoni, *Storia della pleuriti da Ippocrate a Laennec* (Rome: Istituto di Storia della Medicina dell'Università di Roma, 1947); and A. Pazzini and A. Baffoni, *Storia delle malattie* (Rome: Edizioni Clinica Nuova, 1950), p. 255.

two main Arabic authorities. Despars adopted without hesitation the distinction set by Avicenna and Mesue. Savonarola showed some perplexity when listing the different meanings given to *pleura:* "the membrane that covers the ribs" according to Mondino dei Liuzzi, "rib" according to Serapion, "side" in Simon of Genoa and Pietro d'Abano.[61] He also pointed out the other distinctions that some physicians had introduced: they had called *verissima* the pleurisy arising in the muscles of the diaphragm, *magna* the illness affecting the mediastinum. In his view, to increase the number of names was "ignominious" and all these considerations superfluous.

It is hard to estimate the effect on practice that these numerous distinctions between different kinds of pleurisy might have had. If the etiology differed from one author to another, the distribution of symptoms among *vera* and *non vera* followed the almost fixed framework found in the sources. Guaineri was the only one among our three authors to propose, as far as this disease was concerned, some examples drawn from his practical experience, although he rarely specified the kind of pleurisy affecting the patients. The first case that he reported illustrated the possibility that the disease could turn into heart trouble.[62] The rupture of the abscess before its maturity was supposed, because of the abundance of the matter and its bitterness, to bring the inflammation to the heart and provoke a fatal syncope. This happened to two noble ladies, who fainted several times as soon as they became ill. Pain in their left side was so violent that they suffocated and "gave back their soul to angels" within four hours. In the absence of further information on the beginning of these cases, it is hard to identify them in modern terms, but if the main pathognomonic sign, that is, "pain in the side," was the only factor that led Guaineri to diagnose pleurisy, the actual affliction might have been heart disease.

The other cases that he reported had no better end. One of them, which affected one of the duke of Savoy's squires, was presented to illustrate the common rule that "you must not trust the good quality of urine."[63] Called at the seventh day, Guaineri observed great difficulty in breathing, strong fever, weak and difficult expectoration, and livid spit. Whereas the Jewish practitioners who were treating the patient were predicting recovery on the basis of the urine alone, Guaineri delivered a fatal prognosis. At the eleventh day, the patient died. Like most physicians, Guaineri did not mind reporting his colleagues' failures as opposed to his own, imputing them to incompetence and thus keeping intact both his personal skill and the power of medical art in general. Another reported case fits in with this tradition; the lyricism of the story leads me to quote it in full.

> I saw in Evires a very pretty young noble woman, quite robust, to whom some colleague had unfortunately prescribed, at the third or fourth day, one drachma of fetid pills and four grains of diagridium, while pleurisy was severe, accompanied by fever and great difficulty in breathing. When the drugs began to act, the poor thing started vomiting so violently that she fainted. Once I had been called, I prescribed the best remedies I could, but they had no effect and she expired in my arms with pious lamentations. Let such a case be a warning to you, so that you may avoid what happened to this perfidious man, who will incur the torments of Hell for the loss of so angelic a beauty.[64]

[61] Despars, *Commentarium* 3.10.4.2; and Savonarola, *Practica maior* 6.5.13.
[62] Guaineri, *De pleuresi* 6.
[63] *Ibid.*, 7.
[64] *Ibid.*, 12.

From this example, which showed a pretty young woman expiring in a good practitioner's powerless arms, apprentice physicians would learn that resolutive drugs should never be prescribed when fever is strong from the very beginning.

The treatment of pleurisy was indeed a favorite field for controversies. The question of which side should be bled was not discussed in the fifteenth century with the same bitterness as it was in the sixteenth;[65] instead, the most active controversy was about the use of "repercussive" drugs at the beginning of the disease. This prescription was found in Mesue's *Grabadin,* the incontestable authority on the treatment of pleurisy since the thirteenth century. In the fourteenth century, a writer from Montpellier, Bernard de Gordon, inaugurated a debate on this topic. Following the *Regimen in Acute Diseases,* which recommended that side pain be soothed with fomentations, Mesue had prescribed the application of plasters or a sponge soaked with water in order to overheat, and thus cause to evaporate, the matter of the inflammation.[66] He stipulated that this operation be done with care, at the beginning of the disease, so that the overheating should not convert the matter into pus. Bernard de Gordon totally rejected the relevance of this prescription and laughed at Mesue for recommending its application at the very beginning of the disease, "in principio principiante." According to Bernard, it was impossible for physicians to be aware of the very beginning, and even if they were aware of it, they should keep those "repercussive" drugs in reserve for later use.[67] Savonarola thought that Bernard de Gordon had gone too far in ridiculing Mesue, but he considered the Arabic author's prescription inapplicable since patients did not usually consult physicians in the very first hour of their diseases. Despars also advised against this practice, even though Hippocrates and Alexander of Tralles had recommended it as well. Guaineri, who devoted a long passage in *De pleuresi* to this debate, was the only one among our three authors to follow Mesue: according to him, the words *in principio principiante* did not refer to the early moments of the disease but to the first day during which the matter was still slight and flowing in a continuous way.[68] Thus he gave a long list of medicines that would cause evaporation. Although he did not quote it explicitly, it was possibly the *Regimen in Acute Diseases* that led him to adopt Mesue's similar prescription. In that case, he could be called "more Hippocratist" than the other two physicians.

All three authors agreed that bleeding had to be prescribed at the beginning of the disease. On this point Despars was the most intransigent: "I advise practi-

[65] Which side to bleed in pleurisy was a traditional topic for scholastic questions. We find it in, e.g., Taddeo Alderotti's commentary on the *Regimen in Acute Diseases;* see Siraisi, *Taddeo Alderotti* (cit. n. 51), p. 388. Savonarola places a *dubium* on this problem at the end of the chapter on pleurisy: "Utrum si pleuresis sit in dextra parte plus conferat phlebotomia saphene in parte opposita quam in eadem." The debate became more bitter in the sixteenth century, through the impetus given by Pierre Brissot; the supporters of anti-Arabism were in favor of bleeding the diseased side.

[66] Hippocrates, *Régime des maladies aiguës,* ed. R. Joly (Paris: Les Belles Lettres, 1972), App. 33.1, p. 84; and Mesue the Younger, *Antidotarium sive Grabadin medicamentorum compositorum* 1.2.2.

[67] Bernard de Gordon, *Lilium medicine* 4.8 (Lyons, 1550): "Sexto intelligendum sicut dictum est quod repercussiva nunquam competant in pleuresi, etsi aliquo modo competerent sicut voluit Mesue, tunc competerent in initio initiante. Si quis presens esset et Deus dedisset talem gratiam quod principium initians cognosceret, aut quod portarentur in manu vel in bursa medicamina repercussiva, sed quod possunt talia concurrere non videtur. Et ideo summe miror, quomodo bonus Mesue solus dormitavit." And see Demaitre, *Doctor Bernard de Gordon* (cit. n. 2), p. 142.

[68] See Savonarola's *dubium* on this topic: "Dubitatur utrum in principio pleuresis competant repercussiva," in *Practica maior* 6.10.13; Despars, *Commentarium* 3.10.5.2; and Guaineri, *De pleuresi* 9.

tioners in this country to start the treatment of *vera pleurisis* with bloodletting and to spare nobody, from fourteen to sixteen years of age, until seventy for the mighty . . . , I do not except pregnant women, for pleurisy that is allowed to grow without being weakened by phlebotomy kills more quickly and more certainly both the mother and the fetus than a moderate phlebotomy, repeatedly prescribed." Guaineri was more cautious. He considered deplorable the habit of some colleagues of prescribing bloodletting for children under fourteen. He decisively advised people over sixty years of age against it, recommending that its use be adapted to the patient's strength. Savonarola, an unquestioning supporter of bloodletting, nevertheless forbade it in the case of "furious matter," as during the 1440 epidemics.[69]

The treatment of pleurisy also called for adapting certain ancient prescriptions that were found too violent. The reason generally put forward was the weakening of mankind, perverted by gluttony and debauchery. Despars repeatedly blamed as well patients' indiscipline and the difficulty that physicians encountered in commanding obedience.[70] Guaineri did not follow the same path; it seems, on the contrary, that he considered the physicians of antiquity more foolhardy and less subtle than the modern practitioners. Concerning bloodletting, for instance, he preferred the moderate attitude shown by the Arabs, such as Avicenna and Mesue, to the Greeks' intrepidity; he pointed out in particular that in several places Galen prescribed bleeding up to syncope.[71] According to Guaineri phlebotomy was a drastic remedy and only a small amount of blood needed to be drawn each time. Somewhat distancing himself from Galen, he showed in general a certain independence toward authorities. His report of the controversy that Bernard de Gordon had raised over Mesue's use of "repercussive" drugs contains a humorous remark: "But let us leave these good fellows, who were exceptional practitioners, in Paradise, where all disputes subside, and let us come back to the common act of practice." Guaineri had no taste for scholastic discussions; he was searching for truth through, on the one hand, the original meaning of the words and, on the other hand, the comparison of current practices, judging them by their efficacy, whether they were dictated by medical science or by other techniques.

IV. CONCLUSION

This comparison, on a few specific points, of three contemporary physicians who were molded by a similar culture and faced the same problems in their everyday practice reveals a wide range of variation among their different solutions and convictions. The variety of sources, which was always the lot of medieval science, explains this in part. What seems new is that the dialectic approach to

[69] Despars, *Commentarium* 3.10.5.2; Guaineri, *De pleuresi* 10; and Savonarola, *Practica maior* 6.5.13.

[70] See Jacquart, "Le regard d'un médecin" (cit. n. 1), p. 61. Savonarola also conforms to this topos: "Sed hic adverte quod non possumus hodierna die cum ita facilibus cibis procedere ut antique propter ingluviem hominum et debilitatem nature ab eo quod prius fuit": *Practica maior* 6.10.13, in connection with the prescription of barley water to pleuritics.

[71] Guaineri, *De pleuresi* 10. The prescription of bleeding up to fainting already appears in Hippocrates, *Régime* (cit. n. 66), App. 31.2, p. 83. In a general way, Guaineri takes care of his patients' comfort, prescribing a special potion to "a delicate young lady," or reducing the dose of scylla juice and increasing that of honey in some tablets so that they do not "bite" the pleuritic's tongue too much (*De pleuresi* 18).

authorities no longer constitutes a major concern. Greek and Arabic works now offer only a stock of ideas, from which each takes what seems the most suited to his own opinions on specific questions. In a sense the authorities are becoming dead texts. Antonio Guaineri was probably the most independent toward the sources, "leav[ing] these good fellows . . . in Paradise."

The confusion in medical vocabulary, which developed in an anarchical way during the Middle Ages, was probably one reason for keeping the authorities at a distance. "It is not possible to talk properly about cerebral faculties anymore," Jacques Despars noted. The proliferation of technical terms seemed "ignominious" to Michele Savonarola. Like Guaineri, but with less conviction, he resorted to etymology in order to try to find the hidden truth inside the structure of words. In a more philological and erudite manner, Despars, following the example of Simon of Genoa, sought the different senses of words through all the available sources. Some of his investigations were taken up again in the sixteenth century by Berengario da Carpi.

Although their own power within society had increased, learned practitioners of the fifteenth century suffered disturbing competition from other kinds of medicine—empirical, magical, or alchemical. Without being able to justify this policy clearly, Guaineri did not hesitate to integrate into his art, while condemning them, some practices that he found in those medicines. Savonarola, probably out of prudence, was hesitant. Despars firmly condemned those practices in the name of rationality. In spite of their different stands, all three were aware of the crisis in their art. While attributing failures to their colleagues' incompetence or to their patients' disobedience, and sometimes to God's will, they expressed more clearly than their predecessors their difficulties in treating patients successfully. As the particularities of practice became the object of deeper attention, the divergences between authorities (which had long stimulated medical thinking), joined to the confusion of vocabulary, came to be perceived as factors contributing to powerlessness. However, even if their often paradoxical choices contributed to the progressive dissolution of the theoretical grounds on which their learning was founded, fifteenth-century physicians were not ready to question these grounds in the name of their own experience.

Giovanni Argenterio and Sixteenth-Century Medical Innovation

Between Princely Patronage and Academic Controversy

*By Nancy G. Siraisi**

AFTER A CERTAIN AMOUNT of academic infighting, in 1587 Alessandro Massaria became first ordinary professor of practical medicine at the University of Padua. The chair was a prize well worth having, the Paduan medical school being as renowned throughout Europe for instruction in practical medicine as it was for anatomy. Massaria celebrated his success by delivering an oration, during which he reviewed the main contemporary schools of thought in medicine. These, according to Massaria, were three: the adherents of the Greeks, and especially of Galen, among whom Massaria counted himself; the adherents of the Arabs, especially Avicenna; and a third group, those who "are drawn to none of the ancients but in a remarkable way to some more recent doctrine."

In the third category, of which he heartily disapproved, Massaria named two individuals. The second of them was Jean Fernel (1497–1558), who taught at Paris and was the author of an influential and often-reprinted general textbook of medicine, as well as of a work entitled *On the Hidden Causes of Things*. Fernel's unconventionality consisted chiefly in his role as a pioneer in the replacement of teaching by commentary with teaching by means of specially composed modern textbooks, and in idiosyncratic ideas about celestial spirits, innate heat, and disease of the total substance that have been the subject of various recent studies. But Massaria's prime example of an innovator in medicine was Giovanni Argenterio (1513–1572). According to one of his listeners, Massaria inveighed especially against Argenterio, who "had dared to criticize the logical books, *Ethics, Meteorologica,* and entire natural philosophy of Aristotle himself, and at the same time used to hurl his infinite and unbridled audacity at Hippocrates and Galen."[1]

* Department of History, Hunter College and the Graduate School, City University of New York, 695 Park Avenue, New York, New York 10021.

[1] Antonio Riccobono, *De Gymnasio patavino* (Padua, 1598), fol. 71r. On Padua as a center for the teaching of practical medicine see Jerome J. Bylebyl, "The School of Padua: Humanistic Medicine in the Sixteenth Century," in *Health, Medicine, and Mortality in the Sixteenth Century,* ed. Charles Webster (Cambridge: Cambridge Univ. Press, 1979), esp. pp. 347–355. On Fernel's ideas and influence see Charles Sherrington, *The Endeavor of Jean Fernel* (Cambridge: Cambridge Univ. Press, 1946); D. P. Walker, "The Astral Body in Renaissance Medicine," *Journal of the Warburg and Courtauld Institutes,* 1958, *21*:199–233; Jacques Roger, *Jean Fernel et les problèmes de la médecine de la Renaissance* (Les Conférences du Palais de la Découverte, Ser. D, no. 70) (Paris, 1960); Linda

Massaria's oration suggests something of the complexity of developments in medical knowledge that had taken place over the previous century. Between the late fifteenth and the mid-sixteenth centuries, the editorial and translating labors of medical humanists had greatly improved and expanded knowledge of Greek medical texts. The immediate result had been yet further enhancement of Galen's reputation, accompanied by a good deal of denigration of the medieval Arabo-Latin versions of Greek medicine and of Avicenna and other representatives of Arabic medical learning. At the same time, significant new contributions to knowledge began to accumulate through direct investigation of nature in the areas of anatomy and pharmacological botany. Yet the situation was a good deal more ambiguous than the foregoing sketch suggests. Avicenna continued to be studied seriously and valued highly at Padua, which was also especially noted for new anatomy and new botany; Latin scholastic authors of the thirteenth to fifteenth centuries were still read, if only for the sake of refuting them; the repudiation of aspects of Galenic anatomy was highly dependent upon improved access to Galen's anatomical writings; and in the course of the century Galen became as ambiguous a figure as Avicenna—subject to both praise and blame and remade in the different images chosen by his admirers and detractors. As a result, few terms in Renaissance medical writing are more ambiguous than "recentiores," "moderni," "neoterici," and similar expressions.

Massaria's choice of Argenterio rather than, say, Vesalius or Paracelsus as the outstanding example of a recent innovator is a striking reminder of the ambiguities and multiple meanings of these terms. Standard histories of medicine and biographical accounts duly note Argenterio's readiness to criticize Galenic doctrines and to propound alternative views (chiefly on *spiritus* and the causes of disease). Argenterio was among the earliest medical critics of Galen, although he was certainly not a lone pioneer. After his death, Argenterio's ideas continued to play a part in late sixteenth- and early seventeenth-century endeavors to enlarge or replace Galenic explanations of disease. His reputation survived until at least 1647, when one of several treatises attacking him was reprinted for the last time. Yet Argenterio's critique was evidently—at least in the light of historical hindsight—only one, and far from the most important, among the factors contributing to the weakening and slow downfall of Galen's authority.[2]

A. Deer, "Academic Theories of Generation in the Renaissance: The Contemporaries and Successors of Jean Fernel (1497–1558)" (Ph.D. diss., Warburg Institute, Univ. London, 1980); and James J. Bono, "The Languages of Life: Jean Fernel (1497–1558) and *Spiritus* in Pre-Harveian Bio-Medical Thought" (Ph.D. diss., Harvard Univ., 1981). Vivian Nutton informs me that in some sixteenth-century German academic circles Fernel, rather than being denounced or praised as an innovator, was accepted as a solid author in the classical tradition. Tiziana Pesenti, "Galenismo e 'novatio': La scuola medica vicentina e lo Studio do Padova durante il periodo veneto di Galileo (1592–1610)," in *Medicina e biologia nella rivoluzione scientifica,* ed. Lino Conti (Assisi: Edizioni Porziuncola, 1990), pp. 107–147, which reached me after the present article was written, provides further information about late sixteenth-century disputes at Padua over Argenterio's views.

[2] Some of the more informative accounts of Argenterio are to be found in Carlo Bonardi, *Lo studio generale a Mondovi (1560–66)* (Turin, 1895), pp. 77–82; G. Bonino, *Biografia medica piemontese,* Vol. I (Turin, 1824–1828), pp. 222–239; Angelo Fabroni, *Historiae Academiae pisanae,* Vol. II (Pisa, 1792), pp. 254–257; Giammaria Mazzuchelli, *Gli scrittori d'Italia,* Vol. I, Pt. 2 (Brescia, 1753), pp. 1038–1039; F. Mondella, "Giovanni Argenterio," *Dizionario biografico degli italiani,* Vol. IV (Rome, 1962), pp. 114–116; Walter Pagel, *Paracelsus: An Introduction to Philosophical Medicine in the Era of the Renaissance* (Basel/New York: Karger, 1958), pp. 301–304; and Kurt Sprengel, *Histoire de la médecine,* Vol. III (2nd ed., Paris, 1835), pp. 200–207. A perceptive brief analysis of Argenterio's

Inquiry into the actual extent of Argenterio's influence in the later sixteenth and early seventeenth centuries lies beyond the scope of the present article. Instead, it will examine the grounds for the formidable reputation that for Massaria and others made Argenterio the prototype of an innovator in medicine. The source of that reputation lay, obviously and in the first place, in ideas expressed by Argenterio in his own writings; but it was also magnified by the reception initially accorded those ideas in Argenterio's immediate professional and intellectual milieu. In turn, the first outbreak of controversy involving Argenterio allows consideration of criteria for the limits of permissible innovation that appear to have operated in the academic medical community at Pisa, and elsewhere, in the mid-sixteenth century.

I. THE MEDICAL FACULTY OF THE UNIVERSITY OF PISA

Argenterio first emerged as a controversial figure while he was teaching at the University of Pisa from 1543 to 1555. Pisa in the 1540s and 1550s offered notable opportunities for medical teaching, and a situation distinctly favorable to the most active and innovative areas of contemporary academic medical science: the study of Galen and other Greek authors; the cluster of ideas and methods subsumed by historians under the term medical humanism; anatomy; and botanical pharmacology. The small university was for all practical purposes a new one, having been reestablished by the young Duke Cosimo I of Tuscany in 1543.

Cosimo was a generous patron of learning, and he lent his support to projects of renovation promising to lead to innovation. One visitor to Florence who noted the value of Cosimo's patronage with special appreciation was John Caius, the leading medical humanist of his age, who visited Florence in the course of a journey through Italy in search of Greek medical manuscripts.[3] Innovative scholars in both learned professions—law and medicine—benefited from Cosimo's interest and encouragement. Thus, the celebrated edition of the Florentine manuscript of the *Digest* published in Florence in 1553, mistakenly believed to be the original commissioned by Justinian and containing significant variants from the text in general use, was prepared by Cosimo's legal adviser and dedicated to the duke.[4]

The actual organization of the University of Pisa seems to have been largely the work of the duke's adviser Francesco Campana and Campana's colleague Filippo del Migliore, who were charged to spare no effort to secure the best available talent for the Pisan chairs; in the spring of 1542 Del Migliore made a

---

views is to be found in Owsei Temkin, *Galenism: Rise and Decline of a Medical Philosophy* (Ithaca, N.Y./London: Cornell Univ. Press, 1973), pp. 141–144, 149–152. On sixteenth-century pathology see Vivian Nutton, "The Seeds of Disease: An Explanation of Contagion and Infection from the Greeks to the Renaissance," *Medical History,* 1983, *27*:1–34, and his essay in this volume; and Linda Deer Richardson, "The Generation of Disease: Occult Causes and Diseases of the Total Substance," in *The Medical Renaissance of the Sixteenth Century,* ed. A. Wear, R. K. French, and I. M. Lonie (Cambridge: Cambridge Univ. Press, 1985), pp. 175–194. The last attack on Argenterio so far known to me was Caspar Hoffmann, *Pro veritate opellae tres . . . III. Ant-Argenterius* (1647), which I have not seen.

[3] Vivian Nutton, *John Caius and the Manuscripts of Galen* (Cambridge: Cambridge Philological Society, 1987), p. 55.

[4] Hans Erich Troje, *Graeca leguntur* (Cologne/Vienna: Böhlau Verlag, 1971), pp. 41–43; and Anthony Grafton, *Joseph Scaliger: A Study in the History of Classical Scholarship,* Vol. I (Oxford: Clarendon Press, 1983), pp. 29–30, 64–65.

recruiting trip throughout northern Italy for this purpose. The results of their efforts as far as the faculty of medicine was concerned, although short-lived, were striking.

In the first ten years of the university's existence the medical faculty at Pisa included the elderly Matteo Corti, who enjoyed contemporary renown as a pioneer in restoring authentic Galenic medical teaching, and Leonardo Giacchino, a member of a little group of physicians who styled themselves a "new Florentine academy" and who coauthored a collection of *opuscula* attacking Arabo-Latin medicine published at Florence in 1533. Although Leonhart Fuchs, one of the leading exponents of the new botany in Germany, declined Cosimo's invitation to join the medical faculty at Pisa, Luca Ghini accepted and developed for Pisa one of the earliest university botanical gardens. Vesalius accepted an invitation to give the anatomy course in 1544; his dissections and lectures in January and February of that year met with great success. Indeed, Cosimo himself wanted to attend—suggesting that his attention to the medical faculty was not just an aspect of his general patronage of the university but was also a reflection of his personal interest in the more dramatic aspects of contemporary medical teaching. Despite Cosimo's best efforts, Vesalius refused a permanent position at Pisa, but shortly thereafter the university acquired in succession the services of two other accomplished young anatomists: Realdo Colombo from 1546 to 1548, and subsequently Gabriele Falloppio.[5]

Of course, not all the faculty were of the calibre of those named; and as Charles Schmitt observed, the more distinguished tended not to stay at Pisa for very long.[6] Nonetheless, the presence of so many professors of medicine of more than local reputation is striking. Local social or educational elites and patronage at the local level clearly played only a small part in determining the prevailing character of the medical faculty of Pisa in the 1540s. Instead, the significant appointments were the result of (and a tribute to) an ambitious program of recruitment that was not only extraregional but European in scope, a program that backed up informed policy with the financial support and prestige derived from generous princely patronage. The alertness of Cosimo's advisers to current trends and individual reputations in all branches of medical learning ensured that, even if their goal was chiefly to secure the most famous established scholars and most promising beginners for Pisa, they acquired a faculty of medicine that for the first two decades of its existence was notable for the presence of important medical innovators.[7]

[5] On the early history of the medical faculty at Pisa see Fabroni, *Historiae* (cit. n. 2), Vol. II, pp. 247–330 (much of this chapter, however, relates to the later sixteenth and early seventeenth centuries); and Charles B. Schmitt, "The Faculty of Arts at Pisa at the Time of Galileo," in his *Studies in Renaissance Philosophy and Science* (London: Variorum Reprints, 1981), esp. as regards the 1540s and 1550s, pp. 248–255. On Matteo Corti see Vivian Nutton, " 'Qui magni Galeni doctrinam in re medica primus revocavit': Matteo Corti und der Galenismus im medizinischen Unterricht der Renaissance," in *Der Humanismus und die oberen Fakultäten,* ed. Gundolf Keil (Weinheim: Acta Humaniora VCH, 1987), pp. 173–184; and Nancy G. Siraisi, *Avicenna in Renaissance Italy* (Princeton, N.J.: Princeton Univ. Press, 1987), pp. 187–192; on Giacchino, or Giacchini, *ibid.,* pp. 69–70. Regarding Luca Ghini and the Pisa botanic garden see A. G. Keller, "Luca Ghini," in *Dictionary of Scientific Biography,* ed. Charles C. Gillispie, 16 vols. (New York: Scribners, 1970–1980), Vol. V, pp. 383–384; for Vesalius at Pisa, and Cosimo's interest in dissection, C. D. O'Malley, *Andreas Vesalius of Brussels, 1514–1564* (Berkeley/Los Angeles: Univ. California Press, 1964), pp. 199–202.

[6] Schmitt, "Faculty of Arts," pp. 250–251.

[7] The focus of Medici patronage of science upon medicine and related fields (botany, anatomy,

Within this faculty, Giovanni Argenterio was at first appointed—on the advice of Campana and at the behest of Duke Cosimo—to a chair in medical theory, perhaps with the expectation that he would relate theoretical teaching to practical experience (as, for example, Giambattista da Monte seems to have tried to do at about the same time in Padua). Certainly, before coming to Pisa, Argenterio's professional experience and accomplishments consisted entirely of about nine years in medical practice, first at Lyons and then at Antwerp. Any reputation he had at that time must have been primarily as a practitioner, although, as he said of himself, he had also been dedicated to "good letters" from an early age. If the prefaces of the written works he produced later in his career are anything to go by, he was also a diligent cultivator of patrons. It was indeed a patron or client who persuaded Argenterio to come to Italy in order to teach and pursue humanistic and scholarly interests. Within a few months of his arrival, he had acquired a circle of patrician and intellectual acquaintances in both Venice and Florence, and the chair at Pisa.[8]

The extent and place of Argenterio's own medical studies are something of a puzzle. Presumably the early part of his education took place in his native Piedmont. He acquired the degree of master of arts from the University of Paris in June 1534; in about 1535, according to his own account, he began practice in Lyons. This does not leave very much time for any medical studies at Paris after the arts degree, and his name does not appear in the records of the Paris faculty of medicine. Yet Paris in the early 1530s seems a highly probable context for the origins of Argenterio's interest in the critique of Galen. At that time the Paris medical faculty was a major center of Renaissance Galenism. And Parisian teaching helped to form the two leading medical innovators of Argenterio's generation, each of whom—like Argenterio—both drew upon and reacted against the Galenic revival: Vesalius and Fernel.[9]

II. ARGENTERIO AT PISA

During his early years at Pisa, Argenterio's reputation evidently continued to grow, if one may judge by the well-informed medical travelers from abroad who sought him out: Guillaume Rondelet from Montpellier in 1549 and the German Lorenz Gryll, who stayed in Argenterio's house for some months in 1550.

natural history) at the University of Pisa, together with the impact on the university from the 1560s of economic and administrative problems and growing ecclesiastical influence at the Florentine court, is noted in Stefano De Rosa, "Alcuni aspetti della 'committenza' scientifica medicea prima di Galileo," in *Firenze e la Toscana dei Medici nell'Europa del '500,* Vol. II: *Musica e spettàcolo: Scienze dell'uomo e della natura* (Florence: Olschki, 1983), pp. 777–783.

[8] On Da Monte's pedagogical methods at Padua see Siraisi, *Avicenna* (cit. n. 5), pp. 98–103. Argenterio's own account of his decision to go to Italy in 1543, his reasons for the move, and the Italian circle he acquired is found in the preface, dated 1549, to *De generibus, et differentiis symptomatum* in his *Varia opera de re medica* (Florence, 1550), p. 116. Elsewhere he described how he began practice with his brother at Lyons in about 1535 and subsequently moved to Antwerp: commentary on *Aphorisms,* in his *Opera* (Venice, 1606) (hereafter ***Opera* [1606]**), *pars prior,* p. 148.

[9] Argenterio informed the Pisan authorities he had graduated in arts at Paris on 20 June 1534, according to Bonardi, *Studio generale a Mondovì* (cit. n. 2), p. 77, citing Archivio di Stato di Pisa, Università, Atti civili, anni 1548–49–50, parte, fols. 1, 10. I have not seen the document. Argenterio's name does not occur in the *Commentaires de la Faculté de Médecine de l'Université de Paris (1516–60),* ed. Marie-Louise Concasty (Paris: Imprimerie Nationale, 1964)—but neither does that of Vesalius (see also O'Malley, *Vesalius* [cit. n. 5], p. 36).

(Gryll's travels, financed by the Fuggers, took the form of a sort of grand tour from one medical celebrity to another.)[10]

Once translated to an academic environment, Argenterio began to produce medical writings. Probably by 1548 he had already composed the first version of a commentary based on his lectures on the Hippocratic *Aphorisms,* one of the two texts assigned at Pisa for the principal course in medical theory. Also in the mid 1540s he began work on a commentary on the second of these texts, the Galenic *Ars;* this commentary was to occupy him for some twenty years and ultimately contained the final summation of many of his ideas.[11] His transfer to a chair of practical medicine apparently also provided Argenterio the occasion to compose two works on fevers, based on his lectures.

In medieval and Renaissance universities parts of the curriculum in practical medicine were usually taught by means of lectures on an authoritative text, and the important subject of fever was one of these. At Pisa the medical textbooks prescribed by the statutes of 1543—unlike some of the appointments to the medical faculty in the same decade—were, on the whole, conservative. Nothing comparable to the rhetorical denunciation of medieval Arabo-Latin works contained in the contemporary statutes of the University of Tübingen (1538) is to be found in the section on medical textbooks in the Pisa statutes. The books prescribed were for the most part long standard in the medieval universities, although modest, but not insignificant, signs of the times were the absence of Book 1, Part 1, of the *Canon* of Avicenna from the list of works to be taught by the ordinary (that is, senior) professor of medical theory and the presence of lectures on Dioscorides. The prescribed textbook on fevers for the course on *practica* was Avicenna's manual on the subject, Book 4, Part 1, of the *Canon.*[12]

Argenterio's handling of the task of teaching about fever displays an intention to manipulate conservative, in the sense of medieval Arabo-Latin, elements in the statutory curriculum in ways consonant with the goals and standards of medical humanism. For at least two years he delivered lectures on *Canon* 4.1 in which, as he said, he made no attempt to interpret the author, but simply followed Avicenna's ordering of the subject matter for the sake of convenience. In so doing he was adopting tactics used since the 1520s by some medical professors in Italian universities who were confronted with the obligation to teach a traditional Arabo-Latin curriculum. They had devised strategies for introducing modern—in the sense of Renaissance Galenist—elements into their lectures. Matteo Corti was accused of ignoring the actual content of Avicenna's *Canon* in his lectures on that book at Pavia; at Padua in 1528/29 he used his lectures on Book 1, Part 1, of the *Canon* to stress Galen's superiority to Avicenna and to provide

[10] Laurentius Joubertus (Laurent Joubert), *Guilielmi Rondeletii vita, mors, et epitaphia . . . ,* in Joubert's *Operum latinorum tomus primus* (Frankfurt, 1599), p. 152; and Laurentius Gryllus, *Oratio de peregrinatione studii medicinalis ergo suscepta,* printed with his *De sapore dulci et amaro* (Prague, 1566), pp. 5–6.

[11] The commentary on the *Aphorisms* was in existence by 1551 (see below), although Argenterio was still revising it in 1562 (*Opera* [1606], p. 148); it was first published in his *Opera* (1606), *pars prior.* The preface by Bersano Benecchia to Argenterio's *In Artem medicinalem Galeni Commentarii tres* (Mondovi, 1566) describes the work as more than twenty years in the making.

[12] The relevant statutes are printed in *Urkunden zur Geschichte Universität Tübingen aus den Jahren 1476 bis 1550* (Tübingen, 1877), pp. 311–312; and F. Buonamici, "Sull'antico statuto della Università di Pisa: Alcune preliminari notizie storiche," *Annali delle Università Toscane,* 1911, *30*:46–47.

references to works or passages of Galen for every topic covered by Avicenna. In 1544–1546, also at Padua, Giambattista da Monte used his lectures on the *Canon* both as the framework for an attempt to unify the teaching of theory and practice and as the occasion for sharp attacks on some of Avicenna's views.[13]

Argenterio eventually went a step further than either Corti or Da Monte and discarded Avicenna entirely. He informed the *iuvenes studiosissimi* who constituted his audience, "I decided to dismiss this author from our schools." Instead, he lectured on Galen's *De febribus ad Glauconem*. One may assume that both the use Argenterio made of Avicenna and his subsequent substitution of a Galenic text constituted an entirely acceptable form of innovation in a milieu in which Galenism and anti-Arabism still constituted relatively novel and "advanced" tendencies. Nor is it necessary to suppose that Argenterio's announced intention to supplement and fill out topics not fully covered in Galen's treatise would in itself have been controversial.[14]

But the content of the commentary on the *Aphorisms* and of the treatises on fevers also suggests that by the time they were composed Argenterio had already begun to move beyond attacks on the Arabs—a cliché of medical humanism—to the much more radical step of a critique of Galenic pathology. The commentary on the *Aphorisms,* at any rate in the version ultimately published, also evinces a characteristic that became predominant in Argenterio's later works, namely, vehement attacks on individuals among the leading *moderni* of his own day. Controversy with colleagues and attacks on predecessors were of course staple ingredients in medical writing and by no means peculiar to Argenterio. Nonetheless, his attacks on other physicians are distinguished both by their frequency and by their special focus. Thus he seldom bothered to mention medieval Arabic or fourteenth- or fifteenth-century Latin authors by name, even for purposes of refutation. But he was a constant and unremitting critic of the most celebrated contemporary proponents of revived Greek medicine in general and Galen in particular, notably Giovanni Manardo, Martin Akakia, Matteo Corti, and Giambattista da Monte. In addition, long portions of the commentary on the *Ars* are given over to denunciations of an unnamed colleague at Tübingen who is probably to be identified with Leonhart Fuchs, and of that more idiosyncratic modern Jean Fernel.

III. ARGENTERIO'S ATTACK ON GALEN AND MEDICAL LEARNING

For reasons that will become apparent, Argenterio's writings on the Hippocratic *Aphorisms* and on fevers were not published until long after his death; and, as already noted, he delayed publication of his commentary on the *Ars* until late in his life. Consequently, his views first appeared in published form in a collection of treatises on disease theory issued under the title *Varia opera de re medica* in 1550. The choice of topic, physical format, dedications, and means of publication

[13] Nutton, "Qui magni Galeni doctrinam in re medica primus revocavit," pp. 182–183; and Siraisi, *Avicenna* (both cit. n. 5), pp. 189–192, 98–103, 195–202.

[14] Giovanni Argenterio, *De febribus,* pp. 1–65, on p. 2; and his commentary on *De febribus ad Glauconem,* pp. 66–168, on p. 66, in his *Opera* (1606), *pars altera.* The works on fevers were in existence by 1551 (see below and note 37), but were first published posthumously in the edition cited. On the contemporary perception of Galenism as reformatory or innovative up through the 1550s see Vivian Nutton, "Medicine in the Age of Montaigne," in *Montaigne and His Age,* ed. K. Cameron (Exeter: Univ. Exeter Press, 1981), pp. 15–25.

of this work all indicate the author's confident expectation that it would yield yet further rewards in the way of patronage, fame, and advancement. The topic was one bound to attract notice. The massive outbreak of syphilis in Europe that began in the 1490s together with various epidemics of other diseases had introduced a period of especially vigorous medical debate about the nature, transmission, and proper understanding of disease. Argenterio's treatises on the subject appeared in a handsome folio dedicated as a whole to Duke Cosimo; in it each separate treatise also bears a separate dedication to a different Florentine notable. Furthermore, the volume was published by what was in effect a state press: Lorenzo Torrentino (the italianized name under which Laurens Leenaertsz, a printer from Brabant, worked in Florence) held a ducal monopoly of printing in Florence for a number of years and published only works approved by ducal authority. Torrentini's press was an important instrument for officially approved cultural diffusion; he printed major literary and historical works as well as the famous edition of the *Digest* already mentioned.[15]

The front matter of Argenterio's volume immediately plunged the reader into a sweeping critique of contemporary medical learning. The complimentary preface to Cosimo began with an attack on excessive dependence on Galen which in the following preface to the reader broadened into a general condemnation of the main components common to both scholastic and humanist forms of academic education in the arts, philosophy, and medicine: reliance on ancient authority, teaching by commentary, and disputations or other forms of debate about texts.

The grounds on which Argenterio objected to these practices were that the most esteemed ancient authors, notably Galen and Aristotle, were in many instances either factually wrong or internally inconsistent; that exclusive focus on the subject matter they had chosen to treat unnecessarily limited the agenda of inquiry; and that the time, energy, and ingenuity expended in textual exegesis would be better used in the direct investigation of nature. Thus of Galen he remarked, "For because Galen . . . wrote almost infinite volumes, he must have made many ill-considered statements without thinking about them. . . . Hence, our credulity, or rather our stupidity, should be even more condemned when we receive all his books as absolutely true (ignoring the author's own opinion) and attribute equal authority to each one." He further opined that the belief that a few men in antiquity had known everything was the reason why "the disciplines have received no increment over so many centuries, even though outstanding talents were always dedicated to them," and added that no builder built a house according to directions laid down by an ancient authority.[16]

Argenterio characterized the practice of writing commentaries as spending one's whole life in "putting forward explanatory opinions about other people's books, and constructing new senses never thought of by the author," and he went on to inquire:

> Would not the ancients have made greater progress in philosophy if Alexander, Themistius, Simplicius, and so many other most learned men, who strove with one another in interpreting Aristotle and explaining his every word in a recondite manner,

[15] On Torrentino and his press see Roberto Cantagalli, *Cosimo I de' Medici, granduca di Toscana* (Milan: Mursia, 1985), pp. 133–134. Regarding earlier discussions of disease see the article by Danielle Jacquart in this volume; for the sixteenth-century debate see the article by Vivian Nutton.

[16] Giovanni Argenterio, *Ad lectores*, in *Varia opera* (cit. n. 8), pp. 7–19, on pp. 8–9.

> had instead taken up the elaboration and illustration of some part of philosophy treated more negligently by our forefathers? For meanwhile because we examine with superfluous labor the vacuum, the infinite, the three principles, and the opinions of the ancients, we are ignorant of the nature of metals and stones; nor do we know enough of the history of herbs and plants.[17]

As for disputation, he duly noted the opinion that *quaestiones* sharpen wits but commented, "I would not deny it; however, it is permissible to accomplish the same thing on more useful material; for the mathematical disciplines, than which our forefathers judged there to be nothing better for cultivating ingenuity and making it more acute, lie despised." Thus, he concluded, the teaching of all disciplines had been corrupted by the barbarous *quaestionaria doctrina* and by badly conceived views about two or three Greek men—a provocative formulation that united standard and already somewhat shopworn humanist invective against scholastic disputation with a frontal attack on both Aristotle and Galen.[18]

That the attack on Galen was also intended to encompass contemporary enthusiasts for Greek medicine who are not named in the treatises on diseases is clear both from the opening of the preface to Cosimo and from the detailed criticisms of individual *moderni* in the author's other works. From the commentary on the *Ars* it emerges with equal clarity that Argenterio objected not only to particular opinions or textual interpretations espoused by leading contemporaries but also to the methodology and goals of the humanistic approach to Greek medical texts. In its preface he characterized the professors of "the most outstanding academy in the whole world (as the vulgar believe)"—presumably Paris—as grammarians rather than physicians because of their emphasis on Greek philology and Latin rhetoric (throughout the commentary he referred to his medical opponent at Tübingen as *grammaticus Tubingensis*). And he explicitly paralleled humanist philology with scholastic *quaestiones* and reconciliation, asserting that neither provided a valid method for achieving real knowledge in medicine.[19]

The treatises on diseases themselves are linked by their titles to similarly or identically named works of Galen (e.g., *De morborum differentiis, De causis morborum, De causis symptomatum*); at least implicitly, Argenterio thus presumably claimed in the case of Galen, as he had done in his lectures on Avicenna, to retain the earlier author's title and organization, while substituting his own, superior, treatment of the subject. In reality, the works do not amount to anything like a systematic endeavor to discard Galen or to devise an alternative pathology either on the basis of texts of other authors or on that of experience. On the contrary, these treatises are highly dependent upon Galen in a double sense: despite the presence of some innovations or shifts of emphasis claimed as Argenterio's own, they present a pathology that remains, by and large, Galenic; and at the same time, criticism of Galen's opinions on various specific topics constitutes a very large part of their content.[20]

[17] *Ibid.*, p. 10.

[18] *Ibid.*, pp. 17–19.

[19] Giovanni Argenterio, *Tres in Artem medicam Galeni commentarii*, in *Opera*, 3 vols. in 1 (Venice, 1592) (hereafter ***Opera* [1592]**), Vol. I, p. 7.

[20] The titles of the remaining treatises in the collection are *De symptomatis, sive morborum effectibus* (divided into *De generibus, et differentiis symptomatum*, and *De causis symptomatum*); *De temporibus, sive partibus morborum; De signis medicis;* and *De officiis medici* (in editions after the first the title refers to fourteen books on diseases because internal subdivisions of each treatise are

The most striking aspect of these works is, indeed, an attack on Galen of a scale and overtness that seem unprecedented in academic circles. A crude idea of its scope can be gained from a rough comparison between the index entries under "Galen" in Argenterio's collection and an approximately contemporary work famous for—among other things—its significant criticism of Galen: Vesalius's *Fabrica*. The 1555 edition of the *Fabrica,* a volume of 824 pages, has about 300 index entries under Galen's name; a good many of them, but by no means all, indicate that they refer to instances of Vesalius's disagreement with Galen. The index to the 278 pages of the treatises on diseases in the edition of Argenterio's collected works published in 1592 has about 175 such entries, almost every one of which refers to an explicit criticism.[21]

Argenterio's central contention in the treatises on diseases was the ambiguity and inconsistency of Galen's definitions of the terms "disease," "cause" of disease, and "symptom," and the inadequacy of Galenic classifications or enumerations of diseases, causes, and symptoms. In particular, Argenterio returned over and over again to the charge that Galen provided no secure and consistent basis for distinguishing between symptom and disease, for differentiating one *affectus* of the body from another, or for relating particular conditions of the body to particular causes. The specific arguments relating to pathology developed by Argenterio both in the treatises on diseases and throughout his subsequent works all seem to be related to this central cluster of ideas.

Thus Argenterio, along with Fernel, became well known for developing the concept of "disease of the total substance" as an explanatory category for highly specific and recognizable diseases, such as plague and the *morbus gallicus,* that did not fit easily into the standard categories of temperamental imbalance, trauma, or malformation (*mala complexio, solutio continuitatis, mala compositio*). The idea of disease of the total substance was usually justified by an appeal to Galen's own doctrine that, whereas most medicinal substances act through the balance of elementary qualities, the effect of some is due to other, specific, properties, a notion that Arabic and earlier Latin writers had developed into the doctrine of specific form.[22] But Argenterio's interest in disease of the total substance, like his objection to classifying as diseases calcifications formed in the body, such as bladder stones, or anomalies present since birth, such as a sixth finger, was only one aspect of a more general dissatisfaction with contemporary

---

enumerated). Galen wrote treatises *De causis morborum* (*Claudii Galeni Opera omnia,* ed. K. G. Kühn [Leipzig, 1821–1833], hereafter cited by **Kühn** and volume and page nos., Vol. VII, pp. 1–41), *De symptomatum differentiis* (Kühn, Vol. VII, pp. 42–84), *De symptomatum causis* (pp. 85–272), *De morborum temporibus* (pp. 406–439), and *De morborum differentiis* (Vol. VI, pp. 836–880). I have not yet attempted to determine how closely Argenterio followed the internal organization of any of these treatises.

[21] Argenterio, *Opera* (1592), has a separate index for each volume; the treatises on diseases occupy the whole of Vol. II. Temkin, *Galenism* (cit. n. 2), p. 151, notes the plethora of index entries critical of Galen in the Paris 1578 edition of Argenterio's commentary on the *Ars*. Indexes do not, of course, necessarily reflect an author's intentions (since they may be the work of an editor or publisher), but they provide information about the way a work was presented to the public.

[22] See Richardson, "Generation of Disease" (cit. n. 2); and for the background of Galenic and medieval pharmacological theory and the notion of specific form, Michael R. McVaugh, introduction to Arnau de Vilanova, *Aphorismi de gradibus,* Vol. II of *Arnaldi de Villanova Opera medica omnia,* ed. L. García-Ballester, J. A. Paniagua, and M. R. McVaugh (Granada/Barcelona: Univ. Granada, 1975).

acceptance of the terms "disease," "symptom," "accident," or "cause" (of disease) as having secure meanings, when, in his view, the texts from which these meanings were drawn were inconsistent, confused, and uncertain. In his words:

> Galen defines disease in a variety of ways: for sometimes he calls disease that which in the first place injures action. . . . And he is also accustomed to define it in another way, that it is the disposition or *affectus* contrary to nature which in the first place vitiates action. Although the latter definition is received by everyone it does not seem much better than the first to me. . . . Moreover, in *De differentiis symptomatum* he writes that not only disease itself but also symptoms and causes of diseases are unnatural dispositions.[23]

> Therefore he understands by the word "symptom" sometimes whatever is contrary to nature; sometimes that which exists according to nature; sometimes that which follows disease; sometimes that which comes not only from diseases but also from the causes of disease; sometimes he excludes the actions of disease from symptoms; sometimes indeed he also wants symptoms to follow health and a state according to nature; sometimes he wants a symptom to be not only that which occurs in the sick, but also [that which occurs as a result of] nature conquering and expelling what is present; sometimes a symptom is for him whatever is committed by the disease, whether it is the disease itself, or something else; sometimes he decrees a symptom to be the effects which are present in the patient, such as a symptomatic excretion. So now this author teaches nothing consistently and firmly; he changes his mind every time he deals with the subject.[24]

The semantic, philological, and rational character of these objections typifies much of Argenterio's critique, which focused chiefly on pointing out internal inconsistencies in Galen's texts or logical flaws in his arguments. For Argenterio as much as for any of his contemporaries the language of medical controversy was still largely a language of textual criticism. But the underlying disillusionment appears to be that of a practitioner who has made a serious endeavor to use Galenic texts as a guide to pathological understanding in practice. Such at any rate is the impression forcefully conveyed by the episode Argenterio vividly recalled from his sojourn at Pisa. He recounted the story in his commentary on the *Ars:*

> Galen is accustomed to say in *De locis affectis* and everywhere else that some things are ejected from the mouth by spitting only, and he shows these things to indicate a condition of the mouth and palate; others are ejected by excretion or, as the vulgar say, *rascatio,* and these indicate diseases of the throat; others with a light cough, which indicate injuries of the trachea, which is contained in the neck; but others are expelled with a real cough, and these are signs of injury to the lung. And he teaches that ulcers of the lung can be distinguished from ulcerated or bleeding parts just mentioned by the following indication: bloody foaming sputum. In reality, I have learned it can happen otherwise, to my great ill fortune. For Francesca Damiana, my wife, a most beautiful and praiseworthy woman, was twenty years old at the time when I was living with her at Pisa. When the south wind blew (which is very frequent and strong there), as a result of secretions from the head, she began to spit blood in the morning when she was still in bed. As a result, she was profoundly terrified. But since I was instructed in Galen's doctrine, I told her that there was nothing to be

[23] Giovanni Argenterio, *De morbi generibus,* in *Opera* (1592), Vol. II, pp. 1–2. For his views on bladder stones, and disease of the total substance, *ibid.,* pp. 5–6, 12.
[24] Giovanni Argenterio, *De generibus, et differentiis symptomatum, ibid.,* p. 91.

> afraid of, because she brought up blood easily, without a cough, and with only light *rascatio,* and she began to feel better. Subsequently she did not spit up blood, but phlegm and pus, with cough and fever, which took hold right at the start, and from these ills, which began in autumn, at about the end of May she died.[25]

Argenterio refused to allow his wife to be dissected, but he later attended the autopsy of the wife of a colleague who had died after manifesting similar symptoms: "From these things I gather that not everyone who dies of consumption always spits blood, and those who do, do not always eject it with coughing."[26]

In reality, Argenterio's critique of Galen drew—whether he recognized it or not—on some of Galen's own ideas. Galen, too, had stressed the importance of carefully distinguishing between disease and symptom.[27] Argenterio's own contributions in the treatises on diseases usually consisted of attempts to select among, refine, add to, and render more precisely descriptive existing classificatory categories; and he did not hesitate to draw on Galen for his own purposes. His procedure may be illustrated from his discussion of causes: He embarked on a critique of Aristotle's four causes from a Galenic standpoint, taking from Galen the idea that it was necessary for the physician investigating the causes of disease to subdivide the efficient cause into initial, antecedent, and cohesive causes. However, Argenterio's emphasis on the multiplicity of specific, idiosyncratic symptoms and diseases, which necessarily entailed a belief in forms of causation peculiar to specific situations or substances, led him to multiply the types of causation even further, adding two more categories of his own. And he dismissed consideration of final causes of diseases as irrelevant to medicine.[28]

Argenterio's treatises on disease seem closely allied to a genre of Renaissance medical writing that may be termed counter-commentary. Two striking examples of this genre produced in the Italian academic milieu a few years before Argenterio's treatises on diseases were Matteo Corti's commentary on the *Anatomia* of Mondino dei Liuzzi, and the section on humors of Giambattista da Monte's commentary on Book 1, Part 1, of the *Canon* of Avicenna, both notable for the severity with which the commented author is taken to task.[29] Of course, the flexibility of commentary as a form and, again, the pedagogical necessity of continuing to provide lectures on books in a traditional curriculum facilitated the development of commentaries of this type. In the body of the treatises on diseases, as in the general strictures on Aristotle and Galen in the prefatory matter, Argenterio's strategy thus seems to have been to turn against Renaissance Galenism itself a weapon already used by *moderni*—in the sense of Renaissance Galenists—against aspects of medieval Arabo-Latin medicine.

Insistence on the internal inconsistencies of the major medical authority, and on the insecure and fluctuating character of basic definitions in medicine, had, of

[25] Argenterio, *In Artem medicam Galeni, ibid.,* Vol. I, p. 390.

[26] *Ibid.*

[27] E.g., in Galen, *Methodus medendi* 2.1–3 (Kühn, Vol. X, pp. 78–93), and elsewhere.

[28] Giovanni Argenterio, *De causis morborum* 1.2–9, *Opera* (1592), Vol. II, pp. 41–51. Argenterio's discussion *de causis procatarcticis et antecedentibus* (Ch. 8) and *de causa conjuncta* (Ch. 9) indicates that he was among the few who made use of Galen's treatise *De causis procatarcticis;* on this work and its Renaissance fortuna see Nutton, "Seeds of Disease" (cit. n. 2), p. 4.

[29] Aldo Scapini, *L'archiatra mediceo e pontificio Matteo Corti (secolo XVI) e il suo commento all'Anatomia di Mondino di Liuzzi* (Scientia Veterum, 142–143) (Pisa: Giardini, 1970), prints an Italian translation of Corti's commentary on Mondino; and Giambattista da Monte, *In primi libri Canonis Avicennae explanatio* (Venice, 1554).

course, implications for discussions of the status of medical knowledge. Since the thirteenth century the institutional, social, and intellectual priorities of the university milieu had encouraged learned medical writers to try to assimilate at least some aspects of medicine to an Aristotelian model of scientific—that is, certain—knowledge acquired by proceeding from accepted first principles through rational (syllogistic) demonstrations to universal conclusions. Yet such writers, who were themselves practitioners and teachers of practitioners, also recognized that such a model could not easily be used to describe the kind of knowledge underlying most of the therapeutic activities, information, and skills with which they were familiar.

The result was that thirteenth- to fifteenth-century discussions of the status of medical knowledge usually concluded that medicine partook in some way or another of the nature of both *scientia* and *ars;* usually, physiological and pathological theory were the parts of medicine held to partake most fully of the nature of *scientia.* However, such discussions provided few examples of what their authors considered to be certain demonstrations in medicine. Moreover, among important scholastic medical authors, Arnau de Vilanova never fully accepted the assimilation of any part of medicine to *scientia.* And even the very philosophically oriented Pietro d'Abano (d. 1316) readily acknowledged the presence of a conjectural element in some aspects of medical knowledge.[30]

By the time Argenterio began to stress the themes of internal inconsistency in Galen, and the confusing and ambiguous nature of the various Galenic identifications or definitions of "disease," "cause," and "symptom," several additional factors were likely to affect discussions of medical knowledge. One of these was the interest of Renaissance physicians in the history of ancient medical sects (who had differed precisely over the kind of knowledge obtainable in medicine), an interest both fostered and manifested by the late fifteenth- and early sixteenth-century editions or translations of some of the principal ancient works describing and evaluating the sects—Galen's own *De sectis ad introducendos,* the *Introductio* or *Introductorius* attributed to Galen, and the proem to Celsus's *De re medica,* which drew further attention to the idea of medicine as "conjectural." Another, related, factor was awareness of ancient philosophical skepticism.[31]

[30] On Arnau de Vilanova see the article by Michael McVaugh in this volume. For medieval treatments of the theme of certitude of medical knowledge, with special reference to the views of Pietro d'Abano and Gentile da Foligno (d. 1348), see Nancy G. Siraisi, "Medicine, Physiology, and Anatomy in Early Sixteenth-Century Critiques of the Arts and Sciences," in *New Perspectives on Renaissance Thought: Essays in History of Science Education and Philosophy in Memory of Charles B. Schmitt* (London: Duckworth/Istituto Italiano per gli Studi Filosofici, 1990), pp. 214–229. Pietro's remark about conjecture in medicine occurs in his *Conciliator* (Venice, 1565; facs. rpt. Padua: Editrice Antenore, 1985), *differentia* 3, fol. 7r.

[31] *De sectis* (Kühn, Vol. I, pp. 64–105) is Galen's principal account of the ancient medical sects, but there are also scattered passages in his other works. On the influence during the Renaissance of Celsus's account of the sects, and Galen's own writings as a source of information about ancient skepticism, see J.-P. Pittion, "Scepticism and Medicine in the Renaissance," in *Scepticism from the Renaissance to the Enlightenment,* ed. Richard H. Popkin and Charles B. Schmitt (Wolfenbütteler Forschungen, 35) (Wolfenbüttel, 1987). *Introductio seu medicus,* trans. Niccolò da Reggio, was included in early editions of Galen's *Opera.* See Richard J. Durling, "A Chronological Census of Renaissance Editions and Translations of Galen," *J. Warburg Courtauld Inst.,* 1961, *24*:279–280, nos. A1, A4. Argenterio used *De sectis,* Celsus, and the *Introductio* extensively for the account of the sects in his commentary on the *Ars;* he was among those who recognized the *Introductio* as spurious, referring always to "the author of the *Introductorius.*" *Opera* (1592), Vol. I, pp. 29–31. On Renais-

Thus, while I know of no specific evidence that Argenterio was familiar with either Gianfrancesco Pico's *Examen vanitatis doctrinae gentium* (1520) or Henricus Cornelius Agrippa's *De incertitudine et vanitate scientiarum* (1526), both of these works provided models and materials for a drastic reevaluation of the status of medical knowledge. Pico's exercise in radical skepticism devoted much attention to exposing the internal contradictions in a major ancient scientific authority (in this case, Aristotle), and insisted upon the multiplicity of mutually contradictory theories and the absolute lack of any secure knowledge concerning central aspects of human physiology that had been the subject of centuries of philosophical speculation and scholastic debate. The intentions of Agrippa's more popular and rhetorical work remain ambiguous. But it disseminated to a wider circle of readers the theme of the inconsistencies, uncertainties, and mutual contradictions in physiological theories; gave an apparent endorsement to ancient methodism; and assembled in one place a large array of traditional criticisms—heightened with effective rhetorical vituperation—of the efficacy of therapy and the honesty of medical practitioners.[32]

Argenterio's own mature reflections on the subject of the status of medical knowledge are to be found in his commentary on the *Ars*. His arguments are too lengthy to be summarized adequately here, but their main conclusions may be noted. Although he allowed that demonstrations could occasionally take place in medicine, he firmly denied that either medicine or natural philosophy could be classified as *scientia,* that is, as offering certain knowledge (for *scientia* he required that demonstrations always take place, offering the example of geometry). But he also denied that medicine was *ars,* on the grounds that it could not be described as a *habitus* of acting with reason; real arts, according to Argenterio, proceeded according to fixed rules that could be mastered and controlled by the practitioner, but in medicine "the principle of action is often in nature, which can make infinite and complicated motions, so that always new precepts are to be constituted on account of new diseases and new mutations that emerge from nature which one person interprets in one way and another in another." Argenterio was equally decisive in his rejection of the intermediate positions that medicine was partially *scientia* and partially *ars* or both simultaneously. Of the former idea, he remarked, "I add that the part that is called science [usually, physiological and pathological theory] does not have more certain principles . . . than the other."[33]

For the sake of convenience, however, Argenterio allowed that medicine might loosely be termed an art. Indeed, he wondered why it should be a term of abuse to call medicine a mechanical art, since the usual definitions of a mechanical art as involving repair, service, and the use of physical instruments described an honest, and indeed noble, activity when applied to the healing of the human body. He concluded that "lazy philosophers" upheld their own dignity (and inactivity) by affecting to consider such activity inferior to contemplation.[34]

---

sance skepticism in general see Richard H. Popkin, *The History of Scepticism from Erasmus to Spinoza* (Berkeley/Los Angeles: Univ. California Press, 1979).

[32] The problems raised by the medical portions of these works are discussed, and the historiography concerning them reviewed, in Siraisi, "Medicine, Physiology" (cit. n. 30).

[33] Argenterio, *In Artem medicam Galeni,* in *Opera* (1592), Vol. I, pp. 33–42, on pp. 37, 38.

[34] *Ibid.,* p. 42.

On the related question whether medicine was properly described as conjectural, Argenterio noted that Galen had frequently described it as such on a variety of different grounds: the difficulty of recognizing diseases, deciding the proper occasion for action, determining the quantity of remedies, and prognosticating the outcome. Argenterio's own view was that medicine was conjectural, not because of the uncertainties of practitioners faced with specific decisions about diagnosis, treatment, and prognosis, but because it was simply false to suppose that medical teaching embodied precepts that were certain:

> For (as he himself bears witness) he [Galen] can perceive nothing certain in the knowledge and distinction of the pain of colic and nephritis, nor in the cure of diseases of the whole substance, nor in signs which teach crisis to be about to take place through purging of the belly, or the transmutation of diseases, and in many things recovery or death. . . . For we think the art itself, on account of its mode of reasoning, is conjectural in certain things, for example as regards signs and other things that we dealt with before [presumably in the treatises on diseases].[35]

The problem for Argenterio, in short, was not the admitted uncertainties of therapy, but the inadequacy of the whole understanding of disease.

Argenterio's statements about the conjectural nature of all medical knowledge seem to conflict with his repeated demands for more exact and refined classification in pathology, and raise the question whether he considered his own theories any more securely based than those he was trying to replace. But whatever Argenterio's actual views on this point, one possible reading of his treatises on diseases is not merely as an attack on both ancient and modern medical science but also as a repudiation of any possibility of certainty in medical knowledge.

IV. REACTION TO ARGENTERIO

According to Argenterio's pupil Reiner Solenander, the treatises on diseases attracted such widespread attention that booksellers sold out of copies of the first edition. Hostile reaction was immediate, both at Pisa and elsewhere. The same source informs us that "all of France, the whole of Germany, and Italy . . . began to rise up in order to suppress [Argenterio's] labors and industry."[36] In a preface contributed to Argenterio's next publication, which appeared in 1551, Lorenz Gryll wrote that a group of Argenterio's opponents had made a determined effort to drive him out of the schools. Certainly, Giulio Alessandrini (1506–1590), trained at Padua and subsequently for many years an imperial physician at Vienna, connected with influential medical circles in both Italy and Germany, published in 1552 a treatise entitled *Antargenterica*. Again according to Gryll, local opposition at Pisa was sufficient to cause Argenterio to fear the loss of princely favor and to put a stop to his plans to publish the commentaries on the *Aphorisms* and *De febribus ad Glauconem*, together with "not a few other works." Argenterio's dedication of the work in which Gryll's preface occurs, a

[35] *Ibid.*, pp. 42–43.

[36] Reiner Solenander, *Apologia qua Iulio Alexandrino respondetur pro Argenterio* (Florence, 1556), pp. 9–11, 14. On Solenander (1524–1601), a German whose medical education was obtained at Louvain and in Italy, and who subsequently practiced at the baths of Lucca and at the Ospedale Santo Spirito in Rome, see *Biographisches Lexikon der herrvorragenden Ärzte aller Zeiten und Volker,* ed. August Hirsch, 2nd ed. revised, ed. A. Albert et al., Vol. III (Berlin/Vienna: Urban & Schwarzenberg, 1934), p. 334.

bland tract on consultation or collegiality in medicine, to the patriarch of Venice with accompanying fulsome praise of Venetian administration of the University of Padua, may be evidence of an unsuccessful effort to engineer a transfer to the larger and better known university.[37]

The nature of the reaction of the Pisan faculty to Argenterio becomes apparent from a little treatise against his views on putrefaction published by Remigio Meliorati in 1553; the booklet concludes with the "friendly" warning that Argenterio should try like the rest of those who teach at Pisa (men of the highest intellectual and moral integrity) to grow in good arts rather than follow sophist boasting. What chiefly distressed Meliorati, who was a professor of philosophy at Pisa, were Argenterio's departures not from Galen, but from Aristotle. However, he summed up his objections by way of an extended comparison between Argenterio and the ancient methodist physician Thessalus, in which an association between the name of Thessalus and the skeptical views sometimes attributed to the methodists seems apparent.

Thessalus was said by Galen to have claimed that no previous medical sect had established anything with certainty, to have competed in medicine on the basis of oratorical or rhetorical skills, and to have announced a new sect of his own; he was also said to have asserted that he could teach all there was of medicine in six months. According to Meliorati, Argenterio, like Thessalus, had tried to overturn the entire medical tradition handed down from Hippocrates and carried on by generations of learned men; but he exceeded even Thessalus in impudence when he referred to Galen, Hippocrates, and Aristotle as "three Greeklings." Argenterio, declared Meliorati, called himself a methodist (*methodicus*), even though he lacked the command of dialectic necessary for the real study of method (*methodus*).[38]

An explicit association of Argenterio's ideas with ancient skepticism was made a generation later by Giorgio Bertini, author of a work directed against both Paracelsianism and Argenterio. Bertini was unambiguous: he characterized Argenterio as "a new Academic among the physicians," one who seemed knowingly and deliberately to besmirch the purest sources of medicine while "he called back into use the old Academic custom, or the cavilings of the Pyrrhonians, and pronounced doubtingly about certain and true things."[39]

[37] Giovanni Argenterio, *De consultationibus medicis sive (ut vulgus vocat) de collegiandi ratione liber* (Florence, 1551), preface *ad medicinae studiosos* by Lorenz Gryll, pp. 10–12. Giulio Alessandrini (Julius Alexandrinus) *Antargenterica pro Galeno* (Venice, 1552); on Alessandrini see Mazzuchelli, *Scrittori* (cit. n. 2), Vol. I, pp. 449–451.

[38] Remigio Meliorati, *De putredine ad Ioannem Argenterium disputatio* (Florence, n.d.; dedication dated Pisa, 1553), pp. 50–53, on p. 50: "videris ergo Tessali temeritate superare, cum Galenum, Hippocratem, Aristotelem tres Graeculos appellas." Galen's account, and denunciation, of Thessalus appears in *Methodus medendi* 1.2 (cit. n. 27), pp. 7–18, and elsewhere.

[39] Giorgio Bertini, *Medicina libris viginta methodice absoluta. In qua mutuus Graecorum et Arabum consensus legitima veteris medicinae adversus Paracelsistas defensio, vera animadversionum Argenterii in Hippocratem et Galenum confutatio* . . . (Basel, 1587), 8.25, col. 204. In this work only Book 8, Ch. 25, and Book 20, Ch. 52, are specifically directed against Argenterio. "Occurrit primus Ioannes Argenterius novus inter medicos Academicus, qui ad ostentationem sui potius quam ad publicam utilitatem censoria quadam animadversionis severitate, et novae doctrinae usurpatione acumen ingenii in Hippocratem et Galenum convertit, fucataque levissimorum argumentorum facie huius artis speciem tot firmissimis et constantibus ornamentis et luminibus coloratam deformare, prorsusque delere, conatus est. Unde vitiligatoris potius quam interpretis partes agere, mentemque autorum commentariis suis veluti sepia attramenti effusione magis obscurare quam illustrare, sed et limpidis-

Within two years of the publication of Meliorati's treatise, Argenterio left Pisa. He secured an appointment at Naples, whence he mounted an energetic campaign in his own professional and intellectual defense. In an oration delivered at the beginning of the academic year 1555, he paralleled his endeavors with recent advances in anatomy, which he presented as constituting both a model and a justification for the reformulation of other areas of medicine.[40] In the following year he published a new, expanded edition of the treatises on diseases. Despite solicitations from "excellent printers in Lyons, Basel, and Venice," the revised edition first appeared in Florence, again printed by Torrentino, with a new title, *De morbis libri XIV,* but with the same combative preface and dedications to Florentine notables (a Lyons edition followed two years later).[41] Interest in continued good relations with patrons influential in the affairs of the University of Pisa is also implied by the dedicatory preface of Argenterio's work on sleep and waking, also issued at Florence in 1556, addressed to Filippo del Migliore and congratulating him on his role in founding and sustaining the University of Pisa. And Argenterio guided the pen of Reiner Solenander in producing his reply to Alessandrini's *Antargenterica,* to which in 1558 Alessandrini in turn replied with an *Antargenticorum suorum defensio.*[42]

In these treatises, both attacker and defender expended most of their energy on details of Argenterio's anti-Galenic teaching that need not be considered here. More important for our purposes, Solenander took up and developed the theme that Argenterio was only one of a number of moderns who had made new contributions to various areas of medicine. According to Solenander, surgery, anatomy, and pharmacy (especially the preparation of distilled or alchemically prepared medicines) had all improved since the days of Galen. He compared Argenterio not only to Vesalius but also to a series of other *moderni* in medicine and natural history, all of whom had differed from, and improved upon, Galen and other ancients: Georg Agricola, Jean Fernel, Guillaume Rondelet, and Giambattista da Monte. Vesalius especially had criticized Galen in much sharper terms than Argenterio did, so why should Argenterio be condemned, when Vesalius was praised?[43] Yet another version of the same idea crops up in the preface contributed to the 1566 edition of Argenterio's commentary on the *Ars* by another pupil, Bersano Benecchia, who evidently, like Solenander, served as a mouthpiece for Argenterio himself: this time the list was of those who had fruitfully criticized Galen and consisted of Avicenna and Averroes, the Byzantine

---

simos medicinae fontes sciens prudensque conspurcare videatur, dum veterem Academicum morem in usum revocat, et de certis et veris, seu ut Pyrrhonii factitare solebant, dubitanter pronunciat." *Ibid.,* 8.25, col. 204.

[40] *Oratio Ioannis Argenterii Neapoli habita in initio suarum lectionum anno 1555,* unnumbered folio at beginning of *Opera* (1606), *pars prior.*

[41] For the circumstances surrounding the preparation of the edition, Solenander, *Apologia* (cit. n. 36), pp. 14–16. Two separate editions of *De morbis libri XIV,* published at Lyons in 1558, are indicated in *Index aureliensis: Catalogus librorum sedecimo saeculo impressorum* (Bibliotheca Bibliographica Aureliana), Pt. 1, Vol. 2 (Orléans, 1966), nos. 107.231 and 107.232, but these may instead be variant versions of a single edition.

[42] Giulio Alessandrini, *Antargentericorum suorum defensio adversus Galeni calumniatores* (Vienna, 1558). For Solenander's *Apologia,* which Alessandrini claimed (sig. * ii) was virtually dictated by Argenterio, see n. 36 above.

[43] Solenander, *Apologia,* pp. 19–24, 171–172.

compiler Alexander of Tralles, Fernel, Vesalius, Da Monte, and Girolamo Cardano.[44] (This line of defense conveniently overlooked the fact that Fernel and Da Monte had been the objects of some of Argenterio's sharpest attacks.) Moreover, both Solenander and Benecchia asserted that other *moderni,* notably Leonhart Fuchs, made extensive use of Argenterio's ideas without acknowledgment—an accusation Argenterio also made with regard to Fernel.[45]

Alessandrini introduced his response to the attempt to class Argenterio with Vesalius, Agricola, Rondelet, and Fernel with the remark that they would probably be insulted by the comparison. He went on to assert that, as regards the first three at any rate, the parallel was totally invalid. In particular, he explained to his readers that the work of Vesalius (for whom Alessandrini professed high esteem) was not in any way destructive of Galenic medicine. Rather, Vesalian and Galenic anatomy were simply different enterprises: the goal of the former was to investigate human anatomy as thoroughly as possible; that of the latter had been to find out whatever was necessary for medicine, and this Galen was able to learn from the anatomy of apes. Moreover, for simian anatomy Galen's account remained correct. As for Agricola and Rondelet, not only they but other modern authors—Pietro Mattioli on herbs and Conrad Gesner on quadrupeds and birds—had assembled many useful new data on various subjects, just as the new world had yielded valuable and hitherto unknown remedies; but these new particulars fitted into the established and ordered system of medicine. And even Fernel, in Alessandrini's opinion, seldom departed far from Galenic principles when treating of subjects Galen had developed fully, although it had to be admitted that Fernel's ideas about the hidden causes of things strayed beyond the boundaries of "our medicine that is built on reason."[46]

Although his remarks about Galen's anatomy are perhaps somewhat disingenuous, Alessandrini's evaluation of the relation of the moderns he names to Galenic tradition is perceptive; and his own position seems to represent a decidedly open, critical, and accommodating form of adherence to that tradition. Indeed, he himself was active in support of Mattioli's botanical work and engaged in the endeavor to distinguish authentic from falsely attributed works of Galen.[47] Alessandrini was, to be sure, a committed Galenist, but by the mid-sixteenth century

[44] Bersano Benecchia, preface to Argenterio's *In Artem medicinalem Galeni* (cit. n. 11), sig. † iiii v.

[45] *Ibid.*, sig. [† v]r; for the accusation from Solenander, see also n. 36 above. Of Fernel's distinction between *affectus* and *affectio,* Argenterio remarked, "Haec quidem Fernelius in ultima editione suae medicinae tradit, post eaquam nos opus nostrum de morbis edidimus, et illi materiam dedimus scribendi de his rebus, et a nobis dissentiendi, interimque transferendi ex nostris multa." *In Artem medicam Galeni, Opera* (1592), Vol. I, p. 132 [repeated page number].

[46] Alessandrini, *Antargentericorum suorum defensio* (cit. n. 42), fols. 13r–17r. On Fernel: "Nam de Ioanne Fernelio viro comprimis docto nihil amplius dico. Extant illius opera legere volentibus obvia, ex quibus prudens ac diligens lector coniecturam facile facere possit, atque ita demum iudicare quantum inter hunc et illum intersit. Immo vero quantum homo homini praestet, Fernelius dico Argenterio. Illud tamen de Fernelio pronunciare hoc loco non dubitarim, videri mihi hunc in iis, quae Galenus antea tractavit, et ad veram medicinam, quam ratione nitiscimus, hacque tempestate profitemur omnes, pertinent pari semper cum Galeno passu incedere, iisdemque prope inhaerere vestigiis, nulla aut minima, et in minimi fortasse aliquando momenti rebus in diversam sententiam discessione. Quod autem de abditis rerum causis aeditum ab eo opus est per se, utque extra rationalis medicinae limites, ac veluti corpus quoddam positum amplecti possumus ut autem ad medicinae nostrae, quae ratione constat, summam facturum non admittemus, etiam admissum Galenicis tamen adhuc non repugnaturum alterius nimirum generis dicemus" (fols. 16v–17r).

[47] Mazzuchelli, *Scrittori* (cit. n. 2), Vol. I, pp. 449–451.

"Galenism" was a less rigid stance than Argenterio's polemic would suggest. To Alessandrini, the acceptance of Galenic principles as fundamental and of the details of Galen's teaching as in the main correct (he expressly repudiated the belief that Galen never erred, asserting merely that Galen had made fewer mistakes than many other authors[48]) evidently did not seem incompatible with the absorption of large amounts of new data, the emergence or development of new areas of study, and the occasional criticism of Galen's opinion on particular topics.

Argenterio's offense in Alessandrini's eyes was to have attempted to revise or repudiate fundamental principles in a subject area at once central to medicine and fully explicated by Galen. For Alessandrini, pathology—"all diseases, symptoms, *affectus,* together with their causes and diagnoses"—was the core of medicine, and Galen's teaching on the subject his most outstanding achievement; by comparison, anatomical advances, however impressive and desirable in themselves, were ancillary to the main enterprise. Moreover, not only the thoroughness of Galen's own teaching but the long subsequent tradition of study of diseases and symptoms led Alessandrini to think it unlikely that much worthy of note could have escaped the attention of all the highly ingenious men who had devoted themselves to the subject. Accordingly, Argenterio and Solenander seemed to him to have "torn down a strong old building in order to give us a new jerry-built one." Hence Alessandrini, like Bertini, claimed to see Argenterio and his followers as a new Academy and Pyrrhonians who made everything uncertain.[49]

As for Argenterio, his stay at Naples was short. In 1560 he accepted the invitation of Duke Emanuele Filiberto of Savoy to act as his adviser in the establishment of a university at Mondovi, in territories the duke had only recently recovered. Emanuele Filiberto's talents and interests were chiefly military and political rather than intellectual; yet for him, too, patronage of a university was a necessary part of a program of state building.[50] Argenterio spent the rest of his life as a professor at Mondovi and, subsequently, at Turin, where the university was transferred in 1566. Doubtless his Piedmontese origins and connections were responsible for Emanuele Filiberto's invitation or command; moreover, at Mondovi and Turin Argenterio enjoyed the position of a respected member of a local elite, a position he had never attained at Pisa (at Turin he married the bishop's sister).[51] Nonetheless, it seems likely that by the time Argenterio agreed to go to Mondovi, the opposition aroused in the learned medical community by his ideas

[48] Alessandrini, *Antargentericorum suorum defensio,* fols. 13v–14r.

[49] *Ibid.,* fol. 14v; fol. 15v: ". . . et Argenterius tamen facit, factumque Reinerus defendit, nimirum, ut destructa priore ac potiore aedificatione, novam ipsi nobis ac deteriorem structuram faciant." And fols. 111r–v: "Verumtamen minime decere arbitror, hac aetate homines, qua nos sumus propositum ob levicula haec, mutare, et a veteri Academia nostra in novam hanc, omnia nobis incerta prodentem, quaeque Pyrrhoniorum quorundam esse videatur, deflectere."

[50] Argenterio's own account of his role in setting up the university at Mondovi is found in his dedicatory letter to Emanuele Filiberto, in Argenterio, *In Artem medicinalem Galeni* (cit. n. 11). See also Silvio Pivano, "Emanuele Filiberto e le Università di Mondovi e di Torino," in *Studi pubblicati dalla Regia Università di Torino nel IV centenario della nascità di Emanuele Filiberto* (Turin, 1928), pp. 3–7; and Bonardi, *Studio generale a Mondovi* (cit. n. 2).

[51] For the early history of the University of Turin and Argenterio's career there see Mario Chiaudano, "I lettori dell'Università di Torino ai tempi di Emanuele Filiberto (1566–1580)," in *Studi pubblicati dalla Regia Università di Torino,* pp. 35–86. See also the biographical accounts mentioned in n. 2.

had cost him both academic position and princely patronage elsewhere. The cultural program of the Medici court was designed to foster the general renovation of medicine among other university disciplines, not to lend support to one side in a medical controversy.

V. CONCLUSION

Argenterio thus appears to have transgressed the limits of acceptable neotericism at Pisa, and probably also at Naples. Yet what were these limits? His defenders were, after all, correct in pointing out that he was by no means the only contemporary critic of Galen. Nor did he attack ancient medicine wholesale: Massaria appears to have been incorrect in supposing that Argenterio had attacked Hippocrates as well as Galen. Rather, Argenterio asserted that the works of Hippocrates had been written with "divine rather than human ingenuity," noting merely that the *Aphorisms* were not an adequate guide to the whole of medicine.[52] And whatever one may think of his claims to originality or his accusations of plagiarism, it is also true that some of the ideas he claimed as his own became part of the common currency of the mid and later sixteenth-century debate about diseases. Furthermore, his positive theories, which involved mostly the multiplication of classificatory categories, do not in themselves seem particularly radical.

On the whole the academic medical community of the 1540s and 1550s, although certainly predominantly Galenist in one sense or another, was also relatively open and accustomed to a good deal of controversy. Idiosyncratic theories on particular topics, criticisms of ancient authorities on specific points, and energetic disagreements among colleagues were common and, usually, well-tolerated features in medical writing. Taken singly, many of Argenterio's views on particular topics could fit into this context easily enough.

Yet the unity and thoroughness with which the treatises on diseases focused on inconsistencies in Galen, weaknesses of basic definitions in pathology, and errors of celebrated contemporaries made it easy for some of Argenterio's colleagues to read them as a systematic attack from within on the whole enterprise of academic medical learning (not to mention the claims to superior status and emoluments of medical professors and university-educated physicians). Moreover, although by drawing a parallel between himself and Vesalius, Argenterio and those he inspired to defend him implicitly compared the situation as regards pathology with that of anatomy, such a comparison was, of course, quite inexact. Certainly, Argenterio was in no position to provide any very impressive physical demonstrations in support of innovative views. Rather, a reading of the treatises on diseases seems to lead naturally to the conclusions, explicitly drawn out in his commentary on the *Ars*, about the fragile and conjectural nature of physiological and pathological "knowledge." That some contemporary critics linked Argenterio's name with ancient medical methodism, the Academy, or Pyrrhonian skepticism comes as no surprise.

[52] *Opera Iohannis Argenterii* . . . (Hannover, 1610), cols. 764, 770.

Girolamo Mercuriale's *De modo studendi*

*By Richard J. Durling**

GIROLAMO MERCURIALE of Forli (1530–1606) was a celebrity in his day, eagerly courted by rival universities. He taught at Padua (1569–1587), Bologna (1587–1592), and Pisa (1592–1606); was involved in current medical debates, such as that over the Fracastorian interpretation of epidemic diseases (see the article by Vivian Nutton in this volume); and wrote important books on gymnastic medicine (Venice, 1569), skin diseases (Venice, 1572), and pediatrics (Venice, 1583). I shall not linger over his biography here: I refer the reader to the study of Italo Paoletto.[1] I intend to concentrate on Mercuriale's lecture on medical study, possibly to be dated 1570 (the date given by the late Milan manuscript of this work) and hence delivered in Padua. In its relative simplicity it offers an interesting contrast to the similar work of his Austrian colleague Martin Stainpeis, published fifty years earlier.[2]

Mercuriale's address belongs to the select genre of propaedeutic literature aimed at medical students. I have mentioned Stainpeis; similar published collections of lectures designed to direct these beginners' course of reading were Janus Cornarius's *De rectis medicinae studiis amplectendis* (Basel, 1545; NLM 2351), Joannes Heurnius's *Modus studendi eorum qui medicinae operam suam dicarent* (Leiden, 1592; Wellcome 3149), and Johan Georg Schenck a Grafenberg's *De formandis medicinae studiis* (Strasbourg, 1607; Wellcome 5828). The latest and most elaborate in this series is Caspar Bartholin's *Consilium de studio medico inchoando, continuando et absolvendo* (in his *Opuscula . . .* , Copenhagen, 1628; Wellcome 691).[3] I plan eventually to study and compare all these for the light they throw on contemporary medical education.

Mercuriale's *De modo studendi* is a lecture that occurs early in a series he gave at the University of Padua. In his day the lecture course at Padua would have lasted three years and covered both theory and practice. Curiously, Mercuriale here mentions theory only, not practice; there are no hints of the bedside teaching immortalized by Giambattista da Monte and called for by the Paduan statutes.[4] But doubtless the students copied down in the notebooks, or *promptuaria,*

* Institut für Geschichte der Medizin und Pharmazie, Christian-Albrechts-Universität, Kiel, Federal Republic of Germany.

[1] Italo Paoletto, *Girolamo Mercuriale e il suo tempo* (Bologna: Università degli Studia, 1963).

[2] Richard J. Durling, "An Early Manual for the Medical Student and the Newly Fledged Practitioner: Martin Stainpeis' *Liber de modo studendi seu legendi in medicina ([Vienna] 1520)," Clio Medica,* 1970, *5*:7–33.

[3] Figures after "NLM" refer to listings in Richard J. Durling, *Catalogue of Sixteenth-Century Printed Books in the National Library of Medicine* (Bethesda, Md.: NLM, 1967); those after "Wellcome" refer to listings in *A Catalogue of Printed Books in the Wellcome Historical Medical Library* (London: Wellcome Institute for the History of Medicine, 1962–1976).

[4] For details see Jerome J. Bylebyl, "The School of Padua: Humanistic Medicine in the Sixteenth

that Mercuriale advises them to keep the details of cases, cures, symptoms, and the like glimpsed at the local hospital of San Francesco.

Mercuriale's lectures show evidence of the extent to which medical humanism —the recovery of Greco-Roman medicine—had triumphed in Italian universities by the 1570s. It had begun with Niccolò Leoniceno's translations of Galen from the Greek and was continued by Leoniceno's friend Thomas Linacre (1460?–1524) and a host of minor figures, including the tireless Johann Guinther of Andernach (1505–1574). It was the same with Hippocrates: Marco Fabio Calvi (d. 1527) struggled to turn difficult Greek into limpid Latin, and his work was continued by the gifted philologist Janus Cornarius (1500–1558). By 1570 almost all Greek and Latin medical literature had been published, both in the original and in Latin or vernacular translations. Mercuriale is not alone in giving pride of place to the Greeks; he is unusual in that he also stresses the quality of a select group of Arabic writers, partly perhaps because of their undoubted contributions to materia medica (cf. the position of the Parisian Jacques Dubois, or Jacobus Sylvius [1478–1555]).[5] There is none of the hysteria of a Leonhart Fuchs (1501–1566), who would have banned Avicenna from the schools.[6] Mercuriale—as is only right in an address aimed at beginning medical students—eschews controversy.

Mercuriale says nothing of dissection, and nothing of botany or field trips. Yet anatomy had been revolutionized by Andreas Vesalius and his predecessors, and a botanic garden had been opened in Padua in 1546. Both anatomy and botany must have fascinated medical students; of course, botany was not yet an independent discipline but was studied principally for the knowledge of simples.

Nor does he mention his illustrious colleagues, past or present. He surprisingly favors two medieval commentators on Avicenna, Jacques Despars and Gentile da Foligno, but is silent about sixteenth-century Avicennism and contemporary commentators. Yet medieval medical works were studied throughout Renaissance Europe, for example, those of Pietro d'Abano and Arnau de Vilanova, and one should not underestimate the thread of continuity in medical doctrine.

Mercuriale's lecture is found in two (or possibly three) manuscripts, of which one is a hastily written sixteenth-century autograph in Cesena, and the other is an eighteenth-century copy in Milan.[7] The text was posthumously printed—for example, in Venice, 1644—but in a slightly revised form and with many misprints. It is a slight work, occupying only fourteen pages in folio in the Cesena autograph. Unlike Stainpeis, Mercuriale does not present the students with an extensive booklist. Instead, they are given sensible, homespun advice on how to

Century," in *Health, Medicine, and Mortality in the Sixteenth Century,* ed. Charles Webster (Cambridge: Cambridge Univ. Press, 1979), pp. 335–370, on pp. 339n14, 347.

[5] See Gerhard Baader, "Jacques Dubois as a Practitioner," in *The Medical Renaissance of the Sixteenth Century,* ed. A. Wear, R. K. French, and I. M. Lonie (Cambridge: Cambridge Univ. Press, 1985), pp. 146–154.

[6] Galen, *Aliquot opera . . . ,* ed. and trans. Leonhart Fuchs (Paris, 1550), fol. 281r. See Richard J. Durling, "Leonhart Fuchs and His Commentaries on Galen," *Medizinhistorisches Journal,* 1989, *24*:42–47.

[7] Cesena, Biblioteca Malatestiana, Fondo del Comunitativa 166.138, listed in Paul Oskar Kristeller, *Iter Italicum: A Finding List of Uncatalogued or Incompletely Catalogued Humanistic Manuscripts of the Renaissance in Italian and Other Libraries* (London: Warburg Institute, 1965), Vol. I, p. 45; Milan, Biblioteca Ambrosiana S 84 sup., listed *ibid.,* Vol. II, p. 313. Kristeller considers the latter collection to belong to the sixteenth century, but the hand here is surely eighteenth century. A possible third MS, from the sixteenth century, is Gdansk, Akad. Nauk. 2325 (listed *ibid.,* Vol. IV, p. 399).

Girolamo Mercuriale, from an engraving by T. de Bry (ca. 1599). Courtesy of the Wellcome Institute Library, London.

organize their studies. They are to have *norma atque methodus*. They must enjoy a good physique and be young, to support the hard labor of medical study. Their intellect must be keen, their memory tenacious. They must keep good company, and reverence and cherish their teachers. Their reading must be selective and limited to the outstanding writers, Latin, Arabic, and Greek. Of the Latins, Celsus, Scribonius Largus, Pliny, Q. Serenus Sammonicus, Caelius Aurelianus, and Theodorus Priscianus are to be read. Of the Arabs, Avicenna, Averroes, Rhazes, Mesue, and Serapion are recommended. Greek writers are to be studied day and night: Hippocrates, Galen, and Dioscorides are, of course, mentioned, but the philosophers Plato, Aristotle, and the latter's pupil Theophrastus are not overlooked. In addition to these, a wide range of classical reading is recommended, from Homer and Hesiod to Athenaeus and Palladius—this for purposes of recreation. The student must always have to hand Hippocrates, Galen, Celsus, and Avicenna.

Difficult terms are a problem. Students must have ready access to lexica, such as Galen's *De linguis* and those of Erotian, Julius Pollux, and Rufus of Ephesus, as well as Guillaume Budé's commentaries, Joachim Camerarius's *Commentarii utriusque linguae*, "the newly published medical lexicon" (i.e., Henri Estienne's

Dictionarium medicum), and Joannes Gorraeus's (equally new) *Definitionum medicarum*.[8]

Mercuriale wants his pupils to read the ancient authorities without commentaries if possible. An exception is the obscure Hippocrates, in reading whom one should have recourse to Galen. Another exception is Avicenna: Mercuriale's favorite commentators here are Jacques Despars (Jacobus de Partibus) and Gentile da Foligno. Students should avoid compendia and epitomes like the plague: they promise a dangerous shortcut for the young. It is not immediately clear what works Mercuriale had in mind, but there seems to have been a large choice. Indeed, the whole question of what medical texts were available in Padua in the 1570s requires further investigation. Works survive that seem aimed at medical students or newly fledged practitioners by such authors as Cornarius, Fuchs, Alfonso Bertocci, Ognibene Ferrari, Conrad Gesner, Blaise Hollier, Jérôme de Monteux, and Georg Pictorius. Their titles contain words like *compendiaria* or *compendiolum* or phrases like "ex Hippocrate, Galeno & Avicenna"—and they would have been anathema.[9] Mercuriale's advice is instead "Ad fontes!" For

[8] Galen's *De linguis* = *Vocum obsoletarum Hippocratis explanatio* in Rasario's edition of Galen's works (Venice: V. Valgrisi, 1562–1563). Erotian was edited by Bartolomeo Eustachi (Venice: Junta, 1566) (Wellcome 2070). Julius Pollux was first published in Greek in Venice in 1502, at the Aldine Press; excerpts were included in Henri Estienne's *Dictionarium medicum* (Geneva, 1564) (cf. Joannes Gorraeus, *Definitionum medicarum libri XXIII* [Paris, 1564]). Rufus of Ephesus's *De corporis humani partium appellationibus libri tres* was included in an edition of Aretaeus Cappadox, *Libri septem* (Venice: Junta, 1552), in the version of Julius Paulus Crassus. Guillaume Budé's *Commentarii linguae graecae* were first published in Paris by Josse Badius Ascensius in 1529. Camerarius's *Commentarii* had been published in Basel in 1551 by J. Hervagius (Wellcome 1221).

[9] Janus Cornarius, *Universae rei medicae epigraphē* . . . (Basel, 1529, 1534); Leonhart Fuchs, *Compendiaria ac succincta admodum in medendi artem eisagōgē* (Hagenau, 1531; Strasbourg, 1535, etc.); Alfonso Bertocci, *Methodus generalis, & compendiaria, ex Hippocratis, Galeni, & Avicennae placitis deprompta* . . . (Lyons, 1558); Ognibene Ferrari, *De regulis medicinae libri tres, ex Hippocrate, Galeno, & Avicenna* . . . (Brescia, 1566); Conrad Gesner, *Enchiridion rei medicae triplicis* . . . (Zurich, 1555); Blaise Hollier, *Medicae artis theorica* . . . (Strasbourg, 1564); Jérôme de Monteux,

HIERONYMUS MERCURIALIS

De Modo Studendi eorum qui medicinae operam navant

Sicuti illis, qui navigandi professionem sibi toto vitae tempore exercendam proponunt, necesse est rectam aliquam rationem habeant, quae quid bonum, quid malum in ea sit, quomodove a se agendum intelligant, ne in re 5 adeo periculosa et difficili magno et salutis et rerum discrimine fallantur, pariter illi qui artibus aliquibus ex liberalibus veluti medicina est, comparandis ac tum in proprium, tum in aliorum usum exercendis esse dediderunt, summopere curare debent, ut studiorum suorum normam atque methodum aliquam sequantur, ne vel coccorum instar huc atque illuc impulsi 10 a fine sibi proposito frustrentur, vel alio deflexi falsum pro vero et malum pro bono animis imbibant sicque oleum et operam non absque detrimento perdant. Ut itaque huiuscemodi scopulos facile evitare possitis cum viam

This edition and translation are based on the Cesena autograph.

C = Cesena, Biblioteca Malatestiana, Fondo della Comunitativa 166.138, as corrected in Mercuriale's hand (interlineally, unless otherwise specified)
C[1] = *C* before corrections
M = Milan, Biblioteca Ambrosiana S 84 sup.
Ed. pr. = Hieronymus Mercurialis, *De ratione discendi medicinam* (Venice, 1644); copy used: Kiel, Universitätsbibliothek, Qh 1050

example, if the subject is the seat of the soul, students should see what Plato and Aristotle said on the matter; likewise, if the subject is finding medical indications or the nature of some drug, they should see what Hippocrates, Theophrastus, Dioscorides, Aetius, or Alexander of Tralles wrote on these matters.

Authorities should be read in orderly fashion. Fortunately, Hippocrates' and Galen's works are arranged in classes, and those of Avicenna likewise. Galen wrote a book on the order of his works, and Jacques Dubois wrote one on the order in which Hippocrates and Galen were to be read.[10] Particularly important, though commonsensical, is Mercuriale's advice to students to compile *thesauri* and *promptuaria* in which to record commonplaces, such as causes, signs, preservation, cures, drugs, diet, and surgery.

There is really nothing startling about Mercuriale's address, therefore. Conservative, cautious, sober, and civilized, it echoes the 1570s classroom with its emphasis on book learning and note taking. It lacks the sometimes penetrating rigor of Mercuriale's Austrian predecessor Stainpeis, and the authorities recommended are conventional. Few contemporary authors are referred to: Vesalius and Jean-François Fernel, to mention only two, are conspicuously absent. It remains important, however, to study this propaedeutic literature in the light of Renaissance lectures generally. Too few such lectures have been studied in depth. There is abundant evidence tucked away in European libraries. May historians edit this unpublished material!

Compendiolum curatricis scientiae . . . (Lyons, 1556); and Georg Pictorius, *Rei medicae totius compendiosa traditio* . . . (Basel, 1558). Two further such titles, published in Saragossa (NLM 2857) and Krakow (NLM 2860), would hardly have been available in Italy. On Fuchs's textbook see W. P. D. Wightman, *Science and the Renaissance* (Edinburgh: Oliver & Boyd, 1962), pp. 213–217; and Eberhard Stübler, *Leonhart Fuchs: Leben und Werk* (Munich: Münchner Drucke, 1928), pp. 42ff. For Cornarius's *Epigraphē* see *ibid.*, p. 103.

[10] Jacques Dubois (Sylvius), *Ordo, & ordinis ratio in legendis Hippocratis et Galeni libris* (Paris: C. Wechel, 1539; rpt., Paris: A. Gorbinus, 1561) (Wellcome 6155–6156).

GIROLAMO MERCURIALE

On the Method of Studying for Those Who Pursue Medicine with Zeal

Those whose whole life is given over to navigation must necessarily have some correct reasoning as to what is good, what bad in their [art], so that they can understand how to proceed, lest in a matter so dangerous and difficult they may be deceived—at great cost. Like these, those who have dedicated their lives to acquiring liberal arts such as medicine and to exercising them for their own benefit and that of others must take exceptional care to follow some pattern and method in their studies. This, lest like cooks impelled hither and thither they be frustrated of their hopes or deflected elsewhere, or imbibe in their minds falsity for truth, evil for good, and so waste their time, not without detriment. So that you may easily avoid rocks of this sort, and since you have asked me now to point out

---

1 Hieronymus Mercurialis *om. C* 2 *Tit. C:* De Modo . . . navant libellus *M:* De Ratione discendi Medicinam *ed. pr.* 3 Sicuti] Quemadmodum *ed. pr.* illis] his *ed. pr.* 3–4 sibi toto . . . proponunt] *CM:* toto vitae suae tempore exercendam sibi proponunt *ed. pr.* 4 necesse] opus *ed. pr.* est] est ut *M ed. pr.* habeant] ineant *ed. pr.* 11 falsum pro vero] verum pro falso *C* 11–12 malum pro bono] bonum pro malo *C* 12 animis *om.* C^1 13 cum *om.* C^1

aliquam brevem monstrari vobis hoc tempore a me petieritis, eam indicare
conabor, cui si diligenter insistetis, non vereor quin facile concedatur aut 15
labores vestros ad optatum finem perducere, aut saltem quam minimum
dispendii subire. Hoc autem faciam tribus iis accurate explicatis, quae ab
Hippocrate, Platone, Aristotele et Galeno in omnibus rebus feliciter perfi-
ciendis valde requiri debere scriptum est Naturam, Studium, Conversa-
tionem consuetudinemve. De Natura dicam ut corpore et animo esse de- 20
betis. De Studio et labore pauca eritis docendi. Consuetudinem vero
quoniam cum duobus habere potestis, vivis nimirum et mortuis, primo
quibus vivis, et quando, secundo quibus mortuis, quomodo quando et
quantum sit vobis conversandum edocebo atque ita omnem rationem studiorum vestrorum amplectendam cognoscitis. Auspicantes igitur dicimus 25
quod qui medicinae studiis operam navare intendunt corpore vegeto et
sano atque aetate puerili iuvenilive esse debent. Nam cum nihil laude dignum sine continuis laboribus ut cecinit Hesiodus "virtutem posuere Dii
sudore parandam" comparari queat, laborare vero nemo possit, nisi sano
atque valido sit corpore. Propterea ante cetera haec corporis conditio ne- 30
cessaria est, ut non inmerito interrogatus philosophus quidam quid in vita
πρῶτον quid δεύτερον esset semper responderit τὸ ὑγιαίνειν. Sanitas vero,
quamvis ut plurimum ab ortu omnes suam consequantur, tamen fere in
manu cuiusque positum est ut quam habet tueatur vel destruat, quam non
habet magna ex parte sibi nanciscatur, videlicet moderatis exercitationibus 35
et sobrietate vitae, quibus pravas a natura corporis intemperies corrigi sa-
nitatemque et robur acquiri non semel nos Plato in Charmide et Galenus
secundo de tuenda valetudine et quarto nec non libro de cibis boni et mali
suci monuerunt. Animus certe unicuique a natura donatur videlicet ut sit
ingenio perspicaci, ac ad intelligendum facili, ut iudicii acrimonia polleat, 40
ut memoria valeat. Quae omnia sane studiis medicinae operam daturo con-
veniunt. Verum tamen ad animum, ingenium atque memoriam consequen-
dam fere etiam industria nostra atque studium, ut Hippocrates in Prae-
ceptis innuit maximopere conferunt. Quoniam si et Aristotelis et Galeni
sententia animorum nostrorum propensiones ac vires ex temperatura cor- 45
poris ita ortum ducere videntur, ut qualis illa sit, eiusmodi etiam has esse
fere necessarium sit. Quemadmodum in nobis situm est temperaturas conservare corrigere atque renovare, simili pacto non negatur animos, hoc est
ingenia, iudicia, atque memorias quasi nostra manu facere. Neque minimum est quod vobis de aetate adnotavi. Quandoquidem si ulla ars inter 50
humanas habetur quam tempestive addiscere atque exercere oporteat, una
profecto est medicina cui vel mediocriter adipiscendae vix totum hominis
vitae cursum sufficere prodidit Hippocrates, ne veteres in medium addu-
cam, quos a pueritia semper medicae artis studia atque exercitationem in-
choasse Hippocrates Plato et Aristoteles significarunt. At illud fugere vos 55
non oportet non parum ad excellendum in unaquaque arte referre ut quis se
ipsum suasque propensiones cognoscat, quove magis sese a natura animo
et ingenio inclinari animadverterit eo studia sua convertat. Quoniam infi-
ciari nemo potest quin alius legibus, alius medicinae, alius aliis artibus
magis natus videatur, inde fit ut dixerit Aristoteles in libro primo Rheti- 60
corum XXVIII. Problem, homines libentius ea studia atque artes aggredi
licet viles, ad quas nati atque apti sunt quod sperant sese in iis magis excellere debere cuius excellentiae a natura amatores sumus omnes. Erit
summus in medica arte adipiscenda, si, quod monuit Galenus in libro de
constitutione artis, toto vitae tempore unam veritatem sequi, nulli sectae 65
adhaerere, in nullius sententiam aut verba iurare vobis proposueritis. Studium porro ac labore valde etiam requiri in omni disciplina comparanda,
puto vosmet intelligere, quando legisse potestis Virgilianum illud 1. Georg.
labor omnia vincit improbus. Nam et Galenus ubique fere admonet eum qui

14 vobis *om.* C^1 15 *post* quin *add.* vobis C^1 aut] ut *M* 19 scriptum est] scripserunt *C* Studium

some way of life, I shall try to indicate it: if you pay diligent attention, I am confident you will easily conduct your labors to a successful end, or at least incur minimal costs. I shall do this by explaining three things accurately, which Hippocrates, Plato, Aristotle, and Galen declare are urgently required in accomplishing all things, namely Nature, Study, and Association or intercourse. About Nature I shall only say that you are necessarily compounded of body and soul. About Study and labor, you need a little instruction. As to Association, you can associate with two [groups], namely the living and the dead. First I shall teach you with what living [persons] and when; second with which of the dead, how, when, and how much you can converse. Thus you can know all about what course of studies to embrace.

To start with, therefore, we say that those who intend to dedicate themselves to the study of medicine must be possessed of a vigorous body, healthy and young. For since nothing worthy of praise is obtained without continual labor—as Hesiod sang,"Between us and Goodness the gods have placed the sweat of our brows"[a]—no one can labor unless he has a healthy, strong body. Therefore this condition of body is necessary above others, so that not undeservedly a certain philosopher, when asked what is first and what second in life, always replied: "Health." Now as to health, although as a rule all obtain it from the outset, it nonetheless lies in the power of all to keep or destroy what one has, and what one does not have can in large part be acquired by one's own effort, to wit, by moderate exercise and sobriety of life, and by these means bad disorders of the body can be corrected and health and strength acquired, as Plato has told us more than once in his *Charmides* and Galen in Books 2 and 4 of his *De sanitate tuenda* and in his book *De cibis boni et mali succi*. A mind is of course given to each by nature, so that one can be shrewd of temper and quick at learning, so that judgment can be acute and memory excellent. All of these things certainly befit a student of medicine. But to ensure a good intellect and memory our own efforts and industry are absolutely vital, as Hippocrates hinted in his *Praecepta*. Aristotle and Galen held the propensities of our minds and their strengths to originate from the temperament of our bodies, so that as our temperament is, so necessarily are our minds. As it is in our power to conserve, correct, and renovate our temperaments, so it is granted us to mold as it were with our own hands our characters, judgments, and memories.

Nor is what I have pointed out to you about time of life a small matter. If there is any human art we should learn and exercise in due season, it is medicine, for even a moderate knowledge of which a whole life is scarcely sufficient, as Hippocrates said, to say nothing of the ancients, who, Hippocrates, Plato, and Aristotle said, began the study and exercise of the medical art from childhood. But you should not be blind to the fact that no mean role in excelling in any art is played by each one's knowing his own propensities, and recognizing his own inclinations, so that he presses his studies in that direction. For no one can deny that one man is better endowed for law, another for medicine, another for other arts, so that as Aristotle says in the first book of his *Rhetoric*, Section 28, men more willingly embrace the studies and arts, though banausic, for which nature fits them, because they hope to excel—and we all naturally love excellence. A man will be first in medicine if, as Galen says in his book *De constitutione medicinae*, he follows one truth throughout his life, is never a slave to any sect, nor swears an oath of loyalty to anyone. Every discipline involves work, as I think you know, for you may have read that famous remark of Virgil's *Georgics* 1 [146], "labor omnia vincit improbus." For Galen too every-

[a] Hesiod, *Works and Days* 289, translation from *Hesiod, the Homeric Hymns, and Homerica*, ed. and trans. H. G. Evelyn-White (Loeb Classical Series) (Cambridge, Mass.: Harvard Univ. Press, 1936).

---

om. C[1] 20 animo *om. C.* 26 vegeti *C* 27–28 laude dignum *om.* C[1] 26 sanitate *M* 33 suam *om.* C[1] 34 *post* manu *add.* fere *M* positum est] esse positum *M* 38 *post* secundo *add.* libro *M* 42 conveniat *M* 43–44 ut . . . innuit *om.* C[1] 43 *post* in *add.* libro de *M* 46 ita] in *M* 47 sita *M* *post* est *add.* et *ed. pr.* 49–55 Neque . . . significarunt *in margine C* 50 ulla] nulla *ed. pr.* 50–51 inter humanas] in humanis *M* 54 adducam] ducam *M* 60–61 in . . . Problem, *in margine C* 62 sperent *M* magis *M* 63–66 Erit . . . proposueritis *in calce paginae C* 65 nulli] nullae *MSS: corr. ed. pr.*

in medicina excellere cupit φιλόπονον idest laborum amatorem esse de- 70
bere, neque ullis vigilis ullisve curis esse deterrendum. Quod tertium su-
perest et omnium maxime ad studia recte componenda pertinet est in con-
versatione tam vivorum quam mortuorum hoc est librorum positum. Quod
enim ad vivos spectat praeceptum est Theognidis poetae antiquissimi a D.
Platone in Menone relatum 75

> καὶ παρὰ τοῖσιν πῖνε καὶ ἔσθιε, καὶ μετὰ τοῖσιν
> ἵζε, καὶ ἅνδανε τοῖς, ῳν μεγάλη δύναμις.
> ἐσθλῶν μὲν γὰρ ἄπ' ἐσθλα διδάξεαι. ἢν δὲ κακοῖσιν
> συμμίσγης, ἀπολεῖς καὶ τὸν ἐόντα νόον.

Hoc est apud eos bibendum, comedendum, sedendum, iisque esse obe- 80
diendum, quorum magna est virtus, quoniam ut a bonis bona discuntur, sic
si cum malis consuetudo habeatur, praesens bonitas amittitur. Quare in
primis studendum est, ut optimos praeceptores, doctos, ac benevolos vobis
eligatis similiter et amicos, quibuscum tuto et cum fructu continuam con-
versationem habere possitis. Cumque iam vobis electi atque probati fuer- 85
int, vestrum erit illos sequi, diligere, revereri, obedire, semperque aliquid
ab illis de studiis vestris sciscitari, putantes nihil esse quod aeque vobis ad
comparandos bonos habitus ac vestra studia confirmanda et locupletanda
conferat atque assidua cum praeceptoribus et amicis doctis concertatio.
Iam vero intelligatis, qualis vobis esse debeat cum libris ipsis conversatio. 90
Primo non quilibet auctores amplectendi sunt, sed probatissimi quique, et
per multa saecula a viris doctis pertriti, quoniam si secus fiat, duo non
contemnendi errores committuntur. Unus est quod doctrina neque firma
neque probata primo animi imbuuntur, alter quod id temporis quo leguntur
auctores gregarii et populares fere consumitur cum praestaret in opti- 95
morum lectione iugiter insumere. Qui vero sint ii scriptores a vobis amplectendi, facile sciri poteritis tum ex vetustate, quando rari sunt auctores
veteres, quorum lectio magnum fructum non afferat, tum ex consensu pos-
teriorum, qui si aliquos longa temporum serie approbasse atque secuti esse
videbuntur, eos non magnopere aestimare non debebitis. Hos invenio ego 100
partim Latinos, partim Arabas, magna ex parte Graecos extitisse. Qui ex
Latinis medicinae studiosis lectitari debent sunt Celsus, Scribonius Largus,
Plinius, Q. Serenus, Celius Aurelianus, Theodorus Priscianus. Ab huiusmodi namque omnibus si prudenter legantur, non parum ornamenti atque
divitiarum laboribus vestris accedere experiemini. Arabum similiter scrip- 105
torum lectiones maxime proderunt, si ut debent amplectantur, in primis
Avicenna, Averroes, deinde Razes, Mesues, Serapion sunt, quos qui le-
gunt, magnam utilitatem consequuntur, quandoquidem etsi stilo ac ordine
laudari non mereantur, sunt tamen commendandi a rebus multis atque
bonis quorum [*sic*] auctores nobis extiterunt. Ceterum Graeci ii sunt de 110
quibus id Horatianum dicere cogor

> Nocturna versate manu, versate diurna.

Cum enim ipsi tantam hanc nostram artem invenisse, provexisse, aluisse,
ornasse atque locupletasse putentur, apud quos melius quam apud parentes
ipsa quaeramus ac invenire speremus non video quamobrem Hippocratem, 115
Platonem, Aristotelem, Theophrastum, Nicandrum, Dioscoridem, Cas-
sium, Aretaeum, Galenum, Oppianum, Athenaeum, Geoponicos omnes et
Hippiatros, Oribasium, Aetium, Alexandrum, Paullum et ex recentioribus
Actuarium, Damascenum ceteris omnibus anteferre summoque studio ipsorum commentaria pervolvere debetis. Post quos sunt alii auctores praes- 120
tantissimi quorum lectio si vobis non negligetur optime studiis vestris con-
sultum omni tempore experiemini. Atque ii sunt Homerus, Hesiodus,
Lucretius, Virgilius, Horatius, Iuvenalis, Martialis, M. Cato, M. Varro,

where recommends that the man who wishes to excel in medicine be *philoponos,* that is, a lover of toil, and not be deterred by many wakeful nights or cares.

The third matter remaining, and one especially relevant to correct study, is to converse with the living and the dead, that is, through books. As for the living, Theognis, quoted by Plato in his *Meno,* has this precept: "Eat and drink and sit with the mighty, and make yourself agreeable to them; for from the good you will learn what is good, but if you mix with the bad you will lose the intelligence which you already have."[b] Wherefore one's first object is to choose excellent preceptors, learned and benevolent, likewise friends with whom you can safely and fruitfully have a continuous conversation. And when they are chosen and approved, it will be your task to follow them, love them, revere them, obey them, and always ask them about your studies, thinking that nothing conduces so much to your acquiring good habits and confirming and enriching your studies as regular disputation with learned preceptors and friends. First, not every author is to be embraced, but only those most approved, and thoroughly thrashed over by learned men over the ages: if you do otherwise, you commit two considerable errors. One is that your minds are imbued with no firm and approved doctrine; another is that the time in which vulgar, commonplace authors are read is for the most part wasted, when it would be better spent in the reading of the best [authors]. Who those authors are who should be embraced you can easily tell both from their antiquity (since rare are the ancient authors whose reading yields no fruit) and from the consensus of posterity. If posterity has approved and followed certain [authors] over a long period of time, you should value them highly, too. I find these to be partly Latin, partly Arab, and in large part Greek. Those Latin writers to be read include Celsus, Scribonius Largus, Pliny, Q. Serenus, Caelius Aurelianus, and Theodorus Priscianus. You will find no small wealth and ornament accruing from such writers if they are read with prudence. The reading of Arabic writers will also be very beneficial, if they are embraced as they should be; first of all Avicenna, Averroes, then Rhazes, Mesue, Serapion—all very useful. Although they do not deserve praise for their style or arrangement, they are yet to be commended for the many good things they offer. But the Greeks are those about whom I am compelled to quote Horace, "Nocturna versate manu, versate diurna."[c] For since they are thought to have discovered this magnificent art of ours, promoted it, nourished it, adorned and enriched it, and we seek their company and hope to find ourselves in it more than in our parents', I do not see how you cannot prefer Hippocrates, Plato, Aristotle, Theophrastus, Nicander, Dioscorides, Cassius, Aretaeus, Galen, Oppian, Athenaeus, all the *Geoponici* and *Hippiatri,* Oribasius, Aetius, Alexander [of Tralles], and Paul [of Aegina], and of recent writers Actuarius and Damascenus. I say, prefer and read their commentaries as you must. After them, there are other excellent authors you should always be reading. They are Homer, Hesiod, Lucretius, Virgil, Horace, Juvenal, Martial, M. Cato, M. Varro, Columella, Palladius, Vitru-

[b] Here Mercuriale not only quotes the Greek but translates it into Latin: Theognis, *Elegies* 33–36, as quoted in Plato, *Meno,* translation from *The Dialogues of Plato,* trans. Benjamin Jowett, intro. by Raphael Demas (New York: Random House, 1937), Vol. I, p. 375.

[c] "Turn your hand to them day and night." Horace, *Ars Poetica* 269.

75 Memnone *C* 87 nihil] nil *M* 89 assiduam *M* concertationem *M* 91 non *om. M* 93 committantur *M* 94 *post* alter *add.* est *M* 96 sint] sunt *M* 100 debebitis] debetis *M* 102—103 *post* sunt *scr. et del. C:* M. Cato, M. Varro, Virgilius, Horatius, Juvenalis, Vitruvius, Lucretius . . . Seneca, Columella, Palladius 107 Avicennas *C ed. pr.* Averros *C:* Averrhoes *ed. pr.* Rases *ed. pr.* 111 Horatianum *om.* C^1 113 ipsi] qui *M* 117 Athenaeum *om. M.* 120–124 Post . . . Athenaeus *in margine C*

Columella, Palladius, Vitruvius, Herodotus, Strabo, Pausanias, Athenaeus.
Neque miremini quod poetas et historicos vobis proponam, quoniam si 125
Galenum ducem nostrum videatis saepe horum auctorum testimonia citare
comperietis quod sciret ex his medicinae quoque scientiae non paucam
auctoritatem ac lucem afferri. Non putetis tamen me velle ut pari studio
cuilibet horum incumbatis, sed cum praecipuos medicos veluti Hippocratem, Galenum, Celsum, Avicennam in manibus frequenter habebitis, suc- 130
cisivis (ut aiunt) quibusdam horis aliquam operam etiam horum lectioni
dare poteritis, id semper meditantes ut omnia ad medicam disciplinam ornandam ac ditandam atque ad eius scriptores intelligendos spectantia at-
tente colligantur, reliqua vel praetereatis vel minus aestimetis. Atque tot de
auctoribus quos prae aliis ad legendum atque imitandum vobis propo- 135
nendos esse duco. Nunc afferam modum quo in illis non infructuose versari
valeatis. Qui quoniam duobus finibus inservire potest tum ad intelligendum, tum etiam ad memoria quod intellectum est conservandum. Primo
scire operae precium est, qua ratione laborandum sit vobis, ut exactam
scriptorum quos evolvere placebit intelligentiam consequamini. Quod enim 140
verba ac stilum auctoris legendi ante alia probe cognoscere debeatis, ex se
satis manifestum est. Quando velle res cognoscere verbis ac nominibus,
quibus declarantur, ignoratis, vanum omnino putatur. Proinde Lexica pa-
rata habere oportebit, quibus quod non noveritis, prompte invenire ac intelligere valeatis. Huiusmodi sunt praeter Lexica vocata Erotianus, liber 145
Galeni de linguis, Julius Pollux, Ruffus Ephesius, Commentarii Budaei in
omnibus innumera Galeni loca illustrata adinvenietis Camerarius et commentaria doctissima edidit, in quibus dum omnes corporis humani partium
appellationes explicat medicinae studiosis praeclaram viam sternit, Lexicon illud medicum nuper editum et Gorraei exercitationes etiam valde iu- 150
vabunt. Postquam igitur haec suppetent, iam unum vel duos auctores ad
plus proponere debetis, quos statis horis quotidie videatis atque maiori qua
fieri potest sedulitate legatis ita ut nulla sententia, nulla res difficilis prae-
termittatur, quin eius natura, proprietas, ac vis cognoscatur. Quod si interdum contingat locis aliquibus abstrusis et difficilimis occurrere, exposi- 155
tores, si qui erunt, adeundum erit, a quibus si neque ad intelligentiam
satisfiat praeceptores consulatis. Quod si nec ab illis satisfiet vobis no-
tandus erit locus, et ad alia transeundum ea spe ut alia dies, vel alius liber
quod nunc non intelligatur tandem aperiat. Neque enim ea sumus omnes
natura et ingenio nati, quod non liceat interdum aliqua ignorare, quando et 160
Grammaticis hoc laudi adscribere videtur Quintilianus, si aliqua ignorent.
Hoc in loco duo vos admonere volo, unum ut si fieri potest auctores vetus-
tiores absque aliorum commentariis legatis, nimirum qui et confundere iudicia et ingenia ad veritatem indagandam pigriora reddere soleant. Exci-
piam unum Hippocratem, cuius obscuritas atque brevitas eiusmodi est, ut 165
nisi Galeni commentariis aut alicuius alterius probatioris aliqua ex parte
levetur, animos lectorum nimis defatiget, ac nonnunquam deprimat. Excipio etiam Avicennam, cuius si unum interpretem, quem alii praefero, Jacobus de partibus vel Gentilem videre placuerit ad aenigmata illa perci-
pienda non prohibeo, sicuti nec multum laudo probos alioquin scriptores 170
damnari, quod eloquentia res nobis non tradiderint. Etenim quis sana
mente praeditus margaritas vel coeno obvelatas si invenerit, lubentissime
non capiat atque etiam charas habeat? "Ornari res ipsa negat contenta doceri" ut scripsit Celsus, remediis non eloquentia curantur aegri. Alterum
est, quod compendia haec et epitomas, quibus nonnulli sese brevi artem 175
docturos pollicentur, tamquam perniciem summam fugiatis. Scitis quid
dicat Galenus adversus illos qui artem sex mensibus addisci posse praedi-
cebant. Homines certe ut scribit Aristoteles 2° Rhetoricorum breviter addiscere amant, attamen brevitas et suis terminis circumscribitur, quos nisi
attingat, iam in vitium transit. Compendia et epitomae possunt quidem 180

128 ut] aut *M* 129 cum *om. M* 129–130 veluti . . . Avicennam *om.* C^1 131 operam] partem *M*

vius, Herodotus, Strabo, Pausanias, and Athenaeus. And do not be surprised that I propose poets and historians to you, since if you consider Galen, our leader, you will find him frequently citing their testimony, because he knew that they would shed no small authority and light on medical science. Do not think that I wish you to spend the same time on each of these writers, but I want you to have constantly to hand such excellent physicians as Hippocrates, Galen, Celsus, and Avicenna and to spend your spare moments on reading them, always mindful to collect diligently all material that can adorn and enrich the medical discipline in understanding those authors. You are to leave the rest or value them less highly.

So much for the authors whom above others I consider you should read and imitate. Now I shall tell you a method you can employ in reading them fruitfully. For it serves two purposes, both in understanding and in memorizing what has been digested by the intellect. The first thing you should know is how to work, so that you have an exact grasp of the authors you consult. It is clear that you should know thoroughly the words and style of the author to be read. It is absolutely pointless to wish to investigate a matter when ignorant of the words and names in which it is conveyed. Hence you will have to have lexica to hand, from which you can find and understand what you do not know. Such lexica are, besides those expressly called lexica, Erotian, Galen's book *De linguis,* Julius Pollux, Rufus of Ephesus and [Guillaume] Budé's *Commentaries* on innumerable passages in Galen, and, you will find, Camerarius published very learned commentaries in which he explains all the names of the bodily parts and opens up an excellent path for medical students; further, the newly published *Medical Lexicon* and Gorraeus's exercises will also be very helpful. Once these are on hand, you must propose one or two authors at most, which you are to inspect daily at fixed times, and read with the greatest possible attention, so that no idea, no difficult matter may be passed over, but the nature of each, its property, and its meaning must be understood. But if you chance on some abstruse and very difficult passages, you will have to consult the commentators (if there are any), and if they do not help your understanding, you are to consult your teachers. If they cannot help you, the passage must be noted, and you must pass on to the next, in the hope that some other day or another book may finally explain what is not yet understood. For none of us are endowed with such a nature and intellect that we may not be ignorant at times, and therefore [whence] Quintilian even praises grammarians for their occasional ignorance.

At this point I want to give you two bits of advice, one that if possible you read the older authors without the commentaries of others, who are wont to confound the judgment and to render the intellect lazier at investigating the truth. I make one exception in the case of Hippocrates, whose obscurity and brevity is such that unless it is relieved by Galen's commentaries or another excellent writer's, it fatigues the reader's mind and sometimes causes depression. I also exempt Avicenna: my favorite commentator is Jacobus de Partibus, or if you prefer to consult Gentile [da Foligno] to solve his riddles, I do not forbid you. I do not approve highly the blaming of otherwise excellent writers on the grounds that their writings are inelegant. For who in his right mind, if he finds pearls obscured by mud, does not take them and prize them? "The thing itself will not be adorned, but taught," as Celsus wrote, and the sick are healed by remedies, not eloquence. Another recommendation is that you should shun like the plague those compendia and epitomes, with which some promise that they will quickly teach the art of medicine. You know what Galen says against those who hold that medicine can be learned in six months. Certainly, as Aristotle writes in Book 2 of his *Rhetoric,* all like to learn quickly, but there is a limit even to brevity, and if it does not attain its object, it becomes a fault.

---

132 ad *om. M* 134 tot] haec *ed. pr.* 145 Huiusmodi] Cuiusmodi *ed. pr.* 146 Ruffus Ephesius *om.* C^1 146–149 Budaei . . . sternit *in margine C* 149–151 Lexicon . . . iuvabunt *in margine C* 151 Postquam *om. M* *post* haec *add.* ubi *M* 153 potest fieri *M* 157 consuletis *M ed. pr.* ab illis satisfiet] illi prodesse queant *M ed. pr.* 158 alius] alias *M* 159 tandem *om.* C^1 simus *ed. pr.* 160 quod] ut M 161 si aliqua ignorent *om.* C^1 162 vos *om. ed. pr.* 166 aut] vel *ed. pr.* 167–174 Excipio . . . aegri *in margine C* 171–173 Etenim . . . habeat *in calce C* 173 *post* charas *add.* non *ed. pr.* 174 *post* doceri *add.* dicebat ille et *M* scripsit] scribit *M ed. pr.* 177 adversus] contra *M ed. pr.* 178 Homines] Omnes *ed. pr.* 180 et epitomae *om. ed. pr.*

consummatis viris et in legendis auctoribus exercitatissimis ut meminit Galenus 4° de differentiis pulsuum convenire ad ea in memoriam revocanda, quae iamdiu intellecta fere obliterari periclitantur. At iuvenibus innumera propemodum detrimenta pariunt. Primo ex compendiis integras rerum naturas atque auctorum sententias intelligere minime licet. Deinde memoriae (185) atque intelligentiae exercendae occasio prorsus aufertur, quo tyronum ingeniis nil perniciosius inveniri potest. Demum ars ipsa perfecte nunquam comparatur. Quoniam si principio fundamenta debilia ac incerta iaciantur, quicquid ipsis superextruitur infirmum omnino atque incertum evadat, necesse est. Postremo mira semper rerum, argumentorum et aliorum cogni- (190) tionis paupertas in illis observatur, qui per compendia vel potius dispendia medicae artis notitiam nancisci student. Sic ergo gerere vos debetis in auctoribus ipsis quos eligetis, intelligendis. Iam vero ut studia vestra magis extollere atque animos solidis et pluribus cognitionibus ditare valeatis, hanc viam sequamini consulo. Quod si Hippocratem, Galenum, Avicen- (195) nam, Celsum aliumve scriptorum prae manibus habueritis, quicquid apud ipsos discetis, aliorum etiam testimoniis confirmare nitamini. Verbi gratia, si contemplatio de animae sede, de indicationibus inveniendis, de alicuius pharmaci natura vobis occurrat, conamini quid in eadem re Plato aut Aristoteles, quid Hippocrates, quid Theophrastus aut Dioscorides, quid Oriba- (200) sius, Aetius aut Alexander scriptum reliquerint invenire. Quo si eadem illi senserint, magis comprobata sit vobis notitia, sin diversa aut conciliare, aut meliora iudicare, ac sequi valeatis. Si vero aliquid addiderint memoria dignum in vestris thesauris collocetis. Idem etiam in lectitandis poetis atque historicis observabitis, ut si aliquid vel ad materiam medicam, vel ad medi- (205) corum mores, vel ad ipsius medicinae artem pertinere invenietis libris medicorum adiungere eoque ad confirmandam aut ornandam rerum medicarum notitiam opportune utamini, ita tamen quod appareat semper vos ceteris medicos peioribus anteferre. Sed haec fecisse ac rerum plurimarum naturas, auctorumque innumeras sententias novisse parum erit, nisi illa (210) etiam memoria complecti conservareque curetis, ne dum occasio et necessitas exigunt, argumentis et rebus cognitis destituamini. Et quamquam memoria quoddam naturae donum in plurimis esse animadvertatur, potest tamen industria hominum atque diligentia ita efficere, ut tum memoria naturalis conservetur excolaturque, tum si non adest, magna ex parte com- (215) paretur. Ad quod obtinendum quinque studiosis omnibus magnopere utilia esse duco. Unum est, ut aliis curis ac perturbationibus animi vacui atque soluti contineantur. Unde Plato in Timaeo pueros εὐμνημονεύτους esse dicebat, quod eorum animi sint nudi aliis figuris, et ob id quae ipsis tunc temporis imprimantur, facile quoque conserventur. Alterum est ut quae- (220) cunque cogitabitis, quaecunque intelligentia consequimini, iterum atque iterum animis obvolvatis, simulque et tenaciter et cum voluptate imprimi studeatis. Solent enim quae magnam impressionem faciunt perpetuo retineri, similiter et quae cum voluptate capiuntur, perinde ac stomachus si cibos cum delectatione suscipiat arctius complectitur, et melius coquit. Sic (225) nanque Aristoteles in libris Politicorum praecepit pueros statim virtutibus ac rebus honestis dedicari, quod tunc temporis omnia veluti nova, cum admiratione atque voluptate capiant, et subinde firmissime ipsis memoriae haereant. Tertium est, ut in studiis vestris, in lectitandis auctoribus, ordinem aliquem certum et clarum habeatis, cui innixi a recta artis et firma (230) cognitione non deflectetis, dicebat Aristoteles τὰ μαθηματικὰ εὐμνημόνευτα ειναι διὰ τὴν τάξιν mathematicas disciplinas facile memoria comprehendi atque retineri, quoniam ordinem inter se quendam habent. Cuius quasi cathenae cuiuspiam merito et comprehenduntur facile, et comprehensae memoria conservantur. Hippocratis et Galeni Libri iam sunt per (235) classes dispositi, similiter et Avicennae ac aliorum multorum, ut si a principio cuiusque auctoris libros evolvere auspicemini atque ea ratione, quam

182 memoria *M* 185 auctore *M* 187 nil] nihil *ed. pr.* 188 *post* si *add.* in *ed. pr.* 191 *post*

Compendia and epitomes can be, as Galen says in his *De differentiis pulsuum,* Book 4, suitable for highly skilled men, trained in the reading of authors, so that they may recall to mind what they understood long ago but are in danger of almost forgetting. But compendia promote untold hindrances to the young. First, you cannot grasp from compendia the natures of things and the opinions of authors as wholes. Then, the opportunity to exercise memory and intelligence is absolutely lost, and nothing is more harmful to the characters of beginners. Next, the art is never seized in its entirety. For if weak and uncertain foundations are laid at the beginning, every superstructure will necessarily be infirm and unsteady. Finally, a remarkable ignorance of subjects, arguments, and other aspects of knowledge is observed in those who study the medical art through compendia, or rather *dispendia* [= expenses; or wastes of time and effort]. You should therefore conduct yourself with those authors you choose in such a way that you may understand them.

Now, however, that you may be strong enough to exalt your studies and enrich your minds with much solid knowledge, I advise you to follow the following way of life. For if you have Hippocrates, Galen, Avicenna, and Celsus or any other writer to hand, strive to confirm whatever you learn from them with the testimonies of others. For example, if the theme is the seat of the soul, try to find what Plato and Aristotle say about this, or if the theme is medical indications, see what Hippocrates said, or if the theme is the nature of some drug, see what Theophrastus, or Dioscorides, Oribasius, Aetius, or Alexander [Trallianus] have written. If they have agreed, your knowledge is confirmed, but if they differ, you can either reconcile the authorities, or judge what is better and follow that. If they have added something worth remembering, you can put it in your *thesauri* [notebooks]. You will observe the same procedure in reading the poets and historians, so that if you find anything relevant to materia medica, or on the mores of physicians, or on the medical art, you can use it to add to your medical books, to adorn your knowledge of medical things opportunely, provided it is clear that you prefer physicians to their inferiors.

But it will be a slight thing to have done this and to have learned the natures of a great many things and the countless opinions of writers, unless you memorize them, so that when occasion and necessity oblige, you need not be destitute of arguments and knowledge. And although memory is something of a gift of nature in most cases, yet human industry and diligence can be so effective, that the natural memory is conserved and strengthened, and if not present can largely be produced. To obtain it I think five points are of great utility to all students. One is that you should keep your minds free of all other cares and perturbations. Whence Plato in his *Timaeus* said children had a good memory, since their minds are naked of other shapes, because of which they retain impressions that are then stamped upon them. Another is that you should again and again revolve each possible thought and each possible piece of information in your minds and try to impress it on them tenaciously and pleasurably. For those things that make a great impression are usually retained, similarly those things that are received with pleasure, just as the stomach more tightly embraces foods it has received with delight and digests them better. So Aristotle in his books of *Politics* stated that children should be dedicated to virtue and honesty straightaway, because then everything is as new, and they grasp everything with admiration and pleasure, and their memories hold on to it firmly. A third tip is that in your studies, in reading authors, you should have some clear and fixed order, which will ensure that you are not deflected from the proper way of art and from firm knowledge: Aristotle said mathematics is easily remembered because of its order. By means of that order, as of any series [*cathena*], it is comprehended, and having been comprehended it is preserved. The books of Hippocrates and Galen are already arranged in classes, likewise Avicenna's and many others', so that if you start from the beginning to turn the pages of some author and read in the way outlined, you will not be cheated of your hopes. Besides, Galen

---

potius *add.* per *M* 194 eligetis] legetis *M ed. pr.* 195 *post* Galenum *add.* et *M ed. pr.* 199 re *om. ed. pr.* 200 quid . . . Dioscorides *om.* C^1 201 *post* reliquerint *add.* per indices *M ed. pr.* 207 adiungere] adiungatis *M* 212 exigant *M* 218 *post* εὐμνημονεύτους *add.* seu memoria valere *M ed. pr.* esse *om. M* 220 conservent *ed. pr.* 222 imprimere *ed. pr.* 225 coquit] concoquit *ed. pr.* 229 in *post* vestris *om. M* 231 deflectatis *M ed. pr.* 231–232 *verba Graeca desiderantur in M ed. pr.* 233 ordinem inter se quendam] inter se quendam ordinem *M* 236 multorum *om. ed. pr.*

exposuimus, incedatis spes vos non fallat. Praeterea Galenus librum edidit, quo ordinem legendi suos libros non infeliciter nos docuit. Inter recentiores Silvius quoque libellum scripsit, in quo ordinem Hippocratis et Galeni 240 commentaria legendi et cum fructu percipiendi uberrime monstravit. Hos si sequi placuerit, non recuso. Quartum ut non contenti sitis res aliquas semel aut bis lectitasse et percepisse, sed eas animis saepius obversetis atque dum occasio datur, cum amicis atque praeceptoribus conferatis, examinetis, atque disputetis. Ita enim dicebat Aristoteles τὰς μελετὰς τὴν μνημὴν 245 σώζειν meditationes scilicet memoriam conservare, neque aliud significarent Maiores nostri Palladen armatum fingentes quam prudentiam, sapientiam, et bonarum artium cognitionem assiduis concertationibus comparari et comparatam retineri. Quintum et postremum est, ut Thesauros ac promptuaria quaedam vobis faciatis, in quibus quicquid memoria et notitia 250 dignum habueritis per communes locos distributum reponatis atque exigente occasione eo recurratis. Poterunt loci communes esse, caussae, signa, passiones, praeservatio, curatio, pharmaca, diaeta, manuum operatio, ac similia. Sub his capitibus collocare alias species et sub speciebus proprias cuique propositiones disponere, ac propositiones pluribus testi- 255 moniis obsignare, et corroborare, varietates adnotare, conciliationes non praeterire, aliorumque et propria inventa adjicere. Ex his thesauris rerum copiam in primis cumulatissimam facietis, deinde ex descriptione ac iugibus inspectionibus magis cognitiones infigentur. Postremo si aliquid e memoria unquam exciderat, sine multo labore statim revocabitur. Caven- 260 dum est tamen ne id vobis contingat, quod multis ante saeculis hominibus eventurum auguratus est Thamus Aegyptiorum rex, cui (ut scriptum est apud Platonem in Phaedro) cum Theut literarum inventionem quasi magnum memoriae remedium ostendisset, vide (dixit ille) ne potius sint memoriae destruendae aptiores. Quasi significare vellet saepe futurum ut ho- 265 mines literarum monumentis confisi, ea quae scriptis possiderent, minime memoriae mandare, atque continere curarent. Quantum studiis incumbendum sit, nulla certa praescribi a me potest ratio, quoniam alii sunt qui ob corporis nativam imbeccillitatem laboribus studiorum durare nequeunt, alii rursum inveniuntur quibus etiam si tota die libris invigilent, nulla fatigatio, 270 nullumve damnum succedat. Et propterea quisque proprias vires metiri debet, sicque animum et corpus exercere ut neuter neutrum offendat, atque alter alterius statum tueatur. Sunt namque remissiones quaedam in studiis necessariae, quae alacriores animos et ingenia, ac iudicia vividiora reddunt, quas, Domini, si non adhibeatis, necessario ingenii vires, prompti- 275 tudo, ac iudicii acrimonia veluti flatescunt et ut dicebat Poeta ille, "si nunquam cesses tendere, lentus erit," modo tamen, remissio haec non ita alliciat, ac oblectet, ut postmodum in desidiam convertatur. Tempus aptum studiis matutinum est, Auroraque semper contemplationibus et musis amica putata est, quoniam corpore vacuo spiritus magis liberi magisque 280 sicci redduntur. Tales vero quemadmodum Plutarchus et Galenus ex sententia Heracliti dicebant aut animos atque ingenii operationes magis perficiunt. Est et vespertina hora, quando praesertim stomachus cibis non gravatur, studiis apta: pleno ventriculo cibisque adhuc effumantibus ne studeatis hortor, quod non sanitati solum magnopere nocet, verum etiam 285 animis caligatur et torpores quosdam obducit, qui postmodum familiaritatem nescio quam cum illis paullatim contrahentes ingenia hebetant, retardant, simulque iudiciorum puritatem aliquibus temporum intervallis corrumpunt. Multa alia ad amplificandum huiuscemodi tractationem in medium proponere potuissem. At satius duxi me ulterius non progredi, tum 290 quia omnia fere quae ad rem pertinent quaeve alter fusius explicaret, breviter nec obscure comprehensa fuerunt, tum quia unicuique vestrum aliqua addendi vel etiam optima suopte ingenio excogitandi facultatem esse relinquendam iudicavi.

240 Sylvius *M* 242 recusabo *ed. pr.* 245–246 μνημὴν σώζειν *om. M* 246 meditationes scili-

published a work in which he taught us not without success the proper order in which his books were to be read. Among more recent writers [Jacobus] Silvius has also written a book in which he showed very fully the order of reading Hippocrates' and Galen's writings, and how to read them profitably. I do not object if you choose to follow these. A fourth point is that you should not be content to have read something once or twice and to have perceived it, but you should turn your mind to it time and time again, and consult your friends and teachers, examine, and debate. Thus indeed Aristotle used to say, "Meditation conserves memory," nor did our ancients mean anything else when they depicted Pallas armed than that prudence, wisdom, and knowledge of the fine arts are won by assiduous disputation. A fifth and final point is, that you compile notebooks [*thesauri*] and memoranda [*promptuaria*], in which you may have whatever is worth memorizing and noting down arranged topically [by means of commonplaces]: you can store these and come back to them as occasion demands. They can be commonplaces, causes, signs, diseases, prophylactics, cures, drugs, diet, manual operations, and the like. Under these chapter headings place other subtypes, and under the subtypes propositions peculiar to each one, and impress these propositions [on your mind] with more testimonies, and corroborate them, note varieties, do not omit ways of reconciling [conflicting opinions], and add others' and your own discoveries. From these *thesauri* you will first amass a vast collection of materials, and then from writing them down and continually inspecting them your knowledge will increase. Finally, if anything is ever forgotten, it can be recalled immediately without much labor. But you must beware that what Thamus [*sic*], the king of Egypt, foresaw many centuries ago does not happen to you, namely (as Plato wrote in his *Phaedrus*), when Theut showed him his discovery of writing as a great remedy for memory, "See (said he) if it is not more apt to destroy memory." As if he wished to intimate that men, relying on written records, would not memorize what they possessed in writing.

How much study you should put in cannot be laid down by me exactly, since some are unable to bear the physical labor of study because of their native frailty, others again are found who if they spend the whole day at their books never tire and are unharmed. And therefore each person must assess his own strength, and so exercise his body and soul that neither offends the other, and each one preserves the state of the other. Some respite there must be from your studies, to make mind and intellect brisker, and judgments more vivid. If you do not have holidays, gentlemen, necessarily your intellect, promptitude, and sharpness of judgment become as it were flatulent, and as the poet used to say, "If you never stop striving, it will be sluggish"—provided, however, that this interval does not so entice you, and delight you, that it is later turned into laziness. Morning is a fit time for studies, and dawn has always been thought of as a friend to contemplation and the Muses, since the spirits are rendered freer and drier by an empty body. As Plutarch and Galen used to say (on the authority of Heraclitus), such spirits accomplish better the operations of the intellect and soul. The evening is also suitable for studies, especially when the stomach is not burdened with food: I urge you not to study when your stomach is full and the food is still exhaling fumes, because this not only seriously harms health but also obscures the mind and induces torpor, which soon gradually induces some sort of familiarity with that state [*cum illis*, i.e., *torpores*] and makes the intellect grow dull, retards it, and corrupts the purity of judgment over a period of time. I could have proffered many other topics to round out this theme. But I thought it best not to go on further, first because I have touched briefly and not obscurely on things pertinent to the matter, which another would have treated at greater length, and second because I thought it best to leave each of you the opportunity to add something from his own intellect.

---

cet] videlicet meditationes *M* 259 aliquid] quid *ed. pr.* 260 sine] si non *M* 262 Thomus *ed. pr.* 263 Theuth *ed. pr.* 265 vellet *om. C*[1] 268 ratio *post* certa *M* 269 in laboribus *ed. pr.* nequeant *ed. pr.* 270 etiam *om. M ed. pr.* 271 quisque *post* vires *M* mentiri *ed. pr.* *post* corpus *add.* simul *M ed. pr.* 272 ut] et *M* 272–273 atque . . . tueatur *om. ed. pr.* 274 et *om. M ed. pr.* 275 Domini *om. M ed. pr.* 276 flavescunt *ed. pr.* 277 eris *M* 279 -que *om. M ed. pr.* et musis] Musisque *M ed. pr.* 281 Plutarchus et *om. C*[1]*:* Galenus et Plutarchus *ed. pr.* 282 Heraclidis *ed. pr.* animorum *ed. pr.* 285 solum *om. M* 287 paullatim *om. C*[1] 289 huiusmodi *M ed. pr.* 290 me *om. ed. pr.* 292 nec] non *M* 293 suopte ingenio *om. C*[1] 294 *post* iudicavi *add.* Finis. MDlxx. *M*

The Reception of Fracastoro's Theory of Contagion

The Seed That Fell among Thorns?

*By Vivian Nutton**

IN MANY MODERN HANDBOOKS of medical history, the medical Renaissance of the sixteenth century signifies the lighting of the torch of progress after the dark centuries of the Middle Ages. Anatomists and surgeons, the true devotees of experience, at last triumph over the argumentative theoreticians and jargon-ridden physicians. Due prominence is given to the introduction of new drugs from the Americas or from the laboratories of the chemical Paracelsians, but the therapies of the average learned practitioner are dismissed as irrelevant or worse. The dead hand of Galenism is seen as stifling whatever originality there may have been among doctors, who are chided for their willingness to let tradition and book learning overcome any thoughts of innovation or belief in the supreme power of observation.[1] To this general disparagement of sixteenth-century physicians and their theories, there has been one major exception, the Veronese doctor and poet Girolamo Fracastoro (ca. 1478–1553).

Fracastoro's reputation among historians of medicine rests on two works, both relating to the same theme of epidemic disease: his poem *Syphilis sive morbus Gallicus,* of 1530, and the prose treatise in three books, *De contagione et contagiosis morbis et eorum curatione,* of 1546. Within these writings, historians, particularly since the bacteriological revolution of the late nineteenth century, have drawn attention to his theory of contagion, and specifically of "seeds of disease," as an explanation for epidemic diseases such as plague or typhus. They have seen in it a prefiguration of modern germ theory and have contrasted it with the traditional etiology of miasma or bad air.[2] More recently, Carlo Cipolla, in a

* Wellcome Institute for the History of Medicine, 183 Euston Road, London NW1 2BN, England.
I am grateful to Ann Carmichael, John Henderson, Richard Palmer, Christine Stevenson, Andrew Wear, and Alexander Zaharopoulos for their advice and criticism, as well as to audiences in Baltimore, London, and Newcastle who heard and commented on preliminary drafts of this paper.

[1] This is the message of Allen G. Debus, *Man and Nature in the Renaissance* (Cambridge: Cambridge Univ. Press, 1978); and of Sherwin B. Nuland, *Doctors: The Biography of Medicine* (New York: Knopf, 1988), pp. 61–119. Some of this imbalance is corrected by the various essayists in Andrew Wear, Roger K. French, and Iain M. Lonie, eds., *The Medical Renaissance of the Sixteenth Century* (Cambridge: Cambridge Univ. Press, 1985); and by Gerhard Baader, Vivian Nutton, and Gundolf Keil, in *Der Humanismus und die oberen Fakultäten,* ed. Keil, Bernd Moeller, and Winfried Trusen (Weinheim: VCH, 1987).

[2] Fielding H. Garrison, *An Introduction to the History of Medicine,* 4th ed. (Philadelphia/London: Saunders, 1929), p. 233; Arturo Castiglioni, *A History of Medicine,* 2nd ed. (New York: Knopf, 1947), p. 457; F. N. L. Poynter and Kenneth D. Keele, *A Short History of Medicine* (London: Mills & Boon, 1961), p. 89; and Francisco Guerra, *Historia de la medicina* (Madrid: Ediciones Norma, 1982), p. 273.

Girolamo Fracastoro. From the frontispiece to his **Homocentrica** *(1538). Courtesy of the Special Collections Department, Duke University Libraries.*

series of influential publications, has sought to distinguish between a medical, professional, and erroneous miasmatic explanation for plague and an administrative, lay, and correct belief in contagion.[3] In his scenario, Fracastoro's elucidations of the causes (and courses) of such diseases as plague, measles, and typhus represent an isolated attempt to bridge the gap between the two sides by opposing to the rarefied theories of his medical colleagues the robust common sense of the man of affairs. It failed, for, according to Cipolla, this same dichotomy can be traced well into the seventeenth century, and beyond. Most medical theorists continued to prefer a miasmatic to a strictly contagious theory of epidemic disease, and by the early nineteenth century, the name of Fracastoro had come to be more famous for poetry than for pathology. This, if nothing else, proved that here was a man before his time, misunderstood and neglected by lesser writers who did not appreciate the validity of his ideas. Had they been wiser or less blinkered in their Galenism, there would have been a Fracastorian revolution in epidemiology to set alongside the Vesalian revolution in anatomy.

Such a hypothesis, however, needs careful consideration, not least because of

[3] Esp. Carlo Cipolla, *Cristofano and the Plague: A Study in the History of Public Health in the Age of Galileo* (London: Collins, 1973); Cipolla, *Public Health and the Medical Profession in the Renaissance* (Cambridge: Cambridge Univ. Press, 1976); and Cipolla, *Fighting the Plague in Seventeenth-Century Italy* (Madison: Univ. Wisconsin Press, 1981). A different approach is followed by Ann G. Carmichael, *Plague and the Poor in Renaissance Florence* (Cambridge: Cambridge Univ. Press, 1986). The best discussion of sixteenth-century ideas on plague is that of Richard J. Palmer, "The Control of Plague in Venice and Northern Italy, 1348–1600" (Ph.D. diss., Univ. Kent at Canterbury, 1978).

the variety of questions involved. Hence this paper offers only a partial attempt at rectification, for it concentrates on the extent to which Fracastoro's ideas were accepted by other writers on epidemic disease in the second half of the sixteenth century.[4] Acceptance and influence are tricky words, however. It is not enough to document the availability of Fracastoro's writings by listing editions or citations; one must also take into consideration how his contemporaries interpreted his arguments and what they found most striking, unusual, or controversial in them. It will be argued that, far from being neglected, his ideas on the causes of epidemic disease were widely known. However, the stress placed upon them by sixteenth-century doctors was rarely that of a modern medical historian, for they understood his arguments in a different context from that of modern bacteriology. His views ceased to be cited specifically, not because they were too adventurous or too advanced, but because they became part of the common stock of medical knowledge and lost whatever novelty they once possessed.

The reassessment offered here also involves a consideration of sixteenth-century ideas on the causes of epidemics. It will be argued that at the root of modern analyses of Fracastoro or the role of contagion lies a misunderstanding of the hierarchy of causes of disease, and that a modern distinction between miasma and contagion is either anachronistic or overrefined. Despite Cipolla's stimulating advocacy, his neat divisions only obscure the reality of much contemporary medical debate, not least by failing to take into account the ease with which Renaissance authors, both medical and lay, could switch from one term to the other. The word *contagion,* especially if defined as involving indirect as well as direct transmission of harmful material, was far from incompatible with a theory of noxious air. Besides, Fracastoro's own exposition was by no means unambiguous, or incapable of improvement. His opponents were not necessarily wrong in pointing to its weaknesses, or in accommodating many of his arguments and examples to their own purposes. It is only the advent of modern bacteriology and its visible proof of the existence of germs and the like that has enabled the whole debate on causation to pass beyond logical hypotheses about the imperceptible. Failure to appreciate this has led to serious misunderstanding both of Fracastoro and of the other participants in the debate.[5]

At the same time, it is important to state what this paper is not. It is not an investigation into Fracastoro's sources, on which I have written elsewhere, nor, except in passing, is it an exposition of the development of Fracastoro's ideas on contagion over some forty years.[6] To chronicle the complicated relationships

[4] Cf. Charles Singer and Dorothea Singer, "The Scientific Position of Girolamo Fracastoro (1478?–1553) with Especial Reference to the Source, Character and Influence of His Theory of Infection," *Annals of Medical History,* 1917, *1*:1–34. My disagreement with this pioneering article should not obscure my debt to it, for its broad coverage of epidemiological literature was remarkable for its day. Until the present article, no attempt was made to go beyond the Singers' data, nor were the consequences of their investigations followed up.

[5] A similar conclusion, on a related topic in the nineteenth century, has been reached by Nicholas J. Fox in his study of the debate over Listerian antisepsis, "Lister and Humoralism," *History of Science,* 1988, *26*:366–397.

[6] On the sources see Vivian Nutton, "The Seeds of Disease: An Explanation of Contagion and Infection from the Greeks to the Renaissance," *Medical History,* 1983, *27*:1–34; rpt., with the same pagination, in Nutton, *From Democedes to Harvey: Studies in the History of Medicine* (London: Variorum Reprints, 1987) (the present article modifies some of the generalizations made therein). On the development of Fracastoro's ideas see, most recently, Carlo Colombero, "Il problema del contagio nel pensiero medico-filosofico del rinascimento italiano e la soluzione di Fracastoro," *Atti dell'*

between the various drafts of both poem and prose treatise would require at least another paper, if justice were to be done to the subtle modifications of both language and argument introduced over the years.[7] It is enough here to provide a somewhat bland summary of the arguments in the prose treatise as it was published in 1546, since it was this, above all, that offered a comprehensive and accessible explanation for a variety of epidemic diseases.

I. FRACASTORO'S IDEAS ON CONTAGION

The three books of Fracastoro's *De contagione* form only one part of the volume as printed.[8] Between the prefatory dedication to Cardinal Alessandro Farnese and the work itself stands a separate tract on sympathy and antipathy. The exact relationship between the two treatises is never spelled out, but both serve to attack explanations for the working of the universe in terms of "occult" qualities. Far from being hidden from the scientist, the causes of, for example, magnetic attraction or repulsion could be explained logically by the principle of antipathy and sympathy. This same doctrine could also explain why lime and a sponge were both good at soaking up water, and why some bodies were more receptive than others to particular diseases. The action of certain contagions was explicitly compared by Fracastoro to that of poisons or the deadly catablepha, whose gaze was believed to be fatal and whose virulence could be attributed to the antipathy of its "spiritual forms" (*species spirituales*) to the spirits of the human body.[9] In

---

Accademia delle Scienze di Torino, Classe Scientifiche, 1979, *113*:245–283, which is less informative on medicine than the title suggests.

[7] Fracastoro began writing *Syphilis* ca. 1510; a draft, along with a letter on syphilis, was sent to Pietro Bembo in 1525, and some of Bembo's comments were incorporated in the final version of the poem, printed at Verona in 1530 by the da Sabbio brothers. A prose tract on syphilis was begun by 1525 and dedicated to Bembo ca. 1533. The treatise on contagion, in progress in 1534 (T. Porcacchi, ed., *Lettere di XIII uomini illustri* [Venice: G. de' Cavalli, 1565], p. 712), was regarded as finished by 1538 (see below, n. 13), and incorporated some of Fracastoro's observations on the plague of 1534–1535. This first version of *De contagione* was dedicated to Giambattista della Torre, who died in 1538, and would appear to have differed from a draft, beginning "Quid sit contagio . . . ," preserved in Verona, Biblioteca Capitolare, cod. CCLXXV-I. After a further revision ca. 1542, Fracastoro allowed the prose treatise to be printed in 1546. Five years later, in 1551, he commented specifically on the English sweat in a letter to Giambattista Rannusio; see Porcacchi, *Lettere,* pp. 736–742. This complex development has been well chronicled by Francesco Pellegrini in a series of studies that present portions of the various versions: "Frammento inedito di Gerolamo Fracastoro riguardante la pestilenza del 1534–35," *Rivista di Storia delle Scienze Mediche e Naturali,* 4th Ser., 1935, *26*:353–359; *Trattato inedito in prosa di Gerolamo Fracastoro sulla sifilide* (Verona: La Tipografica Veronese, 1939); and *Scritti inediti di Girolamo Fracastoro* (Verona: Valdonega, 1954), pp. 1–70. I have not seen his *La dottrina fracastoriana del "contagium vivum": Origini e primi sviluppi tratti da autografi inediti* (Verona, 1950). These alterations and additions escaped the notice of Geoffrey Eatough, in the introduction to his edition of *Syphilis* (Liverpool: Francis Cairns, 1984). For a recent survey of the manuscripts and their problems see Enrico Peruzzi, "I manoscritti fracastoriani della Biblioteca Capitolare di Verona," *Physis,* 1976, *18*:342–348; and Peruzzi, "Un trattato inedito di farmacologia di Gerolamo Fracastoro," *Rinascimento,* N.S., 1978, *18*:183–187.

[8] Girolamo Fracastoro, *De sympathia et antipathia rerum, liber unus: De contagione et contagiosis morbis et eorum curatione, libri III* (Venice: Junta, 1546). I cite the latter treatise (hereafter **Fracastoro, *De contagione***) by the pages in the edition and English translation by Wilmer Cave Wright (New York/London: G. P. Putnam's, 1930).

[9] Girolamo Fracastoro, *De sympathia,* fols. 7v–11r. Cf. Fracastoro, *De contagione,* pp. 2, 24–28, 38. In Aristotelian terms, occult qualities were qualities not immediately accessible to the senses, and hence possessing an unknown power of action. Besides the study of Colombero (cit. n. 6), see Enrico Peruzzi, "Antioccultismo e filosofia naturale nel *De sympathia et antipathia rerum* di Gerolamo Fracastoro," *Atti e Memorie dell'Accademia Toscana di Scienze e Lettere "La Colombaria,"* N.S., 1980, *31*:42–131. For Fracastoro's definition of occult qualities in terms of unintelligibility see Keith

short, although Fracastoro did not explicitly make the connection, his tract *On Sympathy* provided a general theoretical framework for the discussion of the specific example of contagion and contagious disease.

This discussion involved both theory and practice. The first two books of *De contagione* offered a general theory of contagion, followed by a discussion of a number of different contagious diseases, while the third book dealt with their treatment and cure. The treatise opened with a definition of contagion that was far from unique to Fracastoro. Contagion was, so he stated, "a similar corruption of the substance of a particular combination which passes from one thing to another and is originally caused by the infection of the imperceptible parts."[10] In other words, like could only corrupt like; contagion involved the transmission of a corrupt substance; and infection and corruption began at the most fundamental level of the body's constituents, the various combinations that made up each individual part of the body, beyond the reach of human perception. In that sense, the process of contagion was occult, but it could be revealed by sound reasoning.

More striking than this definition is Fracastoro's division of contagion into three types: by direct contact, by indirect contact through fomites, and at a distance. The first and third types had long been familiar, and Fracastoro was far from alone in incriminating such intermediaries as clothing and wood. But he appears to have been the first to use the word *fomes* as a technical term for the substance deposited on these intermediaries and to describe the transferred infective agent as "seedlets of contagion" (*seminaria contagionis*). Both Galen and his medieval translators had used the phrase "seeds of disease," albeit very occasionally, and Fracastoro himself in his poem *Syphilis* had talked of the seeds of disease "creeping" (*inserpere*) into the body. But it is hard to interpret these allusions and metaphors precisely, even harder to reconstruct a theory of contagion and of seeds solely from within the poem.[11] Such hints as exist were extended by Fracastoro in a prose tract on syphilis, probably written in the early 1530s. In it he oscillated between calling the infective agents of syphilis "seeds" (*semina,* the standard Latin translation for Galen's seeds), "seedlets" (*seminaria*), or simply "first principles" (*principia*) of contagion.[12] His discussion was naturally confined to syphilis, and his account of the general principles behind the seeds was correspondingly limited.

By the late 1530s, however, Fracastoro had elaborated an explanation for the differences between the abilities of the seeds to infect in terms of their various

---

Hutchison, "What Happened to Occult Qualities in the Scientific Revolution?" *Isis,* 1982, *73*:233–253.

[10] Fracastoro, *De contagione,* p. 4. A similar definition of contagion can be found as far back as Isidore of Seville, *Etymologiae* 4.17.

[11] For the terms see Nutton, "Seeds of Disease" (cit. n. 6), pp. 1–5; and Fracastoro, *Syphilis* 2.197. Note also that on two occasions it is the disease that "creeps in" (*serpere*): 2.75 (*labes*), 2.244 (*pestis*). Interpretation is difficult partly because the Lucretian model for the poem makes it uncertain how to understand even so basic a word as *seed.* Eatough, in his edition (cit. n. 7), p. 16, follows standard practice by interpreting the poem from the later book. But if Fracastoro was developing his ideas over many years, it is methodologically unsound to assume that all the details corresponded from the start.

[12] *Semina,* in Fracastoro, *Trattato,* ed. Pellegrini (cit. n. 7), pp. 83–85, 92, 95; *seminaria, ibid.,* pp. 83–85, 87; *principia, ibid.,* pp. 83–84, 87, 91–92. On p. 92 the draft shows an equivocation between *semina* and *principia.* Fracastoro may have returned to *principia* in the letter of 1551; Porcacchi, ed., *Lettere* (cit. n. 7), pp. 739–741. For the date of the tract see Pellegrini in Fracastoro, *Trattato,* p. 5, who also notes that the reference, p. 129, to a cattle plague occurring in 1525 "while we were writing" shows that some parts of the tract were completed earlier.

elemental properties.[13] The combination that made up the seed had to be sufficiently strong to survive the transfer and viscous enough to remain in or on an intermediary, which itself had to be of an appropriate kind for the seed to stick. The seeds that infected at a distance needed both a more powerful combination—or else they would break up in transit—and a greater subtlety, in order to penetrate their host. Some seeds, on adhering to the neighboring humors, might begin at once to propagate and multiply until the whole mass of humors was affected. Others were drawn inward into the body by inspiration of the air and, once deep in the body, brought about a dissolution by exciting, like poison, an antipathetic movement of spirits (or life forces) away from their site. Still other seeds might enter through the smaller blood vessels on the periphery and pass along them to the heart and other major organs.[14] Once in the body, their action depended on their analogies and affinities, for, it was well known, some individuals and some parts of the body were likely to be attacked by some contagions and not by others. Phthisis did not attack the eyes. The type of seed also determined the type of disease. The seeds of syphilis were sharper and harder than those of leprosy and were thus able to penetrate into the deepest parts of the body.[15] Once there, they would initiate putrefaction, the rotting and dissolution of the parts.

The relationship between putrefaction and contagion occupies many pages in Fracastoro's discussion. In Book 1, while accepting contagion as a type of putrefaction, he nevertheless denied that the seeds that produced the contagious putrefaction had themselves to be putrid; their very presence within the host's body might be enough to initiate putrefaction. Unlike Giambattista da Monte, a professor of medicine at Padua and his fellow townsman, Fracastoro did not believe that putrefaction of the humors by itself was enough to cause contagion and plague, for, so he argued, this offered no explanation for the transmissibility of contagious disease. By contrast, his own principle of seeds endowed with a special method of causing putrefaction answered all the questions.[16]

But how were these seeds created? Normally they were engendered and propagated within the host, although, in certain circumstances, they might also be produced in the sky by the heating action of the heavenly bodies or as a result of a particular planetary conjunction. This was an area crying out for further research, but for syphilis, at any rate, Fracastoro was convinced of a sidereal origin.[17] In his opinion, it was a totally new disease, the result of a change in the

[13] See Aloisio Mundella, Letter 3, Nov. 1538, in *Epistolae medicinales* (Lyons: Junta, 1556), p. 319. Mundella talks of the three books as being a "finished book" and comments in passing on the *seminaria,* while criticizing some of Fracastoro's views on the proper treatment for pestilential fever. In a letter of 1541, 23, p. 360, Mundella complained that Fracastoro had broken his promise of 1538 to publish the work.

[14] Fracastoro, *De contagione,* p. 34. Fracastoro apparently distinguishes this form of penetration from both that of seeds analogous to viscous humors being carried along by the blood vessels and that of seeds introduced in the air by the combination of breathing and the dilatation of the arteries. Exactly what he had in mind here is far from clear.

[15] *Ibid.,* pp. 18 (phthisis), 60 (leprosy). In his discussion of specific parts affected, Fracastoro made great use of the traditional pair of phthisis and ophthalmia. It is not immediately obvious whether he thought in terms of seeds of specific diseases, or whether his division of the types of seeds corresponds to that of *classes* of disease and their modes of infection.

[16] *Ibid.,* pp. 40–42, 76–86.

[17] *Ibid.,* pp. 58–60. It should be remembered that Fracastoro enjoyed a wide reputation for his skills as an astronomer. As Norman Howard-Jones pointed out, in his discussion of the generation of the seeds of contagion Fracastoro accepted all the traditional causes—meteorological, planetary, and

disposition of the air, since without such a potent cause, it would be hard to imagine such a slowly acting contagion spreading so quickly. The idea of a new disease or of diseases with a relatively localized spread did not trouble him, for, so he argued, the movements of the planets and the constant changes in the atmosphere lent themselves easily to such transient creations. An aerial cause also helped to explain the changes in the course of syphilis over the years, for with the departure of the original atmospheric conditions, the disease had settled down to its own life cycle. After half a century, it was now in its old age and might in time die or be too weak to propagate itself. But, equally, its disappearance was no guarantee that the same atmospheric conditions would not recur and re-create identical seeds. After all, in certain parts of the world, notably Hispaniola, syphilis was endemic because of the permanence of favorable atmospheric conditions.[18]

Such theoretical questions took up most of the first two books of *De contagione,* and Fracastoro confined to the last book his suggestions for the treatment of the (relatively few) contagious diseases he had identified in Book 2. His general principles involved the extinction of the *seminaria,* stopping them from propagating, expelling them from the body or repelling them by means of antidotes that were "spiritually antipathetic" to them. Alternatively, one might dry the body in order to stop the rot—for example, by administering various powders and preservatives—or one might remove the putrefied substance entirely, by drugs, caustics, or surgery.[19] The precise remedy depended, of course, on both the disease and the patient, but Fracastoro was confident enough to lay down general rules for treatment as well as more specific injunctions against particular contagious diseases. Not surprisingly, given the genesis of the treatise, Fracastoro gave most space to the treatment and cure of syphilis. His theories, however, were cited more often by contemporaries in their discussions of plague, although this perhaps reflects only the relative popularity of plague tracts.[20]

II. THE CONTEXT OF THE RECEPTION: FOMITES, *SEMINARIA,* AND CONTAGION

The first stage in attempting to assess reactions to any medical book in the sixteenth century is to investigate its accessibility. We have no estimates for the number of copies of *De contagione* put out by any printer, but the fact that it went through several printings implies widespread availability. The first edition, published at Venice in April 1546 by the heirs of Lucantonio Giunta, was fol-

mineralogical: "Fracastoro and Henle: A Reappraisal of Their Contribution to the Concept of Communicable Diseases," *Med. Hist.*, 1977, *21*:61–68.

[18] Fracastoro, *De contagione*, pp. 111–116, 151–156. Other observers had equally noted a change in the apparent behavior of the disease over time. Although Fracastoro reports the belief of others that syphilis was brought to Europe by Columbus's men, he does not share it. This is also the impression given by his poem *Syphilis,* for although the Spaniards find in the New World a "contagion never before observed," 3.249, their shipmates who had earlier returned to Europe had discovered the same contagion already raging there, 3.385.

[19] Fracastoro, *De contagione,* pp. 190–204.

[20] Comparative figures are hard to establish, but the proportion could well have exceeded five to one. The library of one German doctor in 1591 contained sixty-eight printed plague tracts and at least one more in manuscript; see Klaus G. König, *Der Nürnberger Stadtarzt Dr. Georg Palma (1543–1591)* (Stuttgart: Gustav Fischer, 1961), pp. 98–102 (books), 92 (MS). For this essay, I have endeavored to read as many tracts as possible from the twenty-five years following the publication of *De contagione.*

lowed in the same year by another Venetian printing, by Girolamo Scoto. In 1550 the Lyons printer Nicolas Bacquenoys put out two printings, the second in an elegant duodecimo at the expense of Guillaume Gazeius. Four years later, Gazeius, in cooperation with a different printer, Jean de Tournes, reissued the book, but in a slightly different and more cramped format. This was the last separate edition of *De contagione* for three centuries, although it was included in the five editions of Fracastoro's *Opera omnia* that appeared between 1555 and the end of the century.[21] Such a publishing history suggests a far from negligible response to his ideas, or at least the expectation that enough buyers would be interested to make reprinting and republication profitable.

A bald listing of editions, however, provides only a starting point for an evaluation of Fracastoro's influence. It cannot show whether and how the book was read or what responses it evoked. Nor does it allow for one of the most obvious features in the process of the reception of Fracastoro's ideas—their deployment at second or third hand by later writers on plague and contagion. That such usage can be traced is the result of Fracastoro's novel terminology. Although the concept of seeds of disease could be found in Galen, and even earlier in Lucretius, and although the Latin phrase *semina morbi* occurs in other medical writers before Fracastoro, the designation of the infective agents as *seminaria* would appear to be his alone. Other authors had used the word, but in the traditional sense of seedbed, and in the prose treatise on syphilis, Fracastoro himself may have employed it in this way when he characterized the hot and moist particles sent off from a rotting apple as the "principle and seedbed [*seminarium*] of putrefaction."[22] Later on in the same tract, however, the word is used interchangeably with *semen* and *principium*. By 1538, however, *seminarium* has replaced *semen* as the technical term. But whatever subtle meaning Fracastoro wished to impart by his choice of *seminarium*, his contemporaries, and all subsequent commentators, have interpreted it as if it were no different from *semen* (seed).

Equally distinctive is his term for the intermediary in the transmission of an infected substance from a sufferer to a previously healthy host—*fomes*. This choice word was a technical term in theology for the minute portion of original sin left behind after baptism, which might, at any moment, burst into the fire of concupiscence when presented with a suitably desirable object. In this context *fomes*, whose literal meaning is woodchip or tinder, is an apt metaphor.[23] It was equally appropriate when used by the Bolognese physician Filippo Beroaldo in his plague tract of 1505. He attributed plague to the inhalation of air that had been made bad through the wrath of God, an unseasonable conjunction of the

[21] Details of the separate printings are most easily found in Wright's edition of *De contagione*, p. 346. I have not seen the second Venetian issue. For the first Lyons issue, by Nicolas Bacquenoys alone, not mentioned by Wright, see H. M. Adams, *Catalogue of Books Printed on the Continent of Europe 1501–1600 in Cambridge Libraries*, Vol. I (Cambridge: Cambridge Univ. Press, 1967), p. 448. For the *Opera omnia* see Leona Baumgartner and John F. Fulton, *A Bibliography of the Poem "Syphilis sive morbus gallicus" by Girolamo Fracastoro of Verona* (New Haven, Conn.: Yale Univ. Press, 1935), pp. 60–64.

[22] Fracastoro, *Trattato*, ed. Pellegrini (cit. n. 7), p. 89. For a few of the other references in which *seminaria* are used interchangeably with *semina* and *principia* see above, n. 12. For the older usage note Jean Fernel, *De abditis rerum causis*, in *Medicina universa* (Frankfurt: Wechel, 1577), p. 25.

[23] For texts see Nutton, "Seeds of Disease" (cit. n. 6), p. 34. Fracastoro used the word in its normal meaning of "tinder" at *Syphilis* 3.163, and this would appear to be the meaning also at 2.257. The sudden appearance of the word at *Trattato*, p. 127, implies that Fracastoro had already adopted it elsewhere as a technical term, probably in a lost section dealing with the principles of contagion.

heavens, foul waters, noxious fumes, or exhalations from the ground. Lawyers might also claim that the cause was the heaps of filth lying around or offensive smells. For Beroaldo, merely to ascribe plague to bad air was not enough, for it was a contagion that crept (*serpere*) around the city and was particularly prone to affect those in closest contact with the sick, especially if they too were out of condition. Then it happened that the disease, once in the patient, acted like tinder (*fomes*), making the evil worse and more potent until it could then be caught by others.[24]

Two points should be noted here. First, the *fomes* is *within* the person becoming ill; it is not an external intermediary. The conjunction of the word *succendere* (to light) also suggests that the metaphorical image corresponds to the traditional one of lighting a fire from a spark, rather than to Fracastoro's technical term. Nonetheless, the presence in Beroaldo's treatise of the words *contagia morbi, serpere,* and *fomes,* as well as of a belief in the creation of new types of disease, helps to provide a context for Fracastoro's ideas, even if it does not prove any influence.

Second, Beroaldo accepted both contagion and an atmospheric explanation for plague; they were not, in his mind, exclusive categories. He was not unusual in this. Although the later Paduan professor Girolamo Mercuriale might be right that Fracastoro was the first to "open men's eyes to the understanding of contagion," it is nonetheless true that many physicians had earlier discussed plague at length in terms of contagion, even if they did not go deeply into questions about the actual process of contagion.[25] Although most of the plague treatises written directly after the Black Death said nothing about contact or contagion, a substantial minority of later authors discussed plague in these terms.[26] Some used "contagion" and "plague" almost as synonyms. In the first decade of the sixteenth century, the Turin physician Pietro Bairo traced this equation back to Isidore of Seville in the sixth century. It was a belief not confined to Italy. When in 1518/19 the University of Oxford was almost closed down by an outbreak of "plague" (*pestis*), both the chancellor and the university authorities described it as a contagion and ascribed its cause to exhalations from the flooded Christ Church meadows.[27] Others were more precise in their explanation of the relationship. In 1529/30 the Leipzig professor Georg Helt wrote a letter to the town clerk of Zwickau, Stephan Roth, rejecting rumors of a plague outbreak at Leipzig. Helt and his colleagues there were sure that what they identified as the English sweat was not plague. Their disease was not contagious; it attacked neither the families of the sick nor the young males, whose spirits were most

[24] Filippo Beroaldo, *De terræmotu et pestilentia* (Bologna: J. de Herberia, 1505), sigs. B iii r, A viii v, B v r, and B vi v: "Fit autem ut morbo cubantis veluti quodam fomite succendere vicium illud protinus exultet atque in aliis concipiatur"; cf. sig. B vii r: "contagio serpebat in civitate."

[25] Girolamo Mercuriale, *De pestilentia* (Venice: P. Meietus, 1577), p. 37 = Mercuriale, *De peste* (Basel: P. Perna, 1577), p. 48.

[26] See, e.g., Dorothea Waley Singer, "Some Plague Tractates," *Proceedings of the Historical Section of the Royal Society of Medicine,* 1915–1916, 9:185–191; Palmer, "Control of Plague" (cit. n. 3), pp. 10–16; and John Henderson, "Epidemics in Renaissance Florence: Medical Theory and Government Response," in *Maladies et société Xe–XVIIIe siècles,* ed. N. Bulst and R. Delort (Paris: CNRS, 1988).

[27] Pietro Bairo, *De peste,* in *De medendis humani corporis malis enchiridion* (Frankfurt: J. Saurius, 1612), p. 680, citing Isidore, *Etymologiae* 4.17; and W. T. Mitchell, ed., *Epistolae academicae 1508–1596* (Oxford: Clarendon Press, 1977–1978), pp. 71–94. For a similar incident at Padua in 1541 see Nutton, "Seeds of Disease" (cit. n. 6), p. 27.

receptive to attack by plague. According to Helt, all outbreaks of plague, and of some other diseases too, were distinguishable by the presence of contagion.[28]

Traces of a vigorous debate in Italy on contagion can also be found in the extensive examination of the whole problem by the new professor of practical medicine at Padua, Giambattista da Monte, in one of his lectures on Rhazes in 1540/41.[29] Da Monte was convinced that for contagion to take place, there must be contact of some sort. This could be either direct or through the air or through such things as clothing, which trapped the putrid vapors. In the process of transmission, the intermediary itself might become putrefied, which marked the action of contagion off from that of poison, where, despite rumors from Turkey of chairs daubed with fatal ointment, the intermediary itself remained unaffected. The main agent of contagion was putrefaction—an opinion Da Monte advocated tenaciously to the end of his life—and the degree of contagion depended primarily on the degree of putrefaction attained. An immense putrefaction, as in plague or phthisis, was contagious; a minor one, as in epilepsy, hardly at all. Apparent exceptions, like ophthalmia and syphilis, could be explained away either by the type of medium involved—the affected eye produced thinner vapors that were more easily inserted into the ambient air—or by the receptivity of the part affected. This could be either naturally high—the eye was a "fine organ" (*membrum rarum*), easily capable of receiving impressions from outside—or enhanced —in syphilis the friction of intercourse heated and opened up the body to noxious humors. Da Monte further distinguished the putrefaction, in his opinion a secondary quality, both from the primary quality, the air, and from any tertiary quality, such as the occult quality found particularly in poisons. Air by itself could not *cause* contagion, although it might lower one's resistance to it. The tertiary quality affected the form of the body, not, like contagious putrefaction, its matter. Its effects were perceptible, but neither the way it worked nor its specific form could be known, and hence it might properly be termed occult. Such a quality might work by direct contact, but its potency was such that, like the poisonous aura of epilepsy, it could act at a distance and without requiring the alteration of any intermediary. This, for Da Monte, was a cogent objection to Averroes' theory that the contagion of plague was the result of the alteration of the intermediate air by a venomous occult quality.

In this and in subsequent lecture courses Da Monte reiterated his view of contagion and putrefaction, even developing it further to suggest that the distinction between contagious diseases related not only to the degree of putrefaction involved but also to the site of the major putrefaction.[30] In ophthalmia this was

[28] O. Seidenschnur, "Ein Beitrag zur Geschichte des englischen Schweisses," *Henschels Janus,* 1846, *1*:173–174. The letter is not included in Otto Clemen, *Georg Helts Briefwechsel* (Leipzig: M. Heinsius, 1907). Although Seidenschnur's transcription is inaccurate, this does not affect the point made here. For the date see Otto Clemen, *Kleine Schriften zur Reformationsgeschichte,* Vol. V (Leipzig: Zentralantiquariat, 1984), p. 97. By contrast, Dr. Simon Stein of Altenburg described the same "new plague" (the sweat) as contagious, to be checked by flight or, failing that, by regulating the six non-naturals: *ibid.,* p. 95. For other German examples of "contagia luis" see *ibid.,* Vol. VI (1985), pp. 118–120 (of 1551).

[29] Giambattista da Monte (Montanus), *In nonum librum Rhasis ad Mansorem . . . expositio, a Valentino Lublino . . . communicata* (Venice: Baltassare Constantino, 1554), fols. 217v–227v. This version of the lectures was later incorporated into the posthumous compendium of his writings, the *Medicina universa,* ed. M. Weindrich (Frankfurt: Wechel, 1587), pp. 381–387.

[30] Giambattista da Monte, *In libros Galeni De arte curandi ad Glauconem explanationes* (Venice: B. Constantino, 1554), fols. 170r–175v = *De febre pestilentiali,* in *Opuscula varia et praeclara*

the eye, in scabies the skin, in plague and pestilential fever the heart (because of their high mortality), and in syphilis the liver.[31] For all these diseases Da Monte claimed to have found an explanation that was neither occult nor "spiritual" and that excluded the belief that contagion could occur as a result of the "impression of a harmful quality from among the intentional qualities" (*propter impressionem malae qualitatis de genere qualitatum intentionalium*).[32] In two versions of this lecture course, this view is ascribed to "many friends" (*multorum ex amicis*); in a third, which comes from Johann Crato von Crafftheim, Da Monte apologized for sticking to his opinion in the face of "many and a friend" (*multorum et amici*), and it is the unnamed friend, "otherwise a man of learning," who is warned by Da Monte not to publish or else face a public contradiction.[33] Was this Fracastoro? At first sight, the answer must be no, for this theory bears little relationship to his. But given that Wellcome MS 568, a dated transcript of the lectures of 1540/41, omits the details of the friend who believed in the deadly power of imagination, the version published by Crato in 1562 looks very much like an afterthought, not least in the clumsiness of its phrasing. Since there was, as we shall see, a certain antipathy between the two men, Da Monte, in the formulation he vouchsafed to Crato in the late 1540s, could merely be altering the text only insofar as was necessary to insert an allusion to his disagreement with Fracastoro on the same topic. The theory of the *multi* was, on this hypothesis, not that of Fracastoro, who was equally opposed to "intentional qualities"; but both theories were alike in being rejected by Da Monte.[34]

Whoever the friend was who had to take second place to truth, Da Monte's lecture makes it abundantly plain that the topic of contagion was controversial even before the publication of *De contagione* in 1546. Indeed, Fracastoro delayed the publication of his own treatise specifically to refute Da Monte's views on putrefaction and plague as they had been given in the 1542 course of lectures on Galen's smaller *Methodus medendi*.[35] Far from being an isolated examination

(Basel: P. Perna, 1558), pp. 379–381. See also Da Monte, *In Aphorismos Hippocratis lectiones* (Venice: B. Constantino, 1553), fol. 142r.

[31] See the extracts cited in Andrew Wear, "Explorations in Renaissance Writings on the Practice of Medicine," in *Medical Renaissance*, ed. Wear, French, and Lonie (cit. n. 1), pp. 141–142. Da Monte in 1543 wrote a specific treatise on syphilis, which was widely printed after his death either separately—e.g., Vienna: M. Zimmermann, 1553; Venice: B. Constantino, 1553–1554—or as part of the *Explicatio locorum medicinae*, edited by Vincenzo Casale (Paris: G. Julianus, 1554). In the *Explicatio* Da Monte's discussion of the role of contagion occupies fols. 210r–214r.

[32] In other words, as the result of some form of imagination or mental activity. In his discussion of syphilis, *Explicatio*, fol. 210r, Da Monte was prepared to ascribe a "venomous quality" to the "*virus*" that easily affected soft flesh, but this was a material cause.

[33] Da Monte, *In nonum librum* (cit. n. 29), fol. 220r = London, Wellcome Institute, MS 568, sig. P vii r (dated to 1540/41); and Da Monte, *In nonum librum Rhasis ad r. Almansorem lectiones . . . emendatae . . . a Johanne Cratone* (Basel: P. Perna, 1562), p. 447 = *Medicina universa* (cit. n. 29), p. 382.

[34] For Fracastoro's view of these "spiritual qualities" see *De contagione*, pp. 24–26. Crato claimed to have worked with Da Monte's authority and approval and hence to have produced more accurate and "emended" versions of the lectures than the variant transcripts passed on to unscrupulous editors. The strangeness of the phrase and the insertion, compared with Crato's own stylish Latin, suggests that it was owed to Da Monte himself.

[35] *Ibid.*, pp. 78–80. Fracastoro claims merely to have provided a version of what Da Monte had said. The order of topics and occasional words correspond to those of the printed lectures: Da Monte, *In libros Galeni* (cit. n. 30), fols. 169r–194r. Since these lectures were never printed in Da Monte's lifetime, Fracastoro's claim that the words were "ab auditoribus notata et litteris mandata" is either a mistaken reference to their appearance in print (as Wright's translation of *De contagione* suggests) or, more likely, a way of distinguishing between student rough notes and more literate presentations of the material, as in Wellcome MS 568. This distinction is made by Gian Matteo Durastante, in his

of the problem, Fracastoro's book was but one contribution to a learned debate throughout Europe on a range of contagious diseases. It offered, as we shall see, one hypothesis among many, and the reactions to it by other scholars were conditioned by their knowledge and understanding of this wider debate. Unlike their modern counterparts, Fracastoro's contemporaries had read more than one book on the subject.

III. FIRST REACTIONS

Fracastoro was a man of many parts. Although until the late nineteenth century his primary reputation was that of a poet, especially as the author of *Syphilis,* he was also highly regarded as a mathematician and a physician.[36] The German poet-physician Petrus Lotichius Secundus described him as one of the glories of the University of Padua and took time off from his studies in Italy to pay his respects at Fracastoro's tomb. Others more prosaic acknowledged his achievements. Giulio Alessandrini of Trento, an erudite, effusive, and somewhat unimaginative scholar, put him sixth (of nineteen) on his list of those who had contributed to his own education and to the recovery of the true Galenic message, a perhaps surprising accolade for one who never held a university chair and who is not today counted as a Galenist. But this was no isolated opinion. When in 1543 Luigi Panizza wished to submit his thesis on venesection for publication, he asked that it be sent for examination either to Antonio Musa Brasavola, the senior professor at Ferrara, to Fracastoro, or to Da Monte at Padua, for their word would be enough.[37] In the event, Brasavola alone acted as assessor, but the inclusion of Fracastoro's name in second place attests his wide reputation as a learned physician.

Panizza's collocation of Fracastoro and Da Monte is of considerable interest for, as we have already seen, their careers and interests intertwined at many points. Both came from Verona, and they may have attended the same patients during the 1520s and 1530s. In 1533 Da Monte became the personal physician of Cardinal de' Medici, whom he followed to Rome.[38] In 1540 he returned north, to

preface to his edition of Giambattista da Monte, *In primi libri Canonis Avicennae primam fen . . . commentaria* (Venice: V. Valgrisi, B. Constantino, 1557), fol. 4v.

[36] The biographical account by W. Parr Greswell, *Memoirs of Angelus Politianus . . . Hieronymus Fracastorius . . . and the Amalthei,* 2nd ed. (London/Manchester: Cadell & Davies, 1805), is almost entirely devoted to his poetry and, in its details, is better documented than the older *De vita, moribus, scriptis Hier. Fracastorii Veronensis* of F. O. Mencke (Leipzig: Breithoff, 1731), which is the foundation of Wright's account in the introduction to *De contagione.* At the beginning of the seventeenth century Bernardino Baldi, in his *Cronaca dei matematici* (Urbino: A. A. Monticelli, 1707), p. 121, emphasized Fracastoro's importance as a mathematician and astronomer and passed over "his other writings and the fact that his excellence as a poet is clearer than daylight." I owe this reference to Judith Field.

[37] Petrus Lotichius Secundus, *Opera omnia* (Heidelberg: G. Voegelin, 1609), *Elegies,* Bk. 3.4 (of 1555), "De Patavii celebritate et suis studiis," p. 76; for his Italian journeys see J. Hagius, *Vita Lotichii Secundi,* in the *Opera omnia,* pp. 59–60. Giulio Alessandrini, *Galeni enantiomaton aliquot liber* (Venice: Junta, 1548), p. 171. Luigi Panizza, *De venae sectione in inflammationibus* (Venice: F. Camozzi, 1561), fol. 100v. Given the flexibility of Galenism, one should not judge Fracastoro not to be a Galenist either on his theory of seeds (which, as will be argued, was perfectly compatible with what Galen had said) or on his attack on Galen's idea of critical days, *De causis criticorum dierum* (Venice, 1538), which aroused the anger of "hard-line" Galenists like Matteo Corti. See Mundella, Letter 3 (cit. n. 13), p. 318.

[38] See Wright, preface to Fracastoro, *De contagione,* p. xvii. One would like to know more about the lawsuits and litigation he had to deal with in the 1530s; see the dedication to Cardinal Ippolito de' Medici prefaced to his translation of Aetius, *Opera omnia* (Basel: H. Froben, 1535), sig. AA2v.

Padua, as professor of medical practice, transferring in 1543 to the chair of medical theory. On his death in 1551, Fracastoro composed an elegant poem in memory of the "second Galen," whom he had always treated with respect, even when they had disagreed. Not least over contagion: Fracastoro had delayed his own book in order to answer the theory of putrefaction given by Da Monte in his Paduan lectures on Galen.[39] His summary of his opponent's views was largely accurate, and his objections carefully and courteously put. Although Fracastoro was prepared to allow a role for putrefaction, in his opinion plague differed from other putrid fevers in three ways: in its course, in its substance or manner of putrefaction, and, above all, in its active principles, the external *seminaria* of plague.

It is not surprising, then, that after this confrontation Da Monte should be among the first to comment on Fracastoro's opinions once they had appeared in print, or that he should reject them, albeit with considerable acerbity. The occasion was a group of four lectures in the middle of his course on the second fen (treatise) of the first book of Avicenna's *Canon*. In his previous year's lectures on *Canon* 1.1, he had discussed pestilential fever, reiterating his theory of putrefaction and his opposition to hidden, occult causes, and he now repeated his views on both pestilential fever and syphilis in the early lectures on *Canon* 1.2. Neither here nor in the possibly contemporaneous lectures on the *Aphorisms* of Hippocrates is there any reference or allusion to Fracastoro.[40] Apparently suddenly, halfway through the course, at the end of lecture 31, Da Monte broke off from a discussion of the various stages and types of fever to announce that the next day he would be dealing with contagion and would have "much to say against our friend from Verona." Indeed he had, for he devoted four lectures to the topic, demolishing to his own satisfaction Fracastoro's ideas on sympathy in lectures 32 and 35, and those on contagion in the intervening pair. The irruption of the subject at a not entirely appropriate point and the vehemence of Da Monte's language, which was somewhat moderated in the summary he gave the next year in commenting on *Canon* 1.4, strongly suggest that these were first reactions to reading *De contagione*.[41]

When did Da Monte light on Fracastoro's book? According to Johann Crato von Crafftheim, a devoted student of Da Monte's at the end of the 1540s, the course on the *Canon* occupied three consecutive academic years, from 1547 to

[39] Fracastoro, *Carmina* (Padua: J. Cominus, 1718), p. 159 (the poem); and Fracastoro, *De contagione*, pp. 76–86. See also his praise of Da Monte in Porcacchi, ed., *Lettere* (cit. n. 7), p. 712. This respect was acknowledged by Crato; see his preface to Da Monte, *Consultationes* (Basel: P. Perna, 1565), sig. A4r. For the crucial role of Da Monte as a Galenist see the encomium by his former pupil Orazio Brunetto, *Lettere* (Venice, 1548), fol. 72v.

[40] Da Monte, *In . . . primam fen* (cit. n. 35), fol. 4v; and Giambattista da Monte, *Lectiones . . . in secundam fen primi Canonis Avicennae* (Venice: V. Valgrisi, B. Constantino, 1557), pp. 63–65 (plague), 383 (syphilis). The latter series was given a year after the former, *ibid.*, p. 1, at Padua, while Da Monte held the first chair of medical theory; see Da Monte, *In . . . primam fen*, preface, fol. 3r. See also Da Monte, *In Aphorismos* (cit. n. 30), Vol. I, fols. 142r–143r, 159r–160r. According to Crato, in his preface to Da Monte, *In nonum librum* (cit. n. 29), pp. ix–xi, this lecture course preceded the lectures on *Canon* 1.1 and was given in 1544, but this is by no means certain, and 1545/46 cannot be excluded.

[41] Da Monte, *In secundam fen*, p. 418; and Giambattista da Monte, *In quartam fen primi Canonis Avicennae lectiones* (Venice: B. Constantino, 1556), fols. 51r–v. The four lectures on the second fen cover pp. 418–428, 428–440, 440–453, and 453–467. Da Monte sophistically makes a connection with the previous lectures, on the *stages* of disease, by talking of contagious disease as a subdivision of disease.

1549.[42] But to set Da Monte's explosive reaction in late 1548, at the earliest, is to imply that, despite ties of origin and acquaintance, links with the Giuntine press in Venice, and two separate issues of *De contagione* in Venice, the volume or its contents remained unknown to him for over two years. This is unlikely. Nor can one believe in the date of 1544/45, mentioned by the aged Orazio Augenio in 1600 during the course of a debate on the establishment of formal lectures on *Canon* 1.2.[43] This is at least one year too early, for Da Monte opened the preceding course on *Canon* 1.1 with an attack on John Caius's unauthorized publication, at Basel in 1544, of his lectures on Galen's *Methodus medendi ad Glauconem*.[44] A possible solution, albeit radical, to this dating problem is to assume that the lectures Crato heard in 1548/49 were a repetition of those of 1545/46, for to judge from a printed summary, Da Monte had repeated his ordinary lectures of 1543 on Galen's *Ars medica* in 1546/47, with slight modifications.[45] The hypothesis of a three-year cycle of repeated ordinary lectures, that is, those on the texts set for examination, would resolve a major difficulty in Crato's chronology, namely, that in the three years 1544–1546 only one course, on the *Aphorisms*, is mentioned. On my suggested schema, as set out in Table 1, the lectures on *Canon* 1.2 and 1.4 would have been given extraordinarily by Da Monte to those students who wished to attend.[46] This proposed explanation and the revised dating allow the extraordinary lectures on *Canon* 1.2 to fit neatly into 1545/46, or rather into the latter part of that academic year,[47] and those on 1.4 to go into the next year.

[42] Crato, preface to Da Monte, *In nonum librum* (cit. n. 29), pp. ix–xi.

[43] Venice, Archivio di Stato, Riformatori dello Studiò di Padova, Filza 419. I am grateful to Richard Palmer for making his transcripts of this and other relevant archival material available to me. For the background to this proposal see Nancy G. Siraisi, *Avicenna in Renaissance Italy* (Princeton, N.J.: Princeton Univ. Press, 1987), pp. 113–119.

[44] Da Monte, *In . . . primam fen* (cit. n. 35), p. 8, referring to John Caius, *De methodo medendi libri duo* (Basel: H. Froben, N. Episcopius, 1544). The preface is dated 15 May 1544, which rules out any course of lectures earlier than 1544/45.

[45] Giambattista da Monte, *Typus trium librorum Artis parvae Galeni* . . . (Venice: Junta, [1546]), is an altered reprint of the summary of the lectures of 1543 and refers to the fuller treatment Da Monte is to give in his ordinary lectures of 1546/47. The date is confirmed by Crato, preface to Da Monte, *In nonum librum* (cit. n. 29), p. xi.

[46] Crato, preface to Da Monte, *In nonum librum,* p. xi, which puts the *Aphorisms* in 1544/45. He also said nothing about the lectures on *De differentiis medicamentorum,* which preceded those on *Canon* 1.4 but in the same academic year (Da Monte, *Explicatio* [cit. n. 31], fols. 26r, 43v), or about the "many extraordinary lectures" that Da Monte gave in 1549; see Archivio dell'Università di Padova, Reg. 660. The relevant documents on the official, ordinary lectures given by Da Monte do not, apparently, survive in the Paduan or Venetian archives. None of the editors of the lecture courses—Johann Crato, Gian Matteo Durastante, Francesco Pegolotti, or Walenty Lublin—was in residence at Padua in the early 1540s, and they made use of transcripts by others. It should be admitted that this hypothetical reconstruction depends on the absence of archival information. If Crato was the "learned and faithful German pupil" who told Da Monte about the German habit of leaving murdered corpses unburied in case the reappearance of their assailant started the blood flowing again and thus identified the murderer (*In secundam fen* [cit. n. 40], p. 467), the argument for the notes edited by Francesco Pegolotti and published in 1557 being of the first presentation would be weakened. But there were many other Germans besides Crato in Da Monte's audience, and nothing to stop Da Monte's repeating his original lecture word for word. Nonetheless, if Crato's dates for the lectures, 1548/49, are preferred, Da Monte's silence for over two years is very curious indeed and requires some explanation.

[47] At first sight, it might seem impossible to fit thirty-one lectures into the period between April 1546, the date of publication of *De contagione,* and the end of term in, usually, early July. But from the many references to "today," "yesterday," or "the day before last" scattered throughout the lectures, it is clear that they were given not on fixed days but as a block, interrupted by feast days and holidays. Hence, assuming that Da Monte got an early sight of the book, the remaining lectures could be squeezed into the term, even if it ended before he could cover all the text of *Canon* 1.2; see *In secundam fen,* p. 832.

Table 1. Lecture schema

	Ordinary	Extraordinary
1543	*Ars medica*	
1544	*Canon* 1.1	
1545	*Aphorisms*	*Canon* 1.2
1546	*Ars medica*	*Canon* 1.4
1547	*Canon* 1.1	
1548	*Aphorisms*	*Canon* 1.2
1549	*Ars medica*	*Canon* 1.4

Only Augenio's dimly remembered date now needs correction, and even here his error is of one year only, not four, and may pardonably be attributed to confusion over the various parts of *Canon* 1.

But even if this proposed redating is not accepted, there can be little doubt that Da Monte's reaction to *De contagione,* whether or not it was delayed until 1548/49, was both vigorous and, it must be acknowledged, deadly accurate. Fracastoro's crime was twofold: he had dared to attack Galen, and he had failed to recognize his own ignorance. His audience knew more about contagion than he, a man of but limited intelligence, and that somewhat shaky, who should have learned to think before opening his mouth in public. The more he wrote, the more he showed his ignorance of Galen and of Aristotle, whose views he had deserted for those of Epicurus and Democritus. His book proved he knew nothing profound about philosophy, and even less about medicine; it was hardly surprising that he got himself into a tangle.[48]

The division between these two scholars was sharpened by their points of agreement.[49] Both spoke the same language, and appealed to the same authorities, of academic discourse; both accepted that contagion involved some form of transfer, direct and indirect; both were prepared to discuss contagion within the broader context of sympathy and antipathy; and both considered putrescence a sign of plague and other contagious diseases. Both were equally convinced of role for individual predisposition and of the importance of the air or atmosphere in epidemics, but the precise weight each gave to these factors differed greatly. In particular, Fracastoro's introduction of the seeds, "those imaginary seedlets," seemed to Da Monte a far from justified attempt to compensate for underrating both air and individual receptivity. Indeed, the seeds themselves were an explanatory luxury. On a Galenic or Aristotelian view, species could well multiply in the air: putrefied air could engender more and deadlier putrefied air, without having recourse to propagating seedlets or Democritean atoms. But even on Fracastoro's own schema, the seeds were an unnecessary intrusion. They served two purposes: to mark off one contagious disease from another, and to permit infection at a distance or through an intermediary. Da Monte repeated his view that the specificity of a contagious disease was determined by the degree and site of the putrefaction; and he claimed that what was transmitted at a distance or by an intermediary was putrid matter, of whatever kind, rather than these strange and purely hypothetical nonputrid seeds whose arrival in the body initiated putrefaction. If, as was apparently agreed, contagion involved putrefaction, Fracastoro's objections to the transfer of putrid matter and the destruction of particular parts of the body through putrefaction were trivial. They were also against the facts of observation. In scabies, putrid skin might be passed on or rot the

[48] Da Monte, *In secundam fen,* pp. 428, 437, 439, 456, 440.

[49] The similarity between the two men's conceptions was noticed by Giovanni Argenterio, who contrasted them with his own and those of Donato Antonio Altomare; see Argenterio, *Opera* (Hanau: Wechel, 1610), p. 2536.

adjacent air, and this air might in turn become trapped in cloth; when the cloth was used again, the vapor released might infect another person. Leprosy, phthisis, and ophthalmia all putrefied the surrounding air and thereby affected those who either were naturally receptive or had become so through overindulgence or, as in syphilis, overactivity. One had only to enter the sickroom of a sufferer from phthisis to smell the foul air and see the putrid vapors around the bed, and it required little imagination to believe that the bedclothes trapped within themselves some of the infected air. As a final tilt at his opponent, Da Monte argued that his own preference for putrefaction as the agent was more consistent with the traditional doctrines of Galen, Aristotle, and the Arabs, as well as being philosophically more satisfying through not involving the creation of another layer of hypothetical entities.[50]

In its own terms, this was a devastating critique. Because he shared so many of Fracastoro's assumptions, Da Monte could put his finger on the great weakness of Fracastoro's theory of seeds. Given that, except for diseases spread by direct contact and perhaps in some instances by fomites, air was involved either in the creation of the seeds or in their transmission, to suppose that additional, invisible entities were involved, rather than putrefied and putrefying air, was to invoke a further unnecessary layer in the chain of causation. As we shall see, almost every author on plague assumed that it was spread by the aid of an agent *X*, but only Fracastoro went so far as to invent another order of living things, the seeds of disease, to be the unknown agent.[51] What he ascribed to seeds, Da Monte attributed to putrefaction arising either in the body or from changes in the upper air, and to the infinite variety of receptivity in man. Such a wide range of possibilities for the cause, degree, and site of the putrefaction easily explained the differences between diseases, and, together with the infinitely varied and variable makeup of each individual, made it no surprise that Galen should have refused to give a definite answer to the question why this, rather than that, person should fall ill.[52] The complexion of the individual was known in its entirety only to God, and mere mortals were inevitably condemned to partial ignorance.

To this neat exercise in logic Da Monte added further arguments in the lectures he gave in 1550 on *Epidemics* 1.3. In them he discussed the causes of fever, rejecting the "hallucinations of the moderns [*novi*]" who believed in poisonous, secret, or occult qualities. He had no more patience with *seminaria,* for this doctrine undervalued the importance of individual receptivity or resistance to infection. Nor was it necessary to posit the continued existence of a seed of infection lurking within the body to account for the persistence or intermittent recurrence of pestilential fever long after the efficient causes of plague or the putrid vapors had been removed. All that was required was the continuing presence, however small, of some corrupted humors within the body.[53] If Fracastoro

[50] Da Monte, *In secundam fen,* pp. 441; 456–458; 438–441; 446, 451; and 441.

[51] Although the seeds undoubtedly had life, to refer to Fracastoro's doctrine as involving a *contagium vivum* may be seriously anachronistic. But at least one of his later followers, Evangelista Quattrami, *Breve trattato intorno alla preservatione et cura della peste* (Rome: V. Accolti, 1586), p. 9, argued that it was precisely its possession of a *living* seed of poison that distinguished plague from putrid or hectic fevers.

[52] Da Monte, *In secundam fen,* pp. 438, 444.

[53] Da Monte, *In tertium primi Epidemiorum sectionem explanationes* (Venice: B. Constantino, 1554), fols. 116v, 107r. Galen himself had used the phrase "seed of disease" in this way in his *In*

accepted that contagious diseases involved some corruption of the body's humors and elements, then the supposition that this process of corruption had not been entirely stayed was a far simpler explanation.

If Da Monte's dialectic is cogent here, his other criticism of "our foolish and stupid friend from Verona" seems very much like special pleading. It focused on what Galen had meant when he talked of changes in the heart during fever. Fracastoro had implied that the heart putrefied like a rotten apple, but, claimed Da Monte, this was impossible, for life could not remain long in so crucial an organ once putrefaction had taken hold, and the sufferer would die before any other part of the body could become affected.[54] What Galen had really meant in *De praecognitione per pulsum,* Book 2, was that the heart shriveled up (*contabescat*), that is, it suffered a related form of putrefaction, not putrefaction itself. Even so, when the heart did become putrefied, as in plague, the putrefaction need not be total, just as an apple might become rotten only in one part and be called rotten even though the rest of it remained edible. While this may well be true, it is hard to identify the Galenic text adduced by Da Monte to distinguish shriveling (or marasmus) from other forms of putrescence. One possibility is the rare case reported by Galen of a conjoint putrefying and hectic fever leading to a withering marasmus in the heart,[55] but equally, some form of misrepresentation may lie behind Da Monte's claim. Renaissance scholars, no less than modern ones, did not always check their references.

The personal animosity so obvious in the lectures on *Canon* 1.2 was somewhat reduced in those given the next year on *Canon* 1.4. Although Da Monte had not changed his mind, the subject matter of his chosen text gave him little opportunity to discuss contagion in detail. Instead he complained about the many misrepresentations and misunderstandings on the part of his audience, which had made the reports of his teaching almost unrecognizable. Hence he was coming under added pressure to ensure an accurate publication of his views on contagion and sympathy. He now condemned Fracastoro for the lesser crime of misunderstanding earlier authorities and for thus being led to "some absurd seedlets" and the "figments of the Epicureans." There is less personal abuse: Fracastoro is "a fellow townsman of mine, of a distinguished family and character, and, generally, possessed of an excellent, almost divine talent [*ingenium*]."[56] Even if we choose to refer this divine talent exclusively to Fracastoro's poetry, we are still some way from the stupid charlatan of the previous year's lectures. The editor of Da Monte's *Medicina universa,* a compilation from all his lecture courses and writ-

---

Hippocratis Epidemiorum libros . . . commentarii 1, *Claudii Galeni Opera omnia,* ed. C. G. Kühn (rpt. Hildesheim: Olms, 1964–1965), Vol. XVIIA, p. 239. See also the parallel discussion in Galen's *In Hippocratis Aphorismos . . . commentarii* 2.12 (Kühn, Vol. XVIIB, p. 468).

[54] Da Monte, *In tertium primi Epidemiorum* (cit. n. 53), fols. 119r–v; cf. Fracastoro, *De contagione,* p. 8.

[55] Galen, *De differentiis febrium* 1.13 (Kühn, Vol. VII, pp. 328–330). Hectic fevers might easily turn into wasting fevers, and at *De praecognitione per pulsum* 3.3 (Kühn, Vol. IX, pp. 342–343) Galen talked of the relationship between putrid and hectic fevers, but without mentioning the word translated as *contabescat*. This combination of half-correct references could account for Da Monte's error.

[56] Da Monte, *In quartam fen* (cit. n. 41), fols. 51r–v. Very few of Da Monte's lecture courses were published in his lifetime, although many copies of student notes were in circulation. For an outsider's complaint about this situation see Pietro Andrea Mattioli, *Opera omnia* (Frankfurt: N. Bassaeus, 1598), p. 180, *Epistula* 4, who hoped for the publication of Da Monte's lectures on drugs and distillation.

ings, carried the process of purification still further. He retained all the argument but carefully (and wisely) omitted the abuse.[57]

Da Monte's students, from Italy, England, Austria, Poland, and Germany, ensured a wide circulation for his views on syphilis and plague.[58] It is an open question whether they, or, less likely still, those of Fracastoro, were included among the "new Italian ideas" drawn upon by Georg Agricola for the plague tract he published in 1554.[59] Agricola was a distinguished Galenist who had settled down as civic physician at Chemnitz after several years in northern Italy. His theory of plague, ostensibly validated by experience and by classical precedent, and starting from the same texts that Da Monte and Fracastoro had used, shows both similarities and substantial divergencies from theirs. Agricola called plague a "creeping contagion" (*serpens contagio*) and ascribed its cause to pestilential air, that is, air containing putrid exhalations. These exhalations, to use a Galenic phrase, he also listed variously as putrid quality, putrefaction, the seeds (*semina*) of plague that arise from putrefaction and are abroad in the air, and (two modern terms) putrid heat and hot poison. This list of apparent synonyms, none of which need be derived from Fracastoro, shows how easily his *seminaria* could be fitted into the schemata of others. The unifying principle of the list was airborne putrefaction, and it only required one man with pestiferous breath for any number of others to become infected, like sheep in a fold with a single mangy sheep. Agricola divided plague into three types, each determined by its degree of putrefaction and contagion. The lowest grade was to be found in those sufferers whose bodies, already prone to putrefaction, had finally succumbed to the results of a poor diet. The other two types were the result of changes in the air. The less contagious was produced by general atmospheric changes, which might cause some bodies to putrefy but which were not seriously dangerous unless they lasted for a long time. By contrast, if the air itself had become pestilential and poisonous, then the plague was severely contagious. It might attack all organs in the body, and its victims might take in and excrete its deadly vapors through invisible pores in the skin as well as through breath. The deadly contagions might not only pass to the patient's nearest and dearest but also contaminate clothes, tableware, furniture, bedclothes, and the like, which, although not themselves putrescent, retained enough of the exhalations and contagious putrefaction to pass on to their later users. Hence prudent states, like Venice, burned clothing, set up lazarettos, and supervised all sales of food and drink.[60] But success in

[57] Da Monte, *Medicina universa* (cit. n. 29), pp. 367–379. In a *consilium* of 1549/50, Da Monte even went so far as to talk of "obstructionem vero ac putredinem, ut Galeni verbis utar *σπέρματα σίπησεως*, id est semina putredinis sequuntur, quae omnia inficiunt extrinseca ut venenum, quemadmodum testatur Galenus": *Consultationes* (cit. n. 39), p. 1005. The fact that Galen did *not* use the precise words Da Monte ascribed to him is a neat indication of how he had come to accept certain features of Fracastoro's work.

[58] See Wear, "Explorations in Renaissance Writings" (ct. n. 31), pp. 141–142.

[59] Georg Agricola, *De peste libri III* (Basel: H. Froben, 1554), p. 5. The most substantial biography of Agricola is that of Helmut Wilsdorf, *Georg Agricola und seine Zeit* (Berlin: Deutscher Verlag der Wissenschaften, 1956); see p. 267 for the context of this tract, the outbreak of plague at Chemnitz. Agricola also wrote a (lost) tract *De putredine; ibid.*, p. 269.

[60] Agricola, *De peste,* pp. 20–21 (synonyms); 22, 93, 48 (types and description); and 62–63 (prudent states). This last is a nice example of the spread of Italian ideas on public health to northern Europe by those who had spent time there in study or travel. See also John Caius, *A Boke or Counseill against the Disease Commonly Called the Sweate or Sweatyng Sicknesse* (London: R. Grafton, 1552), p. 21, praising the dietary regulations of "certein masters of helth in euery citie and toune" in Italy.

controlling plague by these means was not guaranteed, and the contagion of plague might linger for a considerable time in less fortunate areas.[61]

Arguments like this, which accepted some degree of contagion as well as attributing the most significant cause and agency to bad air, can be found in many plague tracts of this period. And given the conventions of the genre, which placed the emphasis more on prevention than on cure, their writers also stressed the importance of individual susceptibility and the need both to build up resistance and to avoid bad air, sewers, cabbage water, dyeworks, tanneries, and rubbish dumps.[62] Like Agricola, several authors noted that the putrefaction of plague could be spread through intermediaries like cloths and clothes, although whether these intermediaries themselves putrefied or merely acted as receptacles for putrid air was a matter for proper academic debate.[63] When writers like Pamfilo Monti or Pamfilo Fiorembene di Fossombrone did use the phrase "seeds of plague" (*semina pestis*), they took their cue from Galen, not Fracastoro, and they employed the phrase in a manner parallel to that of Henri Gibault, professor at Montpellier, who spoke of "poisonous defilements in the air" that produced putrefaction, which might be passed on by contagion to others.[64] Almost without exception, the composers of plague tracts stressed that the causes of plague were manifest and hence explicable in rational terms. True, it was the divine judgment of God that allowed plague to occur, but this "occult" cause was a matter for theologians to consider, not physicians, even though they all acknowledged the power of God and, at times, accepted a specific link between sin and pestilence. No less a man than Janus Cornarius, professor of medicine at Marburg and later Jena, argued that an epidemic in Munster in Westphalia was sent as punishment from God for the heretical activities there of the Anabaptists. But even he then devoted the rest of his tract to explaining how the plague was spread by corrupted air and by contact with plague-infected bodies.[65]

IV. THE SEEDS OF NEW DISEASES?

These authors of plague tracts from the first decade after the publication of *De contagione* are typical in their acceptance of both bad air and contagion, and in

[61] Agricola, *De peste,* p. 63. The adjective does not, I think, refer to *economically* less fortunate areas, although such an interpretation is possible, and modern scholars have shown that, whatever might have been the case with the Black Death, by the fifteenth century plague was becoming primarily a disease of the poor; see Carmichael, *Plague and the Poor* (cit. n. 3), pp. 67–78, 125.

[62] Individual resistance: e.g., Giovanni Argenterio, *Varia opera* (Florence: L. Torrentinus, 1550), p. 11; sewers and cabbage water: Jacques Dalechamps, *De peste* (Lyons: G. Rouille, 1552), p. 19; Prospero Borgarucci, *De peste* (Venice: M. de Maria, 1565), p. 16; tanneries: Johann Guinther, *De pestilentia* (Strasbourg: C. Mylius, 1565), p. 73; and Caius, *Boke against the Sweate* (cit. n. 60), p. 15.

[63] Contrast Jacques Dubois (Sylvius), *Commentarius in Galeni libros de differentiis febrium* (Venice: V. Valgrisi, 1555), p. 77, with Giulio Alessandrini, *In Galeni scripta annotationes* (Basel: P. Perna, 1581), p. 862.

[64] Pamfilo Monti, *Commentarii in Galeni libros de febrium differentiis* (Bologna: A. Giaccarellus, 1550), pp. 190, 193; Pamfilo Fiorembene, *Collectanea de febribus* (Venice: N. Bascarino, 1550), fols. 103r–v; Henri Gibault, *In Claudii Galeni libros de febribus commentarius* (Lyons: G. Rouille, 1562), pp. 182–184, also pp. 110–113, stressing that plague was rightly called contagious. See also the use of the phrase "seeds of disease" by Da Monte, *Consultationes* (cit. n. 39), pp. 1005, 1117, and by his Paduan colleague Vittore Trincavelli, *Opera omnia* (Lyons: Junta & Guittius, 1586), fols. 19v, 146r–155v.

[65] Gibault, *In Galeni libros,* p. 187; Bairo, *De peste* (cit. n. 27), p. 680; Alessandrini, *Annotationes* (cit. n. 63), p. 865; Borgarucci, *De peste* (cit. n. 62), p. 12; Conrad Gesner, *Epistulae medicinales* (Zurich: C. Froschover, 1577), fol. 46r; François Valleriola, *Enarrationes medicinales* (Lyons: S. Gryphius, 1554), p. 269; and Janus Cornarius, *De peste* (Basel: J. Hervagius, 1551), pp. 16–17, 22–23.

their apparent neglect of Fracastoro. His name comes into prominence only at the end of the 1550s, and it may have been the great northern Italian epidemics of 1555–1557 that brought his ideas to a wider notice.[66] So, for example, when François Valleriola of Arles discussed plague in a treatise that was published in 1554, he ascribed its causes to divine punishment, putrid humors, and a harmful occult quality (which he identified with Hippocrates' morbid excretion from the body into the air) and said nothing of Fracastoro. By contrast, four years later, he described at great length Fracastoro's theories—sympathy, contagion, *seminaria,* fomites, and all.[67] His praise was lavish. Physicians everywhere owed Fracastoro a great deal for his work on contagion, and for the manner in which he had treated this whole subject with great precision and learning. Nonetheless, Valleriola found opportunity for dissent. In his opinion, putrefaction, whether carried by the seeds or resulting from their arrival within the body, could not be the crucial factor either in the process of transmission by contagion or in the destruction of the body, once infected. The thin vapors in the spirits were far more suitable carriers, and, through being omnipresent, they could act in all types of contagious diseases. In plague, men might die without a sign of putrefaction, and Valleriola had never come across putrefaction in any case of rabies he had seen. His agent *X* was a morbid excretion or a morbid exhalation of poisonous power, arising always from some corruption of the air, however this corruption might have been caused. It operated by means of an occult sympathy and antipathy to destroy the patient. Once this poison was in the atmosphere, it might be transmitted in a wide range of epidemic and contagious diseases, and its peculiar nature defined its targets. The whooping cough epidemic in France in 1557, a recurrence of that of 1510, killed only small children, since they were unable to expectorate. This aerial poison was not the result of astrological conjunctions but of occult atmospheric changes and might manifest itself in the form of any deadly disease, both old and new. For how could one limit the power of God, should he choose to send down the seeds of plague as a punishment for sin?[68]

This question of the relationship between Fracastoro's seeds and possibly new diseases was specifically raised by Johann Lange (1485–1565) in what is the only substantial critique of Fracastoro from the 1550s. Lange, formerly a professor at Leipzig, had made an Italian tour in the 1520s before settling down for the rest of his long life as court doctor at Heidelberg. His *Epistolae medicinales,* of which the first set was published in 1554, are miniature treatises on a variety of subjects and reveal the substantial classical learning of this monument of German medical humanism.[69] Four letters from this first volume discuss matters also dealt with by

[66] Argenterio, in his 1550 *Varia opera* (cit. n. 62), pp. 11–13, said nothing on Fracastoro, although his ideas were relevant. A dozen or so years later, in his Turin lectures on fevers, he cited him at length (and of equivalent authority to Da Monte and Altomare): *Opera* (cit. n. 49), pp. 2524–2553. Alessandrini, in his *Enantiomaton* of 1548 (cit. n. 37), knows nothing of *seminaria,* in contrast to his later 1581 *Annotationes* (cit. n. 63), pp. 863–865.

[67] Valleriola, *Enarrationes* (cit. n. 65), pp. 267–272; and François Valleriola, *Loci medicinae communes* (Lyons: Gryphius, 1562), p. 359 and the Appendix, Ch. 2, pp. 36–52. The date is established by the reference, p. 45, to the epidemic of 1557 as being "last year."

[68] Valleriola, *Loci medicinae communes,* App., pp. 44, 45, 51–52. At p. 39 Valleriola emphasizes his own practical experience as a physician and contrasts it with the theoretical stance of the (university) philosopher. The vernacular name of the whooping cough was "la cocoluche" (a type of hat or, possibly, "little cocoon"), which he derived from the sufferers' habit of putting their heads underneath the blankets. The word's etymology is still uncertain.

[69] I use the word *monument* advisedly, for by the 1550s Lange was being pointed out as one of the sights of Heidelberg; see Hagius, *Vita Lotichii Secundi* (cit. n. 37), p. 69; and Guglielmo Gratarolo,

Fracastoro: sympathy (Letter 33), scabies and syphilis (16), plague (18), and the British sweat (19). The latter pair of diseases Lange considered at length. He accepted as their cause both contagion and bad air, whether as the result of astral influence, divine anger, or putrid exhalations from caves, stagnant pools, and decaying corpses. In none of these four letters is there a hint that Lange had then read Fracastoro. However, in a letter at the end of the book, 42, Lange adopted the language of Fracastoro in discussing first *morbus Gallicus* (syphilis) and then two more "new" diseases, the British sweat and the *Scherbock,* a disease ravaging the Baltic lands that caused teeth to fall out and ulcers to form on the throat and jaws. For Lange, black bile was the principle *fomes* of the swellings and *seminaria,* and he accepted that these contagions were spread by seeds, even if their original cause might be something as simple as the foul water that the Danes, Norwegians, and Lithuanians were used to drinking. Contrary to what many people believed, these were not new diseases but manifestations of conditions recorded long ago and produced by the same combinations of (principally atmospheric) conditions that had brought them about in classical antiquity. According to Lange, no one could correctly deny that the *morbus Gallicus* had been described in detail in the *Epidemics* of Hippocrates.[70]

He returned to the topic in three consecutive letters, 13–15, in the second volume of the *Epistolae,* which were probably written in 1557. He again repeated his triple division of causes into infected air, a private contagion of a virulent plague (*lues*), and the judgment of God, and he talked happily about the varied *seminaria* of the *Scherbock,* which might possibly now have reached Cologne, the sweat, syphilis, and the apparently new disease of leg cramps that affected the emperor's army. His acquaintance with Fracastoro, "a man of no common erudition," led him to devote a whole letter, 15, to the "new slow fever in Italy," which had broken out there in 1528.[71] His discussion accepted much of what the Veronese physician had written. The virulence of these fevers was due to no occult property, "a vulgar refuge for those ignorant of causation," and he accepted that such diseases might grow weak, age, and die. But he drew the line at believing that putrefaction must in some way be involved, for, like Valleriola, he had not found it in every sufferer, and he too was inclined to think that the "seeds of pestilential plague" acted like a poison, killing without putrefaction.[72] He reserved his greatest scorn, however, for Fracastoro's notion of an infinity of

---

preface to Pietro Pomponazzi, *Opera* (Basel: H. Petri, 1567), sig. a2v. Further biographical details on Lange are given in Vivian Nutton, "John Caius und Johannes Lange: Medizinischer Humanismus zur Zeit Vesals," *NTM,* 1984, *21*:81–87; and in Nutton, "Humanist Surgery," in *Medical Renaissance,* ed. Wear, French, and Lonie (cit. n. 1), pp. 92–96, rpt. in Nutton, *From Democedes to Harvey* (cit. n. 6), with identical pagination.

[70] Johann Lange, *Epistolae medicinales* (Frankfurt: Wechel, 1589), pp. 200–210 = *Epistolae medicinales diversorum authorum* (Lyons: Junta, 1556), pp. 506–507. Letters 33, 16, 18, and 19 occupy, respectively, pp. 149–159, 76–81, 85–91, 91–96 = 1556, pp. 497–499, 486, 487–488, 488–489. Lange also took his explanatory language for syphilis from Fracastoro, p. 203 = 1556, p. 506. Interestingly, he referred in the margin not only to *Epidemics* 3 but also to the account of plague in Lucretius, *De rerum natura* 6, one of Fracastoro's sources.

[71] Lange, *Epistolae medicinales,* p. 619. The second book of letters, which contains Letters 13–15 on pp. 610–621, was not included in the Lyons volume. The date is established by an earlier letter in the same series, 8, p. 591, which refers to the Frankfurt fair of 1556 as "held recently."

[72] "Proprietas occulta . . . quae solet esse vulgare causarum ignorantiae asylum," *ibid.,* 18, p. 627. On diseases growing old see 15, pp. 618, 620; on putrefaction, avowedly summarizing the arguments of 13–15, 23, p. 655.

new diseases. Far from being new creations, the epidemics of the fifteenth and sixteenth centuries were revivals of earlier diseases chronicled by Aetius, Aretaeus, Hippocrates, and the like. The same astral conjunctions, the same meteorological conditions, the same bad diet would easily bring about a recurrence of a plague long thought lost. Its disappearance might have been so long ago that all knowledge of its original name had been lost, but, claimed Lange, playing Fracastoro's game, its "*seminaria*" continued to lurk in the elementary bodies of man, to be revived by bad diet or bad air. Lange congratulated his correspondent, Theoderic Pamphilus, for refusing to be taken in by Fracastoro's theory of an infinity of diseases, each with its own specific form.[73] There was only a limited number of such diseases, and those who thought otherwise were either ignorant of the earlier literature or insufficiently appreciative of the actions of God in preventing the *seminaria* from propagating further.

Lange's arguments, at least in their identification of modern diseases with classical forerunners, had been anticipated by Niccolò Leoniceno half a century before.[74] Their interest lies in the fact that they are unique in concentrating on the question of specific diseases and on the problems involved in positing a multiplicity of new diseases. Fracastoro's apparently new formulation of a relationship between seeds, fomites, air, and contagion causes Lange no difficulty, despite his great reverence for the past, and Fracastoro's vocabulary fits easily into his own classicizing schema. It may be no coincidence that his singular criticism of Fracastoro depends largely on arguments advanced almost a millennium and a half earlier in one of the less familiar writings of Plutarch.[75] The combination of the authority and the obscurity of this classical precedent explains both why Lange, a man of prodigious classical learning, was led to attack Fracastoro on this particular point, and why he appears to have been alone in his concern to deny a possible infinity of new diseases.

V. THE SEEDS OF VENETIAN PLAGUE

The criticisms of Lange and of Da Monte were directed to the theory that underlay Fracastoro's book; neither scholar dealt with the practical recommendations given in Book 3. Modern historians have followed their example and have thereby tended to obscure the immediacy of much of Fracastoro's advice.[76] Yet in this their fault is venial, for the length of time between the conception of Fracastoro's ideas and their appearance in print means that it is difficult to relate them to a specific historical situation, and the publication of *De contagione* as the second part of a book dealing with sympathy and antipathy inevitably emphasized its theoretical component. But the validity of these theoretical claims

[73] *Ibid.,* 13–15, pp. 610–621; 23, p. 655. Theoderic Pamphilus was a Swiss (see 10, p. 596) and the addressee of many letters from Lange. But I cannot identify him further, unless the grecized name is Lange's literary way of referring to his erstwhile colleague at Heidelberg, Thomas Lieber, who is more usually known as Thomas Erastus.

[74] This had been the message of Niccolò Leoniceno's *De epidemia quam morbum Gallicum vocant* (Venice: Aldus Manutius, 1497), the most influential of all the early discussions on syphilis.

[75] Lange, *Epistolae medicinales,* 13, p. 611; 15, p. 619, citing Plutarch, *Symposium* 8.9, on which see Nutton, "Seeds of Disease" (cit. n. 6), pp. 11–13.

[76] One would like to know, for instance, how widely the Diascordium of Fracastoro, *De contagione,* pp. 246–248, was recommended or cited; see Peter Monaw's mention of it in a letter of 1586 to Johann Hermann, in *Johannis Cratonis et aliorum . . . medicorum consilia et epistolae,* Vol. V (Hanau: Wechel, 1619), Ep. 59, p. 587.

could also be judged by their relevance to the actual needs of an epidemic. Here, unless Agricola was relying on Fracastoro in his advice to the inhabitants of Chemnitz in 1553, the debates over the Venetian plague of 1555–1557 may have been the first to bring these novel ideas to bear on a specific outbreak of plague.

Venice and its immediate hinterland had been free from plague for a generation when in March 1555 the first cases of plague were reported from the parish of San Niccolò.[77] The Provveditori alla Sanità took immediate action, recalling some of their officials who had been dealing with an outbreak of plague in Istria and putting the sick into quarantine in the Lazaretto. Although the number of persons affected was not high, there continued to be new cases throughout the year in Venice and in Padua, whither the disease had spread by late April. After a typical remission in the first months of 1556, there was a renewed outbreak in May; and by July many inhabitants of Venice had left for the mainland. But, once again, there was a definite falling off in cases in the second half of the year. From ten deaths a day at the end of August, the rate had dropped to no more than three at the middle of November, and the Provveditori were confident enough in December to pay off two of their physicians. In 1557 there were still a few suspicious deaths, and in July a further exodus of frightened Venetians, but by the end of the year the Health Office was convinced that this time the plague was well and truly beaten. The plague did not recur in 1558, and the superintendent of the Lazaretto could enjoy a well-earned rest.

The Venetian plague of 1555–1557 was, compared with most such outbreaks, remarkably mild. Its spread was slow, and it was for long successfully confined within a few regions of the city. Although it provoked panic among the citizens, who preferred flight to remaining in an infected city, it was deadly only potentially. The Provveditori alla Sanità could congratulate themselves that their procedures had prevented a major disaster. They had taken the best possible medical advice: they had consulted the College of Physicians at Venice and the physicians of the University of Padua, who, at that time, could claim to be the best in the world; and they had received advice from many other quarters.[78]

There were two opposing schools of thought, one blaming the plague on contagion, the other on bad air. But, as Richard Palmer has elegantly shown, the division between the two camps was more one of emphasis than of substance. In their practical recommendations, the miasmatists, like Niccolò Massa, might also advocate quarantine and disinfection, while the contagionists were equally in favor of street cleaning and of purifying bad air. All parties agreed on the need to secure a good supply of food and water, build up individual resistance, and isolate potential sufferers.[79]

What part in this controversy was played by the writings of Fracastoro? Certainly, the importance of contagion as a, or as the, cause of plague is emphasized more than in earlier epidemics, and it is at least plausible to connect this with

[77] For my description of this epidemic, I rely largely upon Palmer, "Control of Plague" (cit. n. 3), esp. pp. 97–110. Other pieces of information can be found scattered through the pages of the exhibition catalogue *Venezia e la peste, 1348–1797* (Venice: Marsilio Editori, 1979).

[78] For a comparison with other towns see also Palmer, "Control of Plague," pp. 27–50; Cipolla, *Public Health and the Medical Profession;* Cipolla, *Fighting the Plague;* Carmichael, *Plague and the Poor* (all cit. n. 3); and Paolo Preto, *Peste e società a Venezia nel 1576* (Venice: Neri Pozza, 1979), pp. 35–43.

[79] Palmer, "Control of Plague," pp. 112–122.

Fracastoro. But the proof is far from conclusive. Not all the contagionists refer to Fracastoro or use his vocabulary. Neither Bernardino Tomitano nor Francesco Frigimelica gives a hint of having read him, and Lodovico Pasini's definition of contagion as a "poisonous aerial vapor that acts through an occult property" runs counter to Fracastoro's doctrines. Although his ideas were familiar to Massa, who considered the plague of 1555 miasmatic in origin but that of 1556 the result of contagion, Massa did not employ them to any recognizable extent in his published tracts.[80]

Three other authors, however, do seem to have assimilated Fracastoro's message. Bassiano Landi, professor of medicine at Padua, conducted an experiment to prove that putrefaction in the air could not be the cause of plague. When the bread, milk, egg, and wine that he had exposed overnight did not go off, he concluded that this was because, even in this notoriously humid atmosphere, there was no seed (*seminarium*) of putrefaction. By contrast, Gianfrancesco Boccalini was convinced that putrefaction of the air was the culprit. The altered air contained the seeds of putrefaction, whose numbers and quality determined the severity of the plague. But while he used the vocabulary of Fracastoro, his interpretation and his authority were entirely Galenic. The second part of his little tract, however, corrects the views of Marsilio Ficino, who wrote earlier on plague, to bring them into harmony with Fracastoro's. Instead of poisonous vapors with an occult property, Boccalini preferred the seeds of pestilence, which operated through known properties. Besides, to think of the air as working through "spiritual qualities," which were active only as long as the original material from which they derived persisted, was to misunderstand the role of fomites and to underestimate the possibility of the seeds of plague being carried at a distance, even across the sea. Boccalini was not a wholehearted believer in quarantine, for, as Fracastoro had said, contagion and infection depended on the degree of viscosity of the seeds, not on their poisonous qualities. A man who had drunk poison, no matter how strong, never infected anyone. Thus only those whose *seminaria* were sufficiently viscous needed to be removed from all human contact.[81]

The most interesting of the plague tracts from a Fracastorian perspective is the little book of plague questions written by the Venetian physician Vettore Bonagente. He openly acknowledged Agricola and Fracastoro as his sources, taking his triple division of plague and his variety of names for the infective agent from the former, and his discussion of fomites, seeds, and the process of contamination from the latter. Yet he too accepted a Galenic precedent for all this and was prepared to discuss learnedly the time of year and the atmospheric conditions

[80] Bernardino Tomitano, *Consiglio sopra la peste di Vinetia l'anno MDLVI* (Padua: G. Perchacino, 1556); Francesco Frigimelica, *Consiglio sopra la pestilentia qui in Padoa dell'anno MDLV* (Padua: G. Perchacino, 1555) (although as a colleague and close friend of Da Monte he is unlikely to have been ignorant of Fracastoro); and Lodovico Pasini, *De pestilentia Patavina anno 1555* (Padua: G. Perchacino, 1556), fol. 6v. For Massa see Palmer, "Control of Plague," pp. 112–115, 118–120. Massa uses Fracastoro's terminology in a contemporary letter: Niccolò Massa, *Epistolae medicinales* (Venice: S. J. Zilletti, 1558), fol. 171r.

[81] Bassiano Landi, *De origine et causa pestis Patavinae anno MDLV* (Venice: B. Constantino, 1555), n.p., cited by Palmer, "Control of Plague," p. 121, n. 4; and Francesco Boccalini, *De causis pestilentiae urbem Venetam opprimentis anno MDLVI* (Venice: G. Jolitus, 1556), pp. 11, 10, 15, 29, 31, 63–64. The plague tract of Marsilio Ficino (1433–1499) was frequently reprinted in the sixteenth century, not least by the Junta press in 1556, along with that by Tommaso del Garbo. The fact that it was written in Italian helps to account for its widespread popularity.

most conducive to plague. On the precise question whether the plague at Padua was the result of contagion or the putrefaction of the air, he inclined to Landi's opinion, although he was not entirely sure that some recent irrigation works had not so tainted the area that pestilential fever of a kind was now endemic.[82]

Bonagente and Boccalini both employ the language and concepts of Fracastoro, but to different effect. Their discussions cover the same range of texts and the same local problems. Neither seems to have any difficulty in resolving what modern scholars have considered the contradiction between contagion and miasma. Their accounts show the extreme flexibility of the Galenist approach, as well as the way in which Fracastoro's ideas were seen by his contemporaries to fit into the standard categories of plague. Above all, both authors believed in Fracastoro's ideas and in their practical application, to plague as well as to other epidemic diseases.

The Venetian plague of 1555–1557 thus provided a forum in which Fracastoro's ideas could be utilized to practical ends. It may be too much to claim that his work set the terms for the argument over contagion, for none of the "miasmatists" attacks an opinion that is obviously his, but the renewed emphasis on contagion may have given his ideas greater publicity. The professors of Padua and the physicians of Venice, in collaboration with the printers of the region, will have made a powerful team, and the dissemination of their writings provided yet another way in which the doctrines of Fracastoro reached a wider world.

By the 1570s his ideas were part of the common currency of Europe. As early as 1556 the Spaniard Francisco Valles had formulated his own ideas on sympathy in partial dependence on Fracastoro, "a modern author of no small learning and acumen." His lectures on the Hippocratic *Epidemics* at the University of Alcalà considered a variety of epidemic diseases and explained their differences of virulence by the differences between their seeds.[83] Fracastoro's doctrines had also crossed the Alps. Whether Giulio Borgarucci discussed with his wealthy London patients the plague of 1563 in terms of seeds of disease and contagion is not known, but his brother's plague tract, which was inspired by Giulio's experiences, clearly did.[84] French physicians from the early 1560s knew of Fracastoro's work, and Johann Guinther of Andernach depended heavily on him for his own plague tract. He accepted that plague was always contagious, that it could be airborne through the "seeds of pestilential lues," and that it could be spread through fomites in clothes and bedding. Even if one lived in clean air, away from the sick, the fomites might still be able to creep into an overheated body and start the plague. Guinther's advice also attests a different Italian influence. One should clean up streams and wells, and shut down the sources of bad smells, such as tanneries and geese farms (a local touch, for Guinther's tract was published at Strasbourg). Since the plague was contagious, strong measures were needed. One should ban all foreigners from entering and follow the Italians in

[82] Vettore Bonagente, *Decem problemata de peste* (Venice: V. Valgrisi, 1556), fols. 4r, 7r, 9r–12v (sources); 3v, 8r–9r, 18v–20r.

[83] Francisco Valles, *Controversiae medicae et philosophicae* (Alcalà: J. Brocarius, 1556), quotation on fols. 70r–v; Valles, *Commentaria illustria in Cl. Galeni libros* (Cologne: F. de Franciscis, J. B. Ciottus, 1594), pp. 625–626. For other early Iberian examples see Amatus Lusitanus, *Curationum medicinalium centuriae quatuor* (Basel: H. Froben, 1556), pp. 196 (written ca. 1553), 384; and Garcias Lopes, *De varia rei medicae lectione* (Antwerp: M. Nutius, 1564), fols. 14v, 31v.

[84] Prospero Borgarucci, *De peste* (cit. n. 62), 1565, but written in late 1563, pp. 13, 23, 42, 49. For Giulio's activities in London see pp. 3, 47.

setting up lazarettos, serviced by public doctors, surgeons, and servants. One should have strict segregation of the convalescent from the healthy. But, as Guinther sadly remarked at the end, whatever practical steps might be taken, flight was still the best remedy.[85]

VI. CRATO AND THE GERMANS

In their survey of writers on plague who accepted Fracastoro's doctrines, Charles and Dorothea Singer said little about Germany, mentioning only Thomas Jordan, whose plague tractate appeared at Frankfurt in 1576.[86] This apparent lack of interest is deceptive, for it hides a vigorous debate within Germany, centered upon Crato von Crafftheim and offering new criticisms of Fracastoro and alternative theories for plague.

Johann Crato von Crafftheim (1519–1585) has already made an appearance as the devoted student, editor, and expositor of Da Monte at Padua.[87] Born into a wealthy family at Breslau, he attempted to study theology at Wittenberg, but was persuaded by Philipp Melanchthon to turn to medicine since his health was not up to the strain of the religious life. With the aid of a bursary from Breslau, he traveled to Padua, where he listened eagerly to all the medical greats of the university. On his return from Italy, he was soon appointed doctor to the emperor Ferdinand, and he continued in imperial service under both Maximilian and Rudolf II until his death in 1585. His interests were broad, and he devoted much of his energies to pursuing them through a voluminous correspondence across most of Europe. As a member of the court he had access to the imperial post system, and he kept in touch with all the latest developments, particularly those at his old medical school.

Plague was one of his special concerns, not just because, as an imperial courtier, he was in a position to offer advice to the government on the precautions to be taken in an epidemic. He had heard Da Monte's denunciation of Fracastoro in his lectures on Avicenna, and his subsequent reading had convinced him of the frivolity and contentiousness of the Italians. The classical authors had much sounder views, even though Hippocrates had not written specifically on plague, and Galen's treatise on that subject was now lost. Nevertheless, it was perfectly clear that Galen had known of the role of contagion, and of the part played in spreading disease by air that had become infected by the seeds of plague.[88]

[85] Guinther, *De pestilentia* (cit. n. 62), 1565, pp. 17–19, 44, 72–75; see also Johann Guinther, *De medicina veteri et nova* (Basel: H. Petri, 1571), p. 123. The Singers provide many references to those who knew of Fracastoro: "Position of Fracastoro" (cit. n. 4), p. 34.

[86] Singer and Singer, "Position of Fracastoro," p. 32, citing Jordan's *Pestis phaenomena* (Frankfurt: Wechel, 1576), and p. 34, n. 55, citing German authors at the end of the century who commented on the "febris hungarica."

[87] See above, under "First Reactions." There is a valuable eulogy of Crato by Michael Dresser, *De curriculo vitae Johannis Cratonis* (Leipzig: J. Steinman, 1587). Crato was the subject of a remarkable biography by J. F. A. Gillet, *Crato von Crafftheim und seine Freunde: Ein Beitrag zur Kirchengeschichte,* 2 vols. (Frankfurt: H. Brönner, 1860–1861), which, despite its subtitle, contains much of value on medicine. In comparison, the shorter notices by J. Graetzer, *Lebensbilder hervorragender schlesischer Ärzte* (Breslau: S. Schottlaender, 1889), pp. 5–19; V. Fossel, "Crato von Krafftheim 1519–1585," in *Studien zur Geschichte der Medizin* (Stuttgart: F. Enke, 1909), pp. 24–25; and Marlene Jantsch, "Crato von Kraftheim," in *Gestalten und Ideen um Paracelsus,* ed. Sepp Domandl (Vienna: Notring, 1972), pp. 99–108, are very jejune.

[88] Crato, *Consilia et epistolae medicinales,* Bk. V: 2, p. 283; Bk. I: 1, p. 180; Bk. IV: 1, p. 249; Bk. V: 2, p. 283; 6, pp. 292–293; cited from the following editions: *Johannis Cratonis . . . et aliorum*

This conviction Crato expressed in over a hundred letters and separate publications, adding his own interpretation of his classical sources.[89] His novelty was to divide plague into two types, public and private. The latter, the result of either bad regimen or bad air, lacked a "morbid exhalation" and was not contagious or dangerous except to the very few who were already predisposed to it. By contrast, public plague derived from air containing lethal exhalations arising from putrescent bodies.[90] These exhalations might also contain seeds of plague, *seminaria,* which multiplied in the air to infect a whole region from small beginnings. The growing number of sufferers would also, by their own tainted breath, add to the corruption of the air, and in turn spread the plague. Both types of plague depended on putrefaction, differing according to its grade, and a private plague could easily change to a public one if its putridity increased. The seeds themselves acted like poison, at times causing death swiftly and unexpectedly, or thickening the humors to form the notorious plague bubo. The sticky seeds might also adhere to clothing and similar intermediaries, and spread disease by some form of contact, although Crato doubted whether this contagion could be identified with putrescence.[91] Here he parted company with his beloved teacher, for he tended to keep the two terms distinct, but this was not a point on which he could be entirely sure. The older authorities did not say enough about contagion, and the excellent physicians with whom he discussed his difficulties muddled matters further by calling anginas and pleurisies plague without distinction.[92]

Crato defended these doctrines against all comers. He criticized the Heidelberg professor Thomas Erastus for a mistaken view on contagion, and openly attacked his own protégé Thomas Jordan for his errors in his description of a mysterious epidemic disease at Brno, the *lues Morava.*[93] To Joachim Camerarius II in Nuremberg he wrote many letters over a period of years, reiterating or refining his own theories. There was no such thing as "putrefaction of the spirits"; when the seed penetrated the heart's substance, it caused an immediate, deadly, and irremediable pestilence called hectic (a notion he derived from reading the book of the Paduan professor Vittore Trincavelli); plague was no disease of the total substance, but neither had it an occult cause.[94] It might indeed be a

---

medicorum consilia et epistolae medicinales, Books I–III (Frankfurt: Wechel, 1591), Book IV (Frankfurt: Wechel, 1593), Book V (Hanau: Wechel, 1619), Books VI–VII (Hanau: Wechel, 1611) (henceforth cited as ***Consilia et epistolae*** or, for letters, ***Epistolae***).

[89] Not least in his plague tract, Johann Crato, *Ordnung der Praeservation; wie man sich zur Zeit der Infektion vorwahren . . . ,* several times reprinted in both Latin and German: see *Consilia et epistolae,* Bk. II, pp. 586–587. I have cited it from the Frankfurt 1585 edition.

[90] Crato, *Ordnung,* sig. D iii r; Crato, *Epistolae,* Bk. I: 9, p. 215 (= Bk. VI: 43, p. 122); Bk. II: 35, p. 384; Bk. IV: 1, p. 413; Bk. V: 16, p. 343; Bk. VII: 4, p. 783; 6, p. 788; 8, p. 794; 12, p. 813.

[91] Crato, *Isagoge* (Venice: V. Valgrisi, 1560), fol. 101r; *Epistolae,* Bk. I: 1, pp. 175–181; 9, pp. 214–215; Bk. II: 3, pp. 235–245; Bk. IV: 1, pp. 405–406; Bk. V: 2, pp. 281–283; Bk. V: 16–17, pp. 342–346; Bk. VII: 1–2, pp. 772–784; 6, pp. 787–789; 8–9, pp. 792–798; 12, pp. 812–816.

[92] Crato, *Ordnung,* sig. D ii r.

[93] Crato, *Epistolae,* Bk. VII: 20, pp. 833–835; cf. Erastus (Thomas Lieber) to Peter Monaw, *ibid.,* Bk. I: 20, p. 312. In Bk. II: 2, pp. 224–232, Crato attacked Thomas Jordan's *Luis novae in Moravia exortae descriptio* (Frankfurt: Wechel, 1580), which publicly acknowledged Crato's patronage on sig. A3. Jordan responded with an appendix, *Censura Cratoniana,* to the second edition (Frankfurt: Wechel, 1583).

[94] Crato, *Epistolae,* Bk. I: 9, pp. 214–216 (= Bk. VI: 43, pp. 121–125); Bk. V: 1–2, pp. 277–283; 6 (of 1558), pp. 292–293; 12 (1561), pp. 328–332; 16 (1564), pp. 342–343 (on Trincavelli); 17 (1564), pp. 345–346; 18 (1565, also on Trincavelli), pp. 350–351; Bk. VII: 1 (1564), pp. 771–773; 2 (1580), pp. 773–775; 3 (1582), pp. 776–780; 4 (1583), pp. 780–785; 6 (1583), pp. 787–789. Camerarius himself

punishment sent from God, but if so, it could not be cured by human agency. The Christian in time of plague should resort to prayer, and free his mind from fear and sadness, lest these emotions weaken the heart and allow it to become more easily affected by poison and attract the pestiferous poison (*virus*) into itself.[95]

Crato's opinion on plague was widely sought. In 1580, Girolamo Mercuriale, professor of medicine at Padua and himself an expert on plague, asked his advice on an epidemic of catarrh that was ravaging northern Italy and other parts of Europe. Mercuriale raised two questions, one on the relationship between contagion and antipathy and sympathy, the other on putrefaction. Fracastoro had talked about sympathy, but Mercuriale thought that this was just a fashionable way of referring to contagion, and that the general public was right to continue to talk of contagion.[96] Mercuriale clearly expected Crato to agree with this populist argument against Fracastoro, but he was less sure of his other proposition. Although he was well aware that the seeds of contagion were usually viscous and sticky, and thus inseparable from putrefaction, his experience in the latest epidemic had suggested to him that dry, acrid vapors breathed out from the sick could not, on being inhaled by others, become moist and sticky and, even though without any putrefaction, attack both body and brain at one and the same time.

Crato's response was a long letter denying Mercuriale's conclusions. The seeds in the air could survive only if they were in some degree putrid, and mouth sores were undoubtedly putrid. True, the catarrhal epidemic, which had begun as early as 1577, was more benign than plague, but this was easily explained by differences in the grade of putrefaction. He too was not satisfied by modern explanations of contagion,[97] but he was convinced that putrefaction was necessary, and that contagious diseases required miasmas in order to spread widely and swiftly. Only the previous August the whole imperial court had gone down with this catarrh because on their journey to the Imperial Diet they had had to pass near infected towns and villages. The spirits and the humors were the vehicle for all contagious diseases, for the seeds of contagion were drawn inward to them from the surrounding atmosphere. Crato continued his criticism in a second, but considerably shorter, letter. In catarrh, it was the phlegm that putrefied, and hence it was the head, the favored site of phlegm, that was most obviously affected. The grade of putrefaction also explained the spread of disease by fomites. But a belief in seeds of contagion needed further justification against those who asked why, if there was a constant production of unhealthy miasmas and if the seeds themselves lasted for some time, there were not more cases of relapses, and more people going down again with the same disease. This Crato explained by a greater tolerance on the part of the patient, and a lesser virulence

wrote a plague tract, *De recta et necessaria ratione preservandi in pestis contagio,* which he published along with his summaries of the plague tracts of Girolamo Donzellini, Giovanni Filippo Ingrassia, and Cesare Rincio, and a catalogue of Italian authors on plague, in his *Synopsis quorundam . . . commentariorum de peste* (Nuremberg: Gerlachin & Montanus, 1583). For Camerarius as a physician see the 1977 Munich dissertation of Karl Gröschel, "Des Camerarius Entwurf einer nürnberger Medizinalordnung Kurtzes und ordentliches Bedencken 1571."

[95] Crato, *Ordnung* (cit. n. 89), sig. A vi r; Crato, *Consilia et epistolae,* Bk. III: 17 (1572), p. 145.

[96] Mercuriale to Crato, *Epistolae,* Bk. II. 3, pp. 232–233; 4, pp. 234–235. The argument from the name is on p. 235.

[97] In the preface to his edition of Da Monte's *Consultationes* (cit. n. 39), sig. A4r., Crato had expressed his doubts about Fracastoro's ideas on sympathy and antipathy and had also called the seeds "imaginary."

in the elderly seeds. Besides, contagion was akin to filth (*foetor*), and it was well known that those who lived all their lives in stinking hovels were unscathed, whereas their visitors were regularly made ill. So in plague, those in frequent converse with the sick were less at risk than those who had lived previously in purer air.[98]

This exchange of letters was continued the next year by Crato's colleague at court Peter Monaw, who reminded Mercuriale how he had sat at his feet and heard him declare in his lectures that putrefaction was always necessary for contagion. Monaw could not see any cogent reason for Mercuriale's change of mind. The topic recurred again in 1583, when Monaw advised his old professor that while plague was indeed spread by *seminaria,* other diseases were transmitted by sympathy, by the imagination impressing itself on the mind and causing various actions or changes to occur.[99] Such queries and criticisms are typical of the letters on plague that make up the seven volumes of the *Epistolarium Cratonianum.* In them writers from many areas of Europe, but chiefly from central Europe and northern Italy, reveal the variety of opinions held and the points that divided their proponents. Was contagion the cause of plague or a concomitant? Were only pernicious diseases contagious? Was the *seminarium* of plague a poison? In what way did plague-infected air contain miasmas or seeds? How exactly did plague kill? What groups were most at risk? And if these happened to be the poor and the craftsmen, what could they do about it, since they had to consort with others in order to live? How long could seeds exist without giving some indication of their power? If plague was the result of a change in the atmosphere, why did some regions apparently escape entirely unscathed? Could plague subsist without an apparent fever? And could physicians, craftsmen in the perceptible (*sensati artifices*), properly believe in occult causes of plague?[100]

When reading this voluminous mass of diverse letters, one cannot fail to be struck by the way in which Fracastoro's concepts and ideas had been accepted over a wide area. Epidemic diseases were being regularly discussed in terms of *seminaria* and fomites, and many of the problems that he had raised were being answered, even before the *lues Morava* so graphically described by Thomas Jordan. In what he had to say about petechial fever and other new diseases, Fracastoro was indeed "most learned," even if it was also accepted that subsequent writers had refined and developed his theories "most accurately." The substantial volume on the phenomena of plague which Jordan put out in 1576 and which

[98] Crato, *Epistolae,* Bk. II: 4, pp. 235–246; 5, pp. 246–248.

[99] *Ibid.,* Bk. V: 49, p. 489 (cf. the contemporaneous letter of Monaw to another Paduan professor, Girolamo Capivaccio, *ibid.*, 48, pp. 484–485, and Mercuriale's reply, *ibid.*, Bk. I: 5, pp. 270–272); and *ibid.,* Bk. V: 50, p. 493, with Mercuriale's reply, Bk. I: 6, p. 284. For Monaw's views on contagion and plague see his letters to Andreas Dudith, *ibid.,* Bk. V: 36–37, pp. 413–422.

[100] Respectively, *ibid.,* Bk. III: 15, p. 300 (Dudith), and Bk. V: 36, p. 418 (Monaw) (cause); Bk. I: 4, p. 270 (Capivaccio), 6, p. 284 (Mercuriale), Bk. V: 48, p. 484 (Monaw and Capivaccio), and Bk. VI: 26, p. 72 (Dudith) (pernicious diseases); Bk. I: 1, p. 175 (Crato), Bk. VI: 27, p. 77 (Dudith), and Bk. VII: 3, pp. 784–787 (Crato to Camerarius) (poison); Bk. V: 36, p. 414, 37, p. 421 (both Monaw to Dudith), and Bk. VII: 8, p. 792 (Crato to Theodor Zwinger) (miasmas); Bk. I: 9, p. 312 (Crato), and Bk. VII: 20, pp. 833–834 (Erastus to Monaw) (plague); Bk. VI: 27, pp. 76–77 (Dudith to Monaw) (groups at risk, and poor); Bk. III: 14, p. 298 (Dudith to Wenceslaus Raphanus) (seeds); Bk. III: 15, p. 300 (Dudith to Raphanus) (regions); Bk. VII: 26, p. 73 (Dudith reporting Alessandro Massaria) (fever); and Bk. VII: 27, p. 77 (Dudith to Monaw) (occult causes). On Andreas Dudith (1533–1589) and his medical learning see Viktor Fossel, "Die *Epistolae medicinales* des Humanisten Andreas Dudith," *Sudhoffs Archiv,* 1913, *6*:34–51.

offered a Fracastorian interpretation of the whole subject was no isolated instance of Italian influence.[101] It was a typical product of Crato's circle in its wish to combine the best of classical learning with Italian ideas and sound observation of local phenomena. Together with the letters and plague *consilia* associated with Crato von Crafftheim and his friends, Jordan's book shows that in Germany, and in many other parts of northern Europe, the message of Fracastoro had penetrated widely by the end of the 1570s, and that his views on plague and epidemic diseases had become part of the normal discourse of the learned physician.

VII. THE SEEDS OF *MORBUS GALLICUS*

If plague and pestilence were the diseases to which Fracastoro's theories were most often applied, it should not be forgotten that it was as an expert on syphilis, or *morbus Gallicus,* that he had first made his name. His poem *Syphilis* had made him famous in the learned world of Europe, and its literary beauty was highly appreciated.[102] Whether the medical ideas that lay behind it and were developed in *De contagione* enjoyed a similar reputation among less poetic experts in syphilis is a question rarely asked. Yet the material for an answer is easily to hand in the three volumes on syphilis compiled by Luigi Luigini and published in 1566–1567, within a generation of the first appearance of Fracastoro's views on seeds and contagion.

As his preface makes clear, Luigini had intended to assemble all previous writings on syphilis, and he incorporated small sections of letters, lectures, *consilia,* and the like as well as long, independent treatises. He included both *Syphilis* and the sections on syphilis, *elephantia,* scabies, and *lepra* from *De contagione,* Book 2, in his collection, but not the relevant observations on treatment from the last book.[103] Elsewhere in Luigini's massive compilation, *Syphilis* is cited only four times. Gabriele Falloppio simply mentions it as the source for the disease name. By contrast, the Frenchman Antoine Lecoq, whose book appeared in 1540, made great use of it, but more for literary effect than for any practical purpose. Even so, his choice of extracts shows how much Fracastoro's poetic formulations were in line with what other physicians had written. Some twenty years later, the Paduan professor Bernardino Tomitano quoted large sections to show the foreign origin of both the disease and its cure. Tomitano was a firm believer in contagion, but his definition of what was passed on by contagion —vapors, fumes, heat, sweat, blood, saliva, and so on, but not *seminaria*—reveals his independence of Fracastoro. Not that there were not followers of Fracastoro around, if one may thus interpret a hint in the syphilis tractate of Prospero Borgarucci, the last author to refer to the poem.[104] All three scholars

[101] Jordan, *Pestis phaenomena* (cit. n. 86), esp. its theoretical section, pp. 67–105; quotations are from p. 238. See also Jordan, *Luis novae descriptio* (cit. n. 93); for a modern German translation, with a sparse commentary, see Tibor Györy, *Der morbus Brunogallicus (1577): Ein Beitrag zur Geschichte der Syphilisepidemien* (Giessen: A. Töpelmann, 1912).

[102] A selection of references is given by Eatough, in his edition of *Syphilis* (cit. n. 7), pp. 204–223. See also H. Hofmann, "La *Syphilis* di Fracastoro: Immaginazione ed erudizione," *Respublica Litterarum,* 1986, *9*:175–186.

[103] Luigi Luigini (Aloysius Luisinus), *De morbo Gallico omnia quae extant apud omnes medicos,* 3 vols. (Venice: J. Ziletus, 1566–1567), Vol. I, pp. 161–172, 173–179. The various texts mentioned in this section are cited from the pages of this edition, although, where possible, I have checked them against the original printing.

[104] Gabriele Falloppio, *De morbo Gallico* (originally, Padua: L. Bertello, 1564), *ibid.,* Vol. I, p. 663;

were highly complimentary to the poet, but, equally, none appears to have taken issue with or developed his ideas. Whether it was not in some way entirely proper to argue with a medical poem or whether, as is more likely, the ideas that seem more interesting to us today were either not noticed or not thought novel is an open question.

To claim that syphilis was a contagious disease was commonplace even before 1546, and later authors who emphasized contagion, like Massa and the Ferrarese professor Antonio Musa Brasavola, did not need Fracastoro's seeds for their explanations.[105] One might think of evaporations from bodies hot with intercourse penetrating the body and, through an affinity with the natural spirits, producing changes within the liver. Like plague, syphilis might have an ultimate cause in a change of atmosphere and be passed on by clothes, cups, and the like. It also needed the appropriate receptivity, which explained why not everyone who had intercourse with the same infected prostitute caught the disease. In short, the same type of explanation was offered for syphilis as for plague, and often by the same authors.[106]

Without a Da Monte to provide an immediate reaction to Fracastoro's later views on the seeds of syphilis, we have to wait until the 1550s to find a mention of the prose work in the venereological literature. In 1551 the Portuguese physician Amatus Lusitanus, who had studied in northern Italy, published in the first volume of his case histories his description of a family tragedy.[107] A father who had thought he was long cured of syphilis infected not only his wife but also their child in the womb. When it was born, its mother was already too ill to feed it herself, and it was handed to a wet nurse. Within a month, she, her husband, two other children she was nursing, and their mothers had all become infected. The unfortunate baby died within a few weeks and was shortly followed by its father, "who had first infected his wife with the hidden *seminaria* of an old illness." The wording here can only derive from Fracastoro and is the earliest instance of his influence in this literature.

Four years later, in 1555, the Polish physician Jósef Struš cited both *De sympathia* and *De contagione* in his explanation of why syphilis was a new disease.[108] It was not because of any novel symptoms or the variety of forms it took, but because of its new quality and its new *seminaria*. It is not until the 1560s, however, that discussions of syphilis appeared using Fracastoro's terminology. Even then, not every author followed a strongly Fracastorian line. Borgarucci's citation of *De contagione* is polite at best, while Antonio Fracanzani, professor at Padua and later at Bologna, although accepting Fracastoro's triple division of infection, declared that in this disease fomites were "very rarely" involved and

---

Antoine Lecoq (Gallus), *De ligno Sancto non permiscendo* (originally, Paris: S. de Colines, 1540), *ibid.*, Vol. I, pp. 396–421; Bernardino Tomitano, *De morbo Gallico, ibid.*, Vol. II, pp. 58–149, on p. 82; and Prospero Borgarucci, *Methodus de morbo Gallico, ibid.*, Vol. II, p. 155.

[105] Niccolò Massa, *De morbo Gallico,* 3rd ed. (Venice: J. Ziletus, 1563), *ibid.*, Vol. I, p. 38; Antonio Musa Brasavola, *De morbo Gallico, ibid.*, Vol. I, p. 567.

[106] See the list of supporters of the theory of contagion given by Falloppio, *De morbo Gallico, ibid.*, Vol. I, p. 670; the whole section, pp. 668–677, is of interest in this respect.

[107] Amatus Lusitanus, *De Gallica scabie* (originally, Florence: L. Torrentino, 1551), *ibid.*, Vol. I, p. 560; for his knowledge of *De contagione* see above, n. 83.

[108] Jósef Struš (Struthius), *De morbi Gallici pulsibus, ex lib. quarto sphygmicae artis* (originally, Basel: J. Oporinus, 1555), in *De morbo Gallico,* Vol. III, p. 96.

preferred an infection by vapors to one by seeds. Leonardo Botallo of Asti, professor at Turin and later a French royal physician, talked at length about *seminaria, fomites,* and what he called the *seminium morbi,* but without developing these ideas further. Among the other authors collected by Luigini, only Alessandro Traiano Petronio, a verbose Bolognese physician, followed Fracastoro in many details. While he acknowledged the role of air in spreading contagion, he devoted three successive chapters to, respectively, the role of *seminaria* of putrefaction; the differences between diseases, including plague; and the importance of fomites in spreading contagious diseases. By placing syphilis in this wider context, Petronio made good use of the doctrines he had taken from Fracastoro.[109]

The massive size of Petronio's treatise offers an explanation for the general neglect of Fracastoro in the other syphilis tracts. It is not just that more tracts were written on plague than on syphilis; it is also that what Fracastoro had to say on the subject of syphilis was less innovative and, apart from his poem, less relevant to any save the most thorough analysis. The doctrine of contagion was far more applicable to plague, measles, and typhus than to syphilis, which was never transmitted at a distance and by fomites only with difficulty. By the 1540s very few authors argued against contagion as the cause of syphilis, and even they accepted contagion at some point in their explanatory process. In such a debate, however much one might approve of Fracastoro's poetry, his ideas on seeds were less effective and could easily be challenged by other, less recondite candidates for the actual agent of transmission. Why accept novelties when the poisons and vapors of older theories were to hand?[110]

VIII. CONCLUSION

This lengthy survey of evidence for the reception of Fracastoro's ideas in the generation after the publication of *De contagione* in 1546 has covered much of learned Europe, from Poland to Portugal. It has focused on two questions: How widely were his theories known and cited? And how did his contemporaries interpret them?

An answer to the first query can be briefly stated. Apart from the assault by Da Monte in his lectures at Padua, there is nothing to show an immediate reaction to the book's publication. By the middle years of the next decade, however, and particularly after the public discussions on the causes and remedies for the Venetian plague epidemic of 1555–1557, several authors were employing his triple division of contagion and his vocabulary of seeds and fomites, first in northern Italy and then further afield. His views were disseminated in his collected works,

[109] Borgarucci, *De morbo Gallico, ibid.,* Vol. II, p. 156; Antonio Fracanzani, *De morbo Gallico* (originally, Padua: C. Gryphius, 1563), *ibid.,* Vol. I, p. 723; Leonardo Botallo, *Luis venereae curandae ratio* (originally, Paris: J. Foucher, 1563), *ibid.,* Vol. III, pp. 10–12; Alessandro Traiano Petronio, *De morbo gallico libri VII* (originally, Venice, 1566), *ibid.,* Vol. II, pp. 1–215, esp. pp. 10–11, and Bk. 1, Chs. 13–15, pp. 26–30.

[110] I have not attempted to cover the influence of Fracastoro's ideas on the theory of epidemic skin diseases or rabies. I have been unable to substantiate the Singers' claim, "Position of Fracastoro" (cit. n. 4), p. 33, that Pietro Andrea Mattioli in his 1554 *Commentary on Dioscorides* applied his ideas to rabies, for neither in this nor in subsequent editions of this book can I find any hint of Fracastoro's teaching on the subject, although there is a substantial section dealing with rabies (as a poison) and its causes.

in reprints of *De contagione,* in published tracts on plague and syphilis, and, as the evidence of the Crato circle shows, in a network of correspondence between European *érudits.* By 1570 at the latest, almost a decade before many of the works cited by the Singers in their pioneering study, and before the disastrous plague of Venice in 1576, they had become commonplace in studies of plague and pestilence.[111] A similar pattern obtains among the smaller number of writings on syphilis. Little reference is made to *De contagione* until the early 1550s, with major discussions using Fracastoro's theories coming only in the 1560s.

This process of influence, however influence is defined, was slower than that of the anatomist Andreas Vesalius, whose impact was almost immediate, but it was no less effective in the end. Yet the comparison with Vesalius may be unfair, for, unlike Fracastoro, who devoted many years to polishing and refining his compositions before letting them be printed, Vesalius was a man in a hurry, talented, ambitious, and, as one of his opponents grudgingly noted, free of all encumbrances, both professional and domestic.[112] Vesalius was also unique in his exploitation of the power of print, and in the infinite care he took over the actual publication of his *Fabrica.* He chose the best artists available, the best printers, and, in Basel, the best distribution network to achieve a swift European diffusion of his ideas. His book was dedicated to none other than the Holy Roman Emperor himself, with an implied plea for his assistance in making its merits known.[113] Compared with Vesalius in his eager self-publicity, Fracastoro was an amateur, and, I suggest, far more typical in his approach to publication and in the way in which, once printed, his ideas were picked up and discussed. Historians of medicine and of printing who have taken the example of Vesalius as the norm have thus been misled in their expectations of what should have happened and have interpreted the lack of an immediate reaction to *De contagione* as indicating disfavor or neglect.[114] On the contrary, the ideas of *De contagione* did become part of the common stock of medical knowledge, albeit relatively slowly, within half a generation of their appearance in print. Further comparison with the reception of other medical authors, including Paracelsus, may confirm whether such a timetable is as typical as I have suggested, but the presence of Fracastoro's ideas in a variety of authors throughout Europe proves that, far from being confined to a handful of northern Italian professors, his *seminaria* bore a substantial fruit.[115]

This is a surprising conclusion only to those who have thought Fracastoro a

[111] Singer and Singer, "Position of Fracastoro," pp. 33–34. For the great plague of Venice see Palmer, "Control of Plague" (cit. n. 3), Ch. 9; and Preto, *Peste e società* (cit. n. 78), which includes a valuable appendix of texts.

[112] Niccolò Massa, *Epistolae medicinales* (Venice: Bindoni & Pasini, 1558), fols. 51v–55v. Vesalius, it is often forgotten, wrote his *Fabrica* when he was only twenty-seven.

[113] Andreas Vesalius, *De humani corporis fabrica* (Basel: J. Oporinus, 1543), preface; translated in Charles Donald O'Malley, *Andreas Vesalius of Brussels* (Berkeley/Los Angeles: Univ. California Press, 1964), pp. 317–324. O'Malley's discussion of the production process of the *Fabrica* and its related epitomes, on pp. 111–138, is classic.

[114] See Elizabeth Eisenstein, *The Printing Press as an Agent of Change,* Vol. II (Cambridge: Cambridge Univ. Press, 1979), pp. 566–574. Eisenstein forgets that, as far as the first century of the printing press is concerned, the great majority of medical books contained only the received wisdom of the past.

[115] Little has been done to answer the question of the spread of Paracelsian printings since Karl Sudhoff's pioneering *Versuch einer Kritik der Echtheit der paracelsischen Schriften,* 2 vols. (Berlin: G. Reimer, 1894, 1899).

prophet without honor. Even more surprising to them, then, will be the discovery that, with the exception of Da Monte's initial reaction, the references to Fracastoro are universally favorable. He is a man of talent, learning, and acumen—terms beloved of any academic—to say nothing of his divine poetic gifts. Although Lange disagreed with his views on new diseases, he appreciated what he had to say, while, once the immediate impact of *De contagione* had passed, even Da Monte reverted to compliments about his fellow townsman. Compared with Vesalius, and still more with the Paracelsians, Fracastoro's theories did not excite controversy or thoughts of medical revolution. He might be accused of falling into Epicurean and atomist heresy, but even this could easily be denied.[116] In short, contemporaries viewed him as one of themselves, a learned Galenist, invoking and interpreting the sound authority of the past in order to deal with the urgent problems of the present.

Such an impression runs counter to the standard modern picture of Fracastoro on two grounds. First, it is argued that, since the physicians of antiquity had no understanding of contagion, Fracastoro could not have based his ideas on their precedents. Second, even if this point is conceded, to advocate a theory of plague largely in terms of seeds of contagion is considered as outright opposition to the standard Galenist position of the sixteenth century. But, as we shall see, neither objection would have carried weight in Fracastoro's own day.

Whether the concept of contagion was known in classical antiquity has long been the subject of ardent debate.[117] True, there is no precise Greek word that can be equated with the Latin *contagio* or *contagium* and their derivatives, nor is there a specific treatise devoted to explaining plague and other epidemic diseases in terms of contagion. But at least as early as the historian Thucydides in the late fifth century B.C., it had been observed that, in certain conditions, those who came into contact or proximity with the sick were liable to become infected themselves.[118] The ease with which a Christian preacher in the third century could recommend the segregation of sinful virgins "like infected sheep or diseased cattle" to avoid their corrupting their companions reveals the familiarity of the practice.[119] Whatever the publicly expressed theory of the medical authors who have survived to us from antiquity, the man in the Roman street was well aware of the phenomenon of contagion.

[116] Cf. Struš's reaction to Da Monte's accusations of Epicureanism, *De morbi Gallici pulsibus* (cit. n. 108), p. 96.

[117] Note, e.g., the long discussion on this point by Mercuriale, *De pestilentia* (cit. n. 25), pp. 7–10. I still stand by the evidence collected by K. F. H. Marx over a century and a half ago, "Origines contagii" (diss., Univ. Göttingen, 1824).

[118] On the importance of Thucydides' evidence see, most recently, A. J. Holladay, "New Developments in the Problem of the Athenian Plague," *Classical Quarterly*, N.S., 1988, *38*:247–250, esp. p. 250. I doubt whether the medical evidence at our disposal is sufficient to exclude a medical acceptance of contagion, as Holladay wants; and, as my paper suggests, contagion and miasma were not thought incompatible.

[119] Cyprian, *De habitu virginum* 17, in *Patrologiae cursus completus, series Latina,* ed. J.-P. Migne, Vol. IV (Paris, 1891), cols. 469–470: "Tamquam contactas oves et morbidas pecudes a sancto et puro grege virginitatis arceri, ne contagio suo caeteras polluant." Another theologian, Origen, talked of a future in which the earth would be shaken by earthquakes and the atmosphere would become pestilential through "taking on a disease-bearing force": "Aer autem vim quandam morbiferam concipiens, pestilens fiat," *Comm. in Matthaeum ser.* 36, in *Patrologiae cursus completus, series Graeca,* ed. Migne, Vol. XIII (Paris, n.d), col. 1649. Origen's exposition is not preserved in the original Greek, and hence his exact phraseology cannot be known. For the *contagio* of sexual urges see Tertullian, *De anima* 38.2.

But it is not necessary to prove the existence of a classical theory to refute the proposition that a contagionist could not also be a Galenist. Fracastoro, in his preface, admitted that Hippocrates, Galen, Paul, and Aetius had all written on contagion. His complaint was that they had not written enough on the subject, or were prepared to accept an explanation in terms of occult properties.[120] His own general system of physiology, despite Da Monte's objections, was Galenic, and his discussion of causes and bodily processes was demonstrably in line with the whole Galenist tradition. Others went even further in aligning him with Galen. The Polish physician Jósef Struš condemned Da Monte for faulty scholarship in accusing Fracastoro of atomism. Was Da Monte not aware that Galen himself had used the phrase "seeds of plague" in *De differentiis febrium* or that there was an exact precedent in Hippocrates' "harmful exhalation" (*νοσερὰ ἀπόκρισις*)? Far from being a renegade, Fracastoro was merely showing his mastery of the classical legacy.[121]

Such a defense of Fracastoro also shows part of the context in which contemporaries considered his ideas on plague and epidemic diseases. Far from being an isolated thinker, he was viewed as merely one of many contributors to a debate about the causes of plague that went back to the Arabs, if not to Galen and Hippocrates. It was a debate whose basic outlines remained the same until the late nineteenth century, and whose structure has been forgotten by modern historians accustomed to thinking in terms of single causes. For Fracastoro and his contemporaries, disease was the result of a multiplicity of causes, which afforded a variety of levels of explanation and action. What follows is a schematic reconstruction of the framework in which the sixteenth-century controversies on plague took place, and its deliberate simplification may make clear why Fracastoro was considered an intelligent, and not particularly revolutionary, Galenist.

At the top of the hierarchy of causes was God, the ruler and predisposer of all. Without his will, to chastise, punish, or warn, there would be no plague or pestilence. Medical intervention would have no immediate effect against divine wrath, which could only be assuaged by prayer and contrition, religious remedies.[122] Such a bleak attribution of all responsibility to God left no place for the physician, and, not surprisingly, such an opinion tended to be modified by the suggestion that God had allowed only the conditions for plague to arise; its continuance might therefore depend on the ability of men to intervene or change the noxious conditions.[123]

Almost as remote as God in the hierarchy of causes were the planets and the stars.[124] On some explanations, plagues were the direct result of planetary forces

[120] Fracastoro, *De contagione,* p. A.

[121] Struš, *De morbi Gallici pulsibus* (cit. n. 108), p. 96.

[122] Rather than append a substantial list of authorities for this and the following points, I have chosen instead to quote Fracastoro and occasionally others. Fracastoro himself does not discuss divine causation but it is regularly mentioned by others; see above, n. 65. Camerarius, in his *Synopsis* of Donzellini (cit. n. 94), sig. C1r, declared that in time of plague one's first duty was prayer and public penance, and then one's medical intervention might prosper.

[123] Note, in particular, the argument of Borgarucci, *De peste* (cit. n. 62), p. 23. A nice example of a clash between a medical and a religious explanation for plague, albeit at a slightly later date, is given by Carlo Cipolla, *Chi ruppe i rastelli a Monte Lupo?* (Bologna: Il Mulino, 1977).

[124] Fracastoro, *De contagione,* pp. 58–60, 64. Note, e.g., the special skills of Jacques Pelletier, *medicus et mathematicus,* in his *De peste compendium* (Basel: J. Oporinus [1560?]). According to Pietro Giacomo Zovelli of Carmagnola, *De pestilenti statu* (Venice: F. Portinari, 1557), fol. 5v, one could refer the cause of plague to the planets, but since the doctor had to measure his actions by what

in conjunction. As such, they might be foreseen by man but could hardly be forestalled.[125] At best, one might either devise a regime that would minimize the effects of the planets or depart from an area likely soon to fall under such a baleful influence.

Both God and the stars, however, were only remote causes, acting indirectly on the individual. Their agent was the air, breathed in by every living thing and thus capable of affecting all and sundry. The air worked in one of two ways. Either its quality was such as to set up a hostile reaction in the bodies of those who breathed it in, although it itself was not precisely bad, or it had been changed from the stuff of life to a deadly poison. These changes might be internal, in the quality of the air, or external, through the admission of something noxious, agent *X*, that was then carried by the air over a wide area.[126]

At the base of the pyramid, there was the individual victim—or indeed the cause of plague. All writers on plague were agreed that, in some instances, bad regimen, in the widest sense of the word, would produce in the individual deleterious changes, what Crato called his "private plague." The body's humors altered for the worse, and the sufferer excreted something noxious as the body tried to restore its balance. This excretion might take the form of breath or vapors, or sloughed-off skin, or whatever else was appropriate to the disease. It might be passed on by contact, or by being carried along in the air. It might even remain, as a little miasma, in clothes, bedding, and goods, or, like fragments of skin, adhere to cups, utensils, and furniture.[127]

Finally, it was universally agreed that, even in the most horrendous epidemic, some escaped. Two reasons were suggested, both incapable of proof or disproof. Either the lucky individual did not come into contact with the noxious agent of plague, or, what was more likely, his or her resistance had been so strengthened that the body was able to maintain its equilibrium and not succumb to the external changes.[128] Only if the body itself changed could one catch the plague, and hence the physician and the writer of plague tracts should offer as much guidance

---

he could sense, he could think of plague only in terms of infected air and such causes of putrefaction as were equally accessible to the senses. See also Borgarucci, *De peste,* p. 23; Alessandro Massaria, *De peste* (Venice: A. Salicatius, 1579), fol. 8v; and Caius, *Boke against the Sweate* (cit. n. 60), p. 13, "the evil disposition by constellation, whiche hath a great power & dominion in al erthly thinges."

[125] Johann Jessen, *Academiae Witebergensis studiosis necnon universis salutem et diuturnam incolumitatem dicit* (Wittenberg: S. Groneberg, 1596), sig. A2r, was not unusual when he talked of the seeds of plague being "shut up" in winter and brought to life again by the arrival of the different climate of spring. The good physician should be aware of the possible recurrence of plague, but even so, whether one caught the plague or not depended on one's own temperament and the propensity of one's humors to putrefaction.

[126] Fracastoro, *De contagione,* pp. 56–58.

[127] On bad regimen see *ibid.*, pp. 60, 190, 223; and Crato, citations in nn. 90, 91. A good example of this hierarchy of causes is given by Guinther, *De pestilentia* (cit. n. 62): Divine vengeance (level 1; p. 8) might so ordain the sun and moon (level 2; p. 8) to poison the air; the air might itself be distempered (level 3), or turned foul by vapors from swamps, latrines, corpses, fogs, and caves; or a wind might bring seeds of pestilence. The air putrefies and acts like a sort of poison (pp. 9–17). Plague is the result of breathing in bad air, aided by the transfer of fomites in clothes and bedding (p. 18). Some plagues are caused by, and always affect those who have, bad bodily humors (level 4), and hence even those who live away from direct contact with the sick may succumb if their bodies are hot and moist (p. 19). Another good summary, incorporating all these causes, can be found in a letter of Monaw to Dudith, Crato, *Epistolae* (cit. n. 88), Bk. V: 37, pp. 419–422.

[128] Note Fracastoro, *De contagione,* p. 60, emphasizing receptivity. This was a cardinal feature of almost all plague tracts.

as possible on prophylaxis, on building up resistance to any possible onslaught. This was in addition to his recommendations for recipes against the plague, once the dread symptoms had appeared.

At the two lowest levels of causation, however, there was much scope for learned debate over the unknown and the unknowable. Was a plague the result of an internal, individual change and then spread by contact, including breathing into the atmosphere, or was it the result solely of noxious air affecting the body? Although such a distinction can be found,[129] it was very much an academic one, for in any schema that involved this hierarchy of causes, both aspects were necessarily taken into consideration, to say nothing of their combination in practice. Hence, as Richard Palmer has argued, the Venetian debates on the plague of 1555 cannot easily be divided into contagionists versus miasmatists *tout court*.[130] Each side made use of both explanations, dividing over the appropriate level of medical intervention. Given that contagion might be airborne, measures would also be needed to change the air, in addition to restricting contact. To the miasmatist, the harmful excretions given off by the sick were most potent to those near them, and hence some form of quarantine might also be useful, on top of the necessary measures to purify the air or clean the streets.

Both sides in this debate accepted the existence of harmful air. It was a legitimate question to ask how and why it had become harmful, or what the various combinations of the planets or the stern judgment of God had actually done to the atmosphere. Here was both a sticking point and a way to reconcile all authorities. Most miasmatists, as well as the contagionists, agreed that it was not enough merely to say that the air was hot and moist, for this was far too common a condition to be a cause of plague. The solution was to posit what I have termed agent *X*, something in the air that would account for the virulence and the sporadic nature of plague. We have already met many of the candidates for agent *X*: putrefaction, putrid air, hot poison, putrid heat, a poisonous occult quality, a morbid excretion, putrid or poisonous vapors, pestiferous odors, and defilements of the air (*inquinamenta aeris*). Seeds, whether of plague, disease, or putrefaction, fit neatly into the same explanatory slot. It is not surprising, then, to find that Fracastoro's contemporaries used his terminology interchangeably with what they had known from Galen, Hippocrates, and their medieval interpreters. For Alessandro Massaria, writing in 1579, whether one chose to describe the agent as putrid or malign particles, vapors, spirits, odors, or whatever was largely irrelevant, when what was obvious was that healthy bodies became infected thereby.[131] Fourteen years earlier, in a letter to Crato, the Swiss physician Conrad Gesner had made a similar point: he could not see why the Hippocratic "morbid excretion" could not be included under the heading of vapor or some such common name.[132] Mercuriale went further, declaring that it was "obvious"

[129] E.g., in Argenterio, *Opera* (cit. n. 49), p. 2536.

[130] Palmer, "Control of Plague" (cit. n. 3), p. 120.

[131] Massaria, *De peste* (cit. n. 124), fol. 14v. He immediately goes on to say that Fracastoro's views on plague are so widely known and accepted that he need not waste paper by copying out what Fracastoro has revealed. Treatises like Francesco Alfani, *Opus de peste, febre pestilentiali et febre maligna* (Naples: O. Salviano, 1577), or Quattrami, *Breve trattato* (cit. n. 51), confirm the truth of Massaria's claim.

[132] Gesner, *Epistulae medicinales* (cit. n. 65), fol. 14r. Crato himself agreed, and in his posthumously printed *Mikrotechnē*, ed. L. Scholtz (Frankfurt: Wechel, 1592), p. 490, identified the *seminaria contagii* of plague with the same Hippocratic words.

that all agreed on the origin of plague and differed only on the terminology: in fact, the atoms of Democritus, the miasmas of Hippocrates, and the seeds of Lucretius and Epicurus amounted to one and the same thing.[133] Finally, the variety of translations offered for the Hippocratic words *νοσερὰ ἀπόκρισις* in *De natura hominis* 9—*"morbosa exhalatio," "morbida excretio,"* and even "seed"—shows the ease with which physicians brought up in the same Galenist tradition could switch from one description to another.[134]

Against such a background, "seeds of disease" was only one possibility for agent *X*, and, save as a metaphor possessed of a decent precedent from classical antiquity,[135] it was far from ideal. No one could *see* the seeds of plague in the same way as they could perceive, by sight or smell, the fetid breath or exhalations from a sufferer from plague. As an explanation for the process of corruption, too, the suggestion of the presence of a possibly nonputrid seed was less immediately informative than that of poison or putrefying matter. Furthermore, by positing as the agent of disease seeds that lived, moved, and multiplied, Fracastoro was laying himself open to the charge of inventing specific beings, a new order of creation. This, as Da Monte and many others argued, was an unnecessary refinement. The existing schema of explanation, with its broad hierarchy of causes, was more than capable of accounting for the phenomena, including the process of contagion, without invoking these novel entities. Besides, the possibility of new diseases, and the explanations offered for or against their novelty, had been the stuff of academic controversy ever since 1492—the classically minded might say ever since Plutarch's day—and those who supported the idea of new diseases then had reached the same conclusions as Fracastoro, without involving seeds. Hence, until the bacteriological and microscopic revolution of the late nineteenth century, there was no convincing reason why one should interpret Fracastoro's seeds of disease literally, and much to be said against it. But a metaphorical interpretation was attractive, for it enabled Fracastoro's formulations of the insights and practices of others to be incorporated easily within traditional guidelines.[136]

Here is the clue to the paradox of Fracastoro. The revival of interest in Fracastoro at the end of the nineteenth century is due to the apparent parallels between his ideas and those of the bacteriologists. But the physicians of the Renaissance, lacking the knowledge, and the technology, of the late nineteenth century, interpreted his ideas very differently, and in a way that may be closer to

[133] Girolamo Mercuriale, *Variae lectiones* 4.6 (Venice: Junta, 1598), fol. 64r (arguing also that Lucretius's description of the plague derived from the Hippocratic *On Breaths*). Girolamo Donzellini, whose plague tract depended heavily on Fracastoro, *Apologia* (Verona: S. a Donnis, 1573; originally 1571), fols. 33r–v, 68r, identified the "noxious exhalation" of plague with the Hippocratic "morbid excretion," the Aristotelian "spiritus corruptus," and the "seminarium pestis" of more recent authors. See also Joachim Camerarius, *Synopsis* (cit. n. 94), sig. B4r–5r.

[134] Respectively, the versions of Marco Fabio Calvi (Rome: F. M. Calvi, 1525), fol. CC; Antonio Brenta, *Aphorismi* (Venice: H. Pincius, 1530), fol. 60v; Thomas Lodge, *A Treatise of the Plague* (London, 1603), sig. B5 (a translation of François Valleriola's *Traicté de la peste* [Lyons: A. Gryphius, 1566], fol. 28r): "The cause, sith Hippocrates, of the generall pestilence . . . is the ayre which we sucke that hathe in itselfe a corrupt and venomous seede."

[135] Cf. the complaint of Dudith to Monaw about those who sought to have the "antiquitatis praeiudicium" on their side before accepting any new discovery or therapy: Crato, *Epistolae,* Bk. VI (1579): 14, p. 44.

[136] Cf. the obviously metaphorical use of the phrase "seeds of disease" by Trincavelli in his lectures on fevers of ca. 1550, *Opera omnia* (cit. n. 64), fol. 19v.

what he had intended. They saw them in a different context, as offering one possibility among many for one of the causes of plague, within the traditional schema of causation. Such an interpretation was not unwise. Fracastoro's explanation involved possible planetary influences, bad air, humoral imbalances, earthquakes, and fleeing rodents, as well as contagion and seeds—in short, all the traditional data of the plague investigator.[137] Indeed, in a letter of 1551 he talked of the English sweat without mentioning the word "seeds," reverting to "principles" (*principii*), and spoke with apparent approval of infection by means of "extremely subtle vapors."[138]

Fracastoro himself never denied that others had preceded him in his enquiries. Although the rhetoric of a dedicatory preface is not always to be trusted, his intention in *De contagione* was not to separate himself from the mainstream Galenic tradition but to carry out more deeply the investigations already begun by others. His contemporaries agreed, welcoming his researches into the process and phenomenon of contagion. Far from being neglected, his ideas and certainly his vocabulary were widely used within a generation, and as the Singers clearly demonstrated, most plague texts from the end of the sixteenth century and the beginning of the seventeenth show signs of his influence, direct or indirect.[139] This is no neglect, no disregard, no isolation, but rather the reverse. Indeed, this apparently universal approbation suggests an alternative explanation for the disappearance of Fracastoro from the pantheon of great physicians until his rediscovery at the end of the last century.[140] Like his *seminaria,* his ideas grew old. Precisely because they were not perceived as radically different but were easily subsumed into the Galenist system of interpretation, they gradually disappeared along with Galenism.[141] Seeds as a metaphor fell out of fashion, and the direct link with their (re)formulator was forgotten. Like the seed that fell among thorns, Fracastoro's ideas were eagerly embraced by the surrounding plants, the hardy perennials of Galenism. It is tempting to continue the metaphor and to argue that thereby all the new growth was quickly choked. But as this paper has argued, it is equally likely that, far from being a healthy seed of grain, the seminal idea of Fracastoro was merely another variety of thorn.

[137] As noted, rightly, by Howard-Jones, "Fracastoro and Henle" (cit. n. 17).

[138] Fracastoro, in *Lettere,* ed. Porcacchi (cit. n. 7), pp. 736–742. In *De contagione,* pp. 96–98, although accepting that the sweat was contagious, he equally failed to mention the word *seed,* talking instead of "vapors that bring with them a subtle contagion that has an analogy with the spirits or the fine foam that floats on the surface of the blood." Such a type of explanation would have been willingly accepted by John Caius: see his *Boke against the Sweate* (cit. n. 60), pp. 13–20.

[139] Singer and Singer, "Position of Fracastoro" (cit. n. 4), pp. 30–34. Their implied linkage of Fracastoro with Girolamo Cardano, however, is not convincing (p. 32).

[140] Wright's introduction to *De contagione,* pp. xxxi–xxxii, shows the late nineteenth-century context; the list of secondary references in the Singers' article also confirms a lack of interest in Fracastoro as a scientist until the 1880s.

[141] Although it would be wrong to assign a date to this disappearance (see Owsei Temkin, *Galenism: Rise and Decline of a Medical Philosophy* [Ithaca, N.Y./London: Cornell Univ. Press, 1973], p. 173), it is hard to trace the medical influence of Fracastoro beyond 1650.

Notes on Contributors

Jerome Bylebyl is a member of the Johns Hopkins University Institute of the History of Medicine. His interests include the early history of physiology and anatomy, and the relation of medicine to philosophy and humanism.

Chiara Crisciani is an associate professor in the Department of Philosophy at Pavia University, Italy. Her research is in the field of late medieval philosophical and scientific thought. She has written a number of articles and books, mainly on the epistemological features of Scholastic medicine and Latin alchemy.

Richard J. Durling is Research Associate at the University of Kiel, Germany. He specializes in the history of ancient, medieval, and Renaissance medicine, with particular interest in translations and commentaries on Galen. He is the editor, with F. Kudlien, of the new series Galenus Latinus.

Eduard Feliu is an independent researcher in Jewish studies and chairman of the Association for the Study of Jews in Catalonia. He has written on the history of the Jews in medieval Catalonia and has translated from both modern and medieval Hebrew.

Lola Ferre is Associate Professor of Hebrew Language in the Department of Semitic Studies of the University of Granada, Spain. She has translated medieval Hebrew natural-philosophical texts into Spanish and is now working on an edition of fourteenth-century Hebrew translations of Latin medical works.

Luis García-Ballester is Professor of the History of Medicine at the Consejo Superior de Investigaciones Científicas. He is now working on the social and intellectual history of medicine in fourteenth-century Spain. He is coeditor (with J. A. Paniagua and M. R. McVaugh) of the critical edition of the biomedical works of Arnau de Vilanova.

Danielle Jacquart is Director of Research at the Centre National de la Recherche Scientific (Paris). Her field of research is medieval medicine, in both social and intellectual aspects. Her books include *Le milieu médical en France du XIIe au XVe siècle* (1981), *Sexuality and Medicine in the Middle Ages* (1988, with Claude Thomasset), and *La médecine arabe et l'Occident médiéval* (1990, with Françoise Micheau).

Mark D. Jordan, Associate Professor of Medieval Studies, University of Notre Dame, is the author of *Ordering Wisdom: The Hierarchy of Philosophical Discourses in Aquinas* (Notre Dame, 1986), and of essays on Aquinas and philosophical rhetoric. He has also published on the Salerno school and the changing place of medicine in the medieval hierarchy of sciences.

Michael McVaugh is Professor of History at the University of North Carolina, Chapel Hill. He is one of the editors of the collected medical works of Arnau de Vilanova and is engaged on a study of medicine and society in the Crown of Aragon during the early fourteenth century.

Vivian Nutton has been a member of the Academic Unit of the Wellcome Institute for the History of Medicine, London, since 1977. A selection of his essays, *From Democedes to Harvey: Studies in the History of Medicine,* was published in 1988.

Nancy G. Siraisi is Distinguished Professor of History at Hunter College and the Graduate School of the City University of New York. She is the author of *Taddeo Alderotti and His Pupils* (Princeton, 1981) and *Avicenna in Renaissance Italy* (Princeton, 1987); she is currently interested in Vesalian anatomy.

Index